"Daddy's Gone to War"

"Daddy's Gone to War"

*The Second World War in the
Lives of America's Children*

WILLIAM M. TUTTLE, JR.

New York   *Oxford*
OXFORD UNIVERSITY PRESS
1993

Oxford University Press

Oxford   New York   Toronto
Delhi   Bombay   Calcutta   Madras   Karachi
Kuala Lumpur   Singapore   Hong Kong   Tokyo
Nairobi   Dar es Salaam   Cape Town
Melbourne   Auckland   Madrid

and associated companies in
Berlin   Ibadan

Published by Oxford University Press, Inc.,
200 Madison Avenue, New York, New York 10016

Oxford is a registered trademark of Oxford University Press

Library of Congress Cataloging-in-Publication Data
Tuttle, William M., 1937– "Daddy's Gone to War": The Second World War
in the lives of America's children
William M. Tuttle, Jr.
p.   cm. Includes bibliographical references and index.
ISBN 0-19-504905-5
1. Children—United States—History—20th century.   2. Children and
war—United States.   3. United States—Social conditions—1933–1945.
4. World War, 1939–1945—Children—United States. I. Title
HQ792.U5T88   1993
305.23′0973—dc20   92-12445

1 3 5 7 9 8 6 4 2

Printed in the United States of America
on acid-free paper

In memory of my parents

Geneva Duvall Tuttle

and

William McCullough Tuttle

PREFACE

Born in 1937, I was a homefront child. My father went into the Army in late 1942 and returned three years later; we had all changed a lot in the interim. I remember many things about wartime: pulling my wagon around the block and collecting bundles of old newspapers, playing war games in the lot next to our house in Detroit, and sitting with my mother in the kitchen listening to the war news on the radio. Both radio and the movies were important parts of my homefront world. Monday through Friday, there were the late-afternoon radio adventure shows in which American heroes pursued enemy spies and saboteurs. And I remember trying never to miss a Saturday matinee at the Norwest, our local movie theater; for three to four hours, the war was a frequent theme in feature films, cartoons, serials, and newsreels.

When we were not playing war in the side lot, we were doing so on the playground of the Peter Vetal School, three blocks from my home. At Vetal, there was a deep division between the middle-class children and the working-class children. In large part, we in the middle class lived on one side of the school, while blue-collar families, including recent arrivals from the southern Appalachians, lived on the other side. I got to know Tommy Fields, whose family had moved to Detroit from Kentucky. We were in the same class, and I visited his house on the other side of the school; I do not think he ever visited mine, but I never thought about it at the time.

Looking back, I see that during the war my little brother George and I lived in a family of women headed by my mother, grandmother, and older sister Susan. Ours was a peaceful household and, from my perspective, a happy one. But I wondered what our lives would be like when my father returned home, and I wondered what he would be like. I did have a V-mail Christmas card that he sent me in 1944 from France, picturing Santa Claus driving a Jeep filled with presents; but I had few memories of him.

In the autumn of 1945, my father re-entered our lives. Forty years old and a major, he arrived sporting both a mustache and the Legion of Merit, which he had earned for his two years as a combat thoracic surgeon in North Africa, Italy, France, and Germany. The more I think about my father's service, particularly in light of what we later learned about the horrors of battle in Vietnam, the more I appreciate the psychic toll which the war took on him.

I was seven when my father came home. Because I had not really known him before he left for the Army, I could not tell how the war had affected him. My father for whom I am named was loud and regaled in storytelling; he liked to laugh, and I enjoyed him when he was having fun. Around him, however, I was usually very shy; I was an outgoing boy and very active, but I think he scared me. We had missed important years together, and we never bridged the gap. My father died in 1962 at the age of fifty-seven. I still believe that, in time, we would have become friends, but we never had a chance to do so. This has been the war's major legacy to me, and it is a sad one.

This book, then, is a personal odyssey as well as a history of a major event in United States history. How I came to write this book, however, is somewhat more complicated. It involves not only the peculiar challenges of writing contemporary history, which I enjoy, but also the recognition that while the writing of history has changed dramatically over the past thirty years, there is still uncharted territory to explore. Some of the theoretical social science concerns that underlie this book are new to the writing of history. So too are some of the topical concerns—not only gender, race, class, and ethnicity, about which we know so much more than we did a few years ago, but also age, or developmental stage, and its function in mediating the effects of history. These topics and issues were not part of my training as an historian in graduate school; yet they are basic to understanding this society.

Like many professional historians trained in the mid-1960s, I learned a great deal of history but received little encouragement to explore the social sciences for their insights into human behavior. My graduate training at the University of Wisconsin was traditional; I studied the historical canon and did my research in manuscript collections, newspapers, and government documents. There was one difference: because I was specializing in twentieth-century United States history, I could interview people who participated in the events I was researching. But what made Wisconsin memorable was the students, who were bold and committed and never missed an opportunity to debate either politics, or scholarship, or both, because we saw them as interrelated.[1]

Having become aware of the discrepancy between America's ideals and its realities in race relations while serving as a training officer in the United States Air Force, I decided that I wanted to study African-American history. (My dear friend in the service, Captain William Woody Farmer, a black B-52 pilot, was killed when his bomber crashed; I think of the good times I had with him, and I think of Captain David Taylor, also African-American, who was my commander and a good friend.) My master's thesis was a history of the Chicago race riot of 1919, and while I read on my own in the sociology of race relations and collective behavior, my teachers seemed to frown upon black history as too narrow a pursuit for a lifelong career of research and writing. I was pleased to resume my studies in African-American history after joining the faculty of the University of Kansas in 1967.

At this time, I also returned to my interest in ordinary voices. Throughout 1968 I talked with black Chicagoans about the 1919 race riot. I interviewed the chief Red Cap at the Illinois Central Station, Chester Wilkins, who told me of the migrant families arriving in the city during the war, and I listened to John Harris, who was fourteen years old and floating on a raft in Lake Michigan with four other black boys on that hot Sunday afternoon in July 1919. John recalled for me almost fifty years later that when the boys' raft floated by the breakwater near Twenty-sixth Street, they saw a white man. The man seemed angry and began throwing rocks at them. Eugene Williams, one of the boys, had just bobbed out of the water when a rock hit him in the head; he slipped back into the water. John Harris dived down to try to save him, but, he remembered, Eugene had "grabbed my right ankle, and, hell, I got scared." Gasping for air, John "shook away from him to come back up, and you could see the blood coming up. . . ." So began the race war that in five days left 38 people dead, 23 blacks and 15 whites.[2]

In conducting interviews and in extrapolating meaning from memory, I had no theoretical underpinnings. I based my approach largely on a detailed knowledge of the historical context. And I was lucky. In 1968, for example, two lovely middle-aged women in Chicago invited me to lunch. They were sisters, Alfreda M. Duster and Ida B. Wells-Barnett, Jr. Over lunch and a review of their mother's then-unpublished memoirs, Mrs. Duster told me about John Harris and arranged for us to talk. I was fortunate too when a single case file number, which I had found in a dusty card file index at the National Archives, led me to the Federal Records Center in Suitland, Maryland, and to literally thousands of typescript pages of verbatim testimony by stockyards workers, both African-American and white, as well as by union officials and representatives of the meat packers, all attesting to the city's troubled race and labor relations.

Later, while doing research on the American homefront during the Second World War, I became curious about the lives of America's children during these years. From my personal experience, I knew that their history consisted of more than the routine of scrap collection drives, Saturday movies, and reciting "The Pledge of Allegiance" every morning in school. Moreover, I sensed that in order to do justice to the children's history, I would need to learn developmental psychology. Research questions occurred to me which seemed to yield best to psychological interpretations. For example, how did family separations affect children? Fifteen million men and women served in the armed forces at one time or another during the war, but historians had not considered the psychological impact of these separations and absences on the homefront girls and boys.

Fortunately, at this juncture the University of Kansas awarded me an Intra-University Professorship for 1982–83, which freed me of all teaching and service obligations so I could devote my attention to developmental psychology. During the year, I read in the literature of Sigmund Freud and psychoanalysis, Jean Piaget and cognition, Arnold Gesell and ontogenesis, B. F. Skinnner and stimulus-response, and several variants of behaviorism, most notably social learning theory. Topics such as moral development, particularly as refined by feminist scholars, captivated me. I also searched for specific psychological articles dealing, for example, with the impact of child care on the sons and daughters of working mothers, or with the effects of father absence.

I was blessed to have as my teachers the psychologists Frances Degen Horo-witz, Aletha Huston, and John C. Wright, who were both inspiring and nurturing.

In fact, throughout the course of this project, psychologists have generously given me guidance and encouragement. In 1983–84, with the support of a National Endowment for the Humanities fellowship, I was an associate fellow at the Stanford Humanities Center. Through the seminars at Stanford's Center for the Study of Youth Development, I met a number of developmentalists, including Sanford M. Dornbusch and Eleanor E. Maccoby, who helped me conceptually in more ways than they realized. Above all, it was two distinguished older psychologists who especially encouraged me: Lois Meek Stolz and Robert R. Sears. I miss them, but I was most fortunate to have benefited from their knowledge, wisdom, and kindness. In 1986–87, I was a research associate in the Institute of Human Development at the University of California, Berkeley, where the psychologist Paul H. Mussen was my sponsor; he also became a friend.

By early 1990, with the assistance of a grant from the Interpretive Research Division of the NEH, I had finished writing a 600-page history of America's homefront children. It was a detailed study of social change in children's lives as illuminated by insights from the social sciences, including psychology, sociology, and anthropology. But I concluded that what I had written was essentially one-dimensional and boring. What was missing was authenticity—that is, the voices of the homefront children themselves.

To rectify this, I devised an approach which, I hoped, would give me the first-person testimony that was lacking. I wrote to the hundred largest newspapers in the United States as well as to several dozen African-American, Hispanic-American, and Jewish-American publications. "Dear Editor," I wrote, "I am a professor of history . . . and the book I am now writing may be of interest to some of your readers." It "deals with American children's experiences on the homefront during the Second World War. . . . The child's perspective seldom appears in history. Yet for many Americans now in their fifties, their wartime experiences were of crucial importance not only then, but have remained so throughout their lives. . . . I would like to hear from people . . . who have stories to tell about their lives on the homefront. . . ."

With this appeal, the letters began arriving at the rate of thirty to forty, even fifty, a day. In all, I heard from more than 2500 people. In addressing an amazing variety of topics ranging from air raids and the atomic bomb to V-J Day and War Bonds, and from anti-Semitism and the Boy Scouts to scrap rubber and spies, these letters changed my book entirely. For one thing, they expanded my conception of what was happening in the children's lives. Their letters introduced new topics, many of which strongly reinforced my belief that gender is the most important variable in children's lives. As I had suspected, their letters also showed that chronological age, or developmental stage, is the most overlooked variable in the historical study of children's lives. "Questions about age and aging through the life course," a German scholar has reminded us, "belong on the agenda of general social history along with questions about gender, race, and class."[3]

At the same time, I realized that these letters, while diverse and numerous, would not allow me to make quantitative judgments. Although I had learned much about the richness of children's experiences, I could not say specifically how many children shared these experiences, or whether they were typical or not. While I might hint at typicality in some cases, my objective was to document the range of children's experiences and, using the children's recollections, to try to understand the meaning of the war years in their lives. Thus, my goal was not—and, in all modesty, could not be—cliometrics, but rather "thick description," in the words of Clifford Geertz, that is, a dense amount of ordinary information about people and their cultures. "We must . . . descend into detail," Geertz has written, "past the misleading tags, the metaphysical types, past the empty similarities to grasp firmly the essential character of not only the various cultures but the various sorts of individuals within each culture if we wish to encounter humanity face to face."[4]

I certainly have not attained the goal described by Geertz. But I have tried, partly by tapping people's memories. In studying certain subjects, there is no substitute for the voices of the people themselves. The homefront children's memories comprise one of the three major research components of my book. First, there is the historical dimension, based on the facts concerning the history of social change during these years; second, there are the insights to be derived from studies in the social sciences; last, there are the homefront children's rich letters. Because these children have never been studied systematically before, an appreciation of their lives can come through research that synthesizes traditional historical sources, social scientific studies, and people's memories.

At the same time, historians should not look to oral history, retrospective letters, or other life narratives as a way of "proving" anything. That is not their, or our, purpose. While historians seek to explain change, we should not be so presumptuous as to think that we can elucidate natural laws. On the other hand, there is no doubt that it is people's stories—whether gathered through interviews, diaries, or letters—that enable scholars to peel away the layers of anonymity that have hidden America's plain folk from view.

A final issue is the reliability of people's memories. Specifically, how should an historian assess these letters? Clearly, this is a self-selected survey, the contributors to which had self-serving reasons for writing. However, I found very few letters which, on the surface, were fabrications. I did question a letter I received from an inmate at the Missouri State Prison, who confided that he had "discovered the cause and the cure for cancer." On the other hand, I found hundreds of letters that were utterly compelling, written without guile, and in which I could detect no motive to try to deceive me. Historians always have to assess their evidence, whether it comes from personal letters in manuscript collections, reports in governmental archives, or articles in newspapers or magazines. And the guide whose advice I have followed in assessing these letters is Sir Isaiah Berlin, who believes that the historian is an artist whose challenge is to convey reality. "History, and other accounts of human life," Berlin writes, "are at times spoken of as being akin to art. What is usually meant is that writing about human life depends to a large extent on descriptive skill, style, lucidity, choice of examples, distribution of emphasis, vividness of characterization, and the like. But there is a profounder sense in which the historian's activity is an artistic one. Historical explanation is to a large degree arrangement of the discovered facts in patterns which satisfy us because they accord with life—the variety of human experience and activity—as we know it and can imagine it."[5]

I would like to acknowledge the generous support I have received from the National Endowment for the Humanities, which awarded me a Fellowship for Independent Study and Research and a Projects Research Grant. I also wish to thank the American Historical Association for an Albert J. Beveridge Grant, the University of Kansas for a Faculty Development Grant, and the Hall Center for the Humanities at the University of Kansas for a Research Fellowship. I also want to acknowledge the scholars who served on my NEH grant advisory board: William H. Chafe, Carl N. Degler, Glen H. Elder, Jr., Frances Degen Horowitz, David M. Katzman, Paul H. Mussen, Charles E. Rosenberg, Barbara G. Rosenkrantz, Robert R. Sears, and Richard L. Schiefelbusch.

The University of Kansas has supported this project in a number of ways, first by awarding me an Intra-University Professorship for 1982–83. I wish to thank Deannell Reece Tacha for her constant encouragement; Daniel H. Bays, chairperson of the Department of History, and staff members Jan D. Emerson, Ellen L. Garber, Sandee Kennedy, and Terri Rockhold for countless favors; and the University of Kansas Libraries, which provided me with an office and remarkable cooperation from its Reference, Interlibrary, and Government Documents departments and its Copying Services. I also acknowledge my deep gratitude to the personnel associated with two research institutions at the University of Kansas: Richard L. Schiefelbusch and Paul Diedrich of the Bureau of Child Research, and Andrew P. Debicki, Theodore Wilson, Janet S. Crow, and Chad Krause of the Hall Center for the Humanities.

While working on this book, I was fortunate to be in residence at the Stanford Humanities Center and at Berkeley's Institute for Human Development. I wish to thank Ian Watt, director of the Stanford Humanities Center, and Mort Sosna, associate director. I very much enjoyed my year at the Center, particularly the intellectual and social interactions with the other fellows. In addition, Marilyn Yalom, deputy director of Stanford's Center for Research on Women, was a most gracious host in Palo Alto, as was Alberta E. Siegel of the Stanford Medical School. At Berkeley, Guy E. Swanson, director of the Institute for Human Development, arranged for my appointment as a research associate. I also want to acknowledge the kindness in the Bay Area of Jolene Babyak, Cornelia and Lawrence Levine, Rhoda and Leon Litwack, Glenna Matthews, and George Duvall Tuttle.

An important phase in my education began in 1984, when I joined a panel sponsored by the Social Science Research Council's Subcommittee on Child Development in Life-Span Perspectives to explore "Historical Perspectives on Child Development." The other participants were Orville G. Brim, Jr., Glen Elder, John Modell, Ross D. Parke, and Peter N. Stearns. In subsequent meetings, our group expanded to include Joan Jacobs Brumberg, Tamara Hareven, Jay Mechling, William Kessen, Steven Schlossman, Sheldon H. White, Viviana A. Zelizer, Michael Zuckerman, and other developmentalists and historians; in 1993, the group's efforts were published as a book entitled *Children in Time and Place: The Intersection of Developmental and Historical Perspectives*. I especially want to thank Glen Elder and John Modell for helping me to conceptualize the chapter I contributed.

Many individuals have assisted me, and they have done so in a variety of ways. I offer thanks to Sandra L. Albrecht, Karen Tucker Anderson, Nanette C. Auerhahn, the late Roger Barker, Surendra Bhana, John Bodnar, Joseph Boskin, Sue Brand, Urie Bronfenbrenner, Hopeton Brown, W. Gregg Buckner, Gene A. Budig, Peter N. Carroll, Lyn K. Carlsmith, John A. Clausen, Alan Clive, Hamilton Cravens, Thomas Cripps, Ralph L. Crowder, Susan Tuttle Darrow, Dennis E. Domer, Lynn Dumenil, Randy Edmonds, Scott Ellsworth, David Farber, Robert Fyne, Linda Gordon, Richard Gwin, Al Habegger, F. Allan Hanson, Louise M. Hanson, Tamara Hareven, Susan Hartmann, Norma Jean Hemphill, N. Ray Hiner, Jim Hoy, Sheila B. Kamerman, Sharyn Brooks Katzman, Michael E. Lamb, Annette Lawson, Judy Barrett Litoff, Sharon R. Lowenstein, Arthur F. McClure, David Macleod, J. Fred MacDonald, Elaine Tyler May, Renee D. Melchiorre, Robert M. Mennel, Sonya Michel, Sarah Minden, the late Theodore M. Newcomb, Norma Norman, Jill Quadagno, Saralinda Rhodes, Debra Rhoton, Beth Scalet, Howard Shorr, Ann Schofield, Judith Sealander, Arlene Skolnick, David Clayton Smith, Judith Stein, Gary Tebor, David Thomas, Beatrice A. Wright, the late Herbert F. Wright, Lorene Wright, Lawrence S. Wrightsman, Jr., Norman R. Yetman, and Edward F. Zigler.

I also am deeply grateful to my friends who have read all or substantial portions of this book in manuscript: Beth L. Bailey, William Chafe, James Gilbert, David Katzman, Burton W. Peretti, and Marilyn Yalom. Their sensitive critiques guided me as I made substantial revisions in the final drafts of the manuscript. Sheldon Meyer of Oxford University Press also made important suggestions for the final version. More than that, he was patient and encouraging even as my concept for this book became increasingly ambitious, causing my original deadline itself to become history.

I owe a great deal to these friends. Friendship often has a political as well as an emotional dimension. Hannah Arendt has written that "for the Greeks the essence of friendship consisted in discourse. They held that only the constant interchange of talk united citizens in a *polis*. In discourse the political importance of friendship, and the humaneness peculiar to it, were made manifest.... However much we are affected by the things of the world, however deeply they may stir and stimulate us, they become human for us only when we can discuss them with our fellows. . . . We humanize what is going on in the world and in ourselves only by speaking of it, and in the course of speaking of it we learn to be human."[6]

At one time or another, my children have all helped me with this book; I want to thank William M. Tuttle III, Catharine D. Tuttle, and Andrew S. Tuttle. When they were little children, I knew next to nothing about fathering; I know more now, but they have grown up and gone.

Finally, I want to thank three people—and one dog, sweet Sam—who have buoyed my spirits throughout the writing of this book. First, David Manners Katzman, who is not only brilliant, but ethical and caring as well; he has been my steadiest intellectual and professional advocate. I cannot imagine a better friend. Second, Kathryn Nemeth Kretschmer, whom I first met in 1970. A beautiful red-haired woman from a small Kansas town, Kathy sat quietly in the back row and got the top grade in the class. We met again in 1987, and she has enriched my life ever since. Last, in memory of my mother Geneva Duvall Tuttle, who died last night at the age of eighty-three. Mother kept our family together while my father was overseas during the war; she never let us down. I adored her. To her and to my father, I lovingly dedicate this book.

Lawrence, Kansas   W.M.T.
March 10, 1992

CONTENTS

The childhood shows the man
As morning shows the day.

—John Milton,
Paradise Regained (1670)

It became difficult to remember a time
when we were not at war.
It seemed that it has always been with us
and would always be with us.

—Barbara Wells, born in 1934

The past is never dead. It's not even past.

—Gavin Stevens
in William Faulkner,
Intruder in the Dust (1948)

"Daddy's Gone to War"

CHAPTER 1

Pearl Harbor: Fears and Nightmares

S UNDAY, DECEMBER 7, 1941, began as it usually did for eleven-year-
old Jackie Smith: with church. Standing in front of her family's quarters
at the Pearl Harbor naval base, she was waiting to walk with her family to
the base chapel. Jackie noticed airplanes flying overhead, but airplanes were nothing unusual around the Army and Navy installations on Oahu. She thought she
was witnessing a simulated dogfight between American airplanes, but then "all
of a sudden flames were shooting up," and it "looked like the whole island was
on fire." Her father ran inside to telephone the base, but the operator told him to
get off the line because Pearl Harbor was under attack. By then, airplanes were
flying at treetop level; Jackie could see the rising sun insignia on the wings. The
family dashed inside and hid under tables. Jackie ran upstairs with her father to
grab the mattresses off the beds to provide other hiding places from the bombs
and machine-gun bullets. Through the second-story window, she saw not only
the airplanes but even the faces of the pilots. "I could almost touch them, it
seemed."[1]

Pearl Harbor, air raid drills, blackouts, and nightmares mark the memories of
America's homefront children. Numerous men and women who were children
then remember precisely where they were and what they were doing on December
7, when they heard the news. For them, time stopped at that moment in what
psychologists called flashbulb memory, the freeze-framing of an exceptionally
emotional event down to the most incidental detail. The indelible memory—for
these homefront children as for their elders—was of that precise moment when
radio reports of the Japanese attack at Pearl Harbor stunned the nation into
silence.

Stephanie Carlson, ten years old, also was close to the action that day. Living
on Oahu, she remembered Pearl Harbor in a poem written soon after the attack:

> Dec. 7 started like any other quiet Sunday in Hawaii. . . .
> Then things began to happen.

> We heard the rumble of guns firing.
> There were black airplanes in the sky.
> We thought it was a drill. . . .
> The telephone rang.
> My father answered it.
> He was called to his destroyer because the Japanese were
>   attacking Pearl Harbor.
> We could stand on the porch and see some of the fighting.
> It looked like a Fourth of July celebration. . . .
> We saw three Japanese airplanes fly right over our house.
> They dropped two bombs which just whistled down.
> It looked just like it was going to fall on you. . . .[2]

While most children remained in Hawaii, Stephanie Carlson and other military dependents boarded ships and left for safety on the mainland. As they headed to the United States, the children sang a song they had just learned: "Let's remember Pearl Harbor as we go to meet the foe. Let's remember Pearl Harbor as we do the Alamo."[3]

What began at Pearl Harbor not only reverberated throughout the United States, but also forever changed the lives of America's homefront children. Most of these girls and boys lived thousands of miles from Pearl Harbor, but on December 7, as they witnessed their parents' fearful reactions, they knew the Japanese attack portended grave danger. Deeply affected were those children who had siblings stationed at Pearl Harbor. Barbara Wells, a seven-year-old, lived in Camp Wood, Texas; her older brother Dub, age eighteen, was in the Navy stationed at Pearl Harbor. Her mother was in "a state of shock," listening to the radio. It was several days before the family learned that Dub was safe. "While we were still awaiting word from him," Barbara recalled, "his birthday presents arrived for Mama's December 18th birthday." In the package were a gold-and-black striped fountain pen and a pearl stationery box. "I remember how she cried when she opened the package, not knowing if that would be the last present from her adored oldest son."[4]

Fathers also cried in the aftermath of Pearl Harbor, and the children were stunned to witness such emotion from these men. A ten-year-old boy entered the living room to see his father sitting in front of the floor-model RCA radio, "its single green tuning eye beaming out in the darkened room. Dad was bent over, his head in his hands . . . his shoulders were faintly shaking as the announcer rattled on. . . . I stood there for a while, feeling shaken; it was the first time I had seen my father cry."[5]

On December 7, girls and boys in the United States witnessed various kinds of adult behavior they had not seen before. Enraged mothers and fathers shouted and screamed, and some even struck their children. All day long, they talked about "war" and its ugly and tragic possibilities. One ten-year-old girl could take it no more. "War, war, war! That's all I hear. I'm tired of hearing about war." At that point, her mother slapped her across the mouth. Another girl, nine-year-old Patty Neal, was sitting with her family around the radio when the news arrived. Patty was "chattering away," her mother asked her to be quiet, but, Patty recalled, "I kept on talking, and my mother, who NEVER spanked me, slapped me, and said, 'Patty, you will remember this day.'"[6]

Children listened as their fathers and other men swore and raged at the Japanese. Racism fed the stereotypes that portrayed the Japanese as duplicitous plotters, hiding behind steel-rimmed glasses and toothy grins. And Americans everywhere vowed to avenge the sneak attack. "Why those dirty sons-a-bitches," screamed one man. Another man, deep in drink, repeated over and over, "I'm gonna get me a machine gun and kill every one of those slant eyed sons-of-bitches I can find."[7]

Whatever untoward behavior the children observed that memorable day, the effect was generally the same: it deepened their fears. Marian Hickman was six years old; December 7 was her uncle's birthday, and there was a party at her St. Louis home. Including grandparents, aunts, uncles, and cousins, about twenty-five people were gathered there. After a delicious and bountiful meal, the men played Rook in the living room; the women chatted in the kitchen and washed the dishes. The radio was on. Suddenly everything stopped. "No more happy talk in the kitchen. The card game was never finished." And Marian vividly remembered the "fear that went through that room. . . . The adults whom I adored and trusted were afraid! If they were afraid, how could I be safe and protected? It was overwhelming. . . ."[8]

The next day in grade schools across the country, girls and boys gathered to talk about the war and to comment on their parents' behavior. "Fear of what was to come permeated the air," recalled one of the homefront children. Some were afraid they would be strafed by airplanes or killed by bombs on the way to school. Many gathered at all-school assemblies to hear President Franklin D. Roosevelt's nationwide radio address denouncing the Japanese for having made December 7 a "day of infamy." A mother wrote the Children's Bureau for advice, explaining that in her daughter's class after Pearl Harbor, the radio was "in constant use" for three days, and it was "more than her nerves could stand. . . ." Naturally, the children had many questions: "Would the Japs come over here and bomb us next?" "Where would they bomb if they came?" "Would they come *even before Christmas?*" "What are we going to do?" asked a ten-year-old girl in rural Indiana. "I am afraid."[9]

Throughout the United States following Pearl Harbor, children feared that enemy bombs might rain down on them too. Their anxieties deepened as they participated in air raid drills and blackouts. Some, afraid of bombs, ran to and from school. After running home at 3:30 each day, the excited first words of a second-grader to her mother were always the same: "We didn't get bombed today." Other children ran at the urging of panicked adults. An eight-year-old girl at P.S. 18 in Yonkers was "let out of school." She started to run home, her heart pounded, and she was "scared to death ... sure a bomb would get us."[10]

Reportedly, coastal children felt more vulnerable to bombing raids than did children living inland. "I have promised my boys and girl," explained a Baltimore mother, "that I will take them to their grandmother in the Middle West, where we will be safe from seacoast bombings." But it was evident to midwestern parents and teachers that their children also felt deeply endangered. In the aftermath of Pearl Harbor, a grade-school boy in Janesville, Wisconsin, dreaded the nights with their blackouts, for they were "punctuated with fear. *All* the lights had to be turned out when the siren went off." The boy's terror was so real that he imagined he could hear the airplanes "in the darkness overhead. . . . We thought about being bombed. We thought about dying. We thought about losing each other."[11]

Dr. Joseph C. Solomon, a San Francisco psychiatrist who studied children's behavior during the blackouts, observed that "the abrupt entry" of America into war "acted as a source of sudden and unexpected danger." In San Francisco, children's fears were compounded by the sirens that blasted on the night of the American war declaration. Some children vomited, others cried. After New York City's first air raid alarm, a boy who had been trying to overcome difficulties with stuttering began to have "many dreams in which he tried to escape bombing by making off on his bicycle," and his stuttering grew worse.[12]

Children experienced deep fears during air raid drills and blackouts. Sirens screamed both during the school day and at night. During the blackouts, pitch darkness was required. Heavy black or dark green drapes or black tarpaper covered all windows; no illumination at all, including flashlights or matches, was permitted. Air raid wardens in Civil Defense helmets knocked on doors to chastise violators.[13]

Children did not need to be warned, however. One reason was that many boys and girls did not really understand that these were drills and not actual attacks; these children tended to interpret warnings literally. After an air raid warden had announced at one house that he could see the glow of a flashlight, a four-year-old girl inside was "frightened to death" that because of this infraction the enemy would bomb their home. Another little girl living in Eureka, California, was afraid that "light would seep out" and the entire town would be shelled by Japanese submarines. "I lived in fear," she said, "that I might inadvertently touch a drape and cause it to shift."[14]

Adding to children's anxieties were the special responsibilities that some took on during the blackouts. A ten-year-old girl in upstate New York cared for her Italian-born grandparents—a grandfather paralyzed on one side, a grandmother with a weak heart. Neither spoke English. "When those sirens screeched, I would pray silently that it was just a drill again." If not, she would have to take her ailing grandparents from their second-floor flat to the shelter in the basement. "Quietly

... I tried to comfort these two gentle people in Italian, reassuring them ... while deep inside I was terrified myself."[15]

Parents, of course, had their duties during the air raids. Many fathers and mothers served as air raid wardens. Although children took pride in their parents' participation, the absence of a father or mother during the blackouts heightened their fears. As a homefront girl from Wausau, Wisconsin, remembered: "When Dad, wearing his warden arm band, exited the house to make his patrol, the last lit lamp was turned off. Eerie darkness reigned everywhere. Even the street lamps were dark. Howling sirens screamed a cruel warning. The whole scene filled my imagination of a real attack. I was very frightened and cried."[16]

Children also panicked during air raid drills in the schools. One girl, hearing the sirens, marched with her classmates to the basement and sat against the wall. "Right at that time, I heard an airplane and started to scream: I knew for sure we were going to be bombed." Indeed, even though these drills occurred during daylight and not darkness, the terror could be greater, for at school the children were separated from their mothers and fathers. "I was always afraid," confessed a homefront girl, "that we would get bombed when I was at school and not at home with my family. That terror stuck with me for many, many years."[17]

School officials tended to be inattentive to the children's emotional distress. The model for civil defense in the schools was a twenty-four-page manual issued by New York City's board of education and reprinted by the National Education Association, which sent it to superintendents throughout the country. The manual did not address the children's fears; its main concern was the school building itself, and how to prevent "fires, breaks in gas and water lines, and interruptions in communications...." During the actual drills, children either went out into the hallway or down to the basement and sat against the wall, or they stayed in the classroom and crouched under their desks. Girls had another imperative to follow as well; even in the midst of the danger, they had to be ladylike. "At school when the alarm sounded," a homefront girl recalled, "we filed into the halls and squatted near the walls, the girls of course making sure their skirts were tucked around them modestly...." Sometimes the children played games like "Twenty Questions" and "Who Am I?" A different set of rules governed the responses of children who were playing in the schoolyard when the sirens wailed. School children in Detroit were told to lie in the street next to the curb.[18]

Preschoolers also were fearful. Some little children, though not really comprehending what was happening, experienced the terror of air raid drills and blackouts almost by osmosis. They felt vulnerable, even if still in highchairs or playpens. Born in 1941, Gerry Lunderville remembered being at home and hearing "the sound of the air raid drills. That eerie shrill will forever be imbedded in my mind. The darkness was foreboding. I may not have been aware that there was a war going on, but I certainly knew there was danger lurking...."[19]

Some preschoolers, however, seemed inspired by the sirens. In nursery schools, simulated air raids became a center of playground activity. According to teachers surveyed for a 1942 article in *Mental Health* magazine, "the artificial, yet realistic whistling of bombs, the falling down dead, and the imitated drone of bombers have effectively taken the place of the hearty 'Hi-ho, Silver.'" Significantly too, there were school-age children who enjoyed the excitement of the air raid drills. A homefront girl in Detroit recalled enjoying the blackouts so long as she did not think about "the real reason" for having them. When "there would be a siren, we would turn off every light in the house, go out on the porch, and see nothing but stars, a moon, and searchlights. We loved it. . . ."[20]

In almost every case of a child's lack of fear, the reason was the same: Mother. One girl born in 1940 would join her mother under the dining-room table; her mother would bring a flashlight and a book. "I looked forward to those special times when she would read to me under the table." Other mothers led their children in song: "My sister and I enjoyed that and thought blackouts were so much fun!" recalled a homefront girl living near Los Angeles. Fathers helped too, but seemingly far less often. Four sisters shared a bedroom in Waukesha, Wisconsin. One of the girls remembered that "we'd be scared to death" when the siren would sound. "My dad would come into our room and sit between our beds and calmly tell us stories of growing up on the farm . . . stories which we still remember vividly. He was very comforting and reassuring to us."[21]

Children living near centers of defense production felt a special vulnerability. Girls and boys in the San Francisco Bay Area had to wear I.D. tags stamped with numbers corresponding to those on lists held by Civil Defense, in case they became evacuees or casualties. Children in Washington, D.C., wore copper "dog tags," and those in New York City were not only given identification numbers and tags, but were fingerprinted as well. Educators and doctors, as well as such organizations as the Children's Bureau and the National Congress of Parents and Teachers, advocated the establishment of a national program of registration and identification. Children in war-boom areas also were given tetanus shots and inoculations. Some communities had special requirements; because of the nearby chemical plants, children in Richmond, California, put Kleenex balls in their mouths for protection during air raid drills. Finally, families devised their own evacuation plans. One girl's parents "talked very frankly to us about the severity of our location—Detroit." The family had a plan "in which we would pitch in and help Mom load the car and head up to my grandmother's," a hundred miles away. "This 'plan' was discussed frequently and was a real part of our lives."[22]

In Hawaii, where the fear of another Japanese attack persisted, children received gas masks and had periodic tear-gas tests. Richard Chalmers, who was eleven when Pearl Harbor was attacked, recalled that an Army officer would close off a classroom, fill it with tear gas, and check the fit of the masks by walking the students through the room. "Usually, just as we got ready to leave the room, they would ask us to open our gas masks so we would know what the smell of tear gas was like," and their eyes would immediately fill with tears.[23]

Surprisingly, children living in rural areas often feared enemy attack as much as those living in defense-production areas. In fact, the blackout experience of a four-year-old girl living on a farm in remote northeastern Kansas sounded very much like that of her urban counterparts: "We turned out the lights, pulled the shades and listened intently to the old ornate radio. . . . I expected to hear planes overhead any minute." Another girl, Arvenia Welch, born in 1935, lived in Appalachia. Her father, an itinerant worker who took his family with him to jobs in southeastern Kentucky, southwestern Virginia, and northeastern Tennessee, was very serious about observing blackouts and had his family hang dark shades over the windows. "Looking back," Arvenia mused, "it seems ridiculous to be so cautious considering we lived in the hills and used kerosene lamps for light. . . ."[24]

A tragedy in May 1945 proved that some rural children were in danger. Indeed, it was in the countryside that the only mainland American children died as a result of enemy action. On May 5, 1945, six Americans, five of them children between the ages of eleven and fourteen, lost their lives in the explosion of a Japanese bomb in the uplands of south-central Oregon.

This tragedy occurred at the end of the war, when Germany was collapsing and Japan was struggling to stave off defeat. Beginning in late 1944, Japan had released 6000 balloons armed with bombs, expecting them to ride the S-shaped "jet stream" that swings northeastward and then southeast and finally eastward across the United States and Canada. Once over enemy territory, built-in timers would cause the balloons to drop and the bombs to explode, thus igniting forest fires and causing other destruction. At speeds of up to 300 miles per hour, the bombs floated from Japan thousands of miles across the Pacific toward unsuspecting residents in North America in little more than a day.[25]

Saturday morning, May 5, was sunny and clear in Bly, Oregon, a lumbering and ranching community of 750 people at the foot of Gearhart Mountain. This was such a beautiful spring morning that Elsye Mitchell, five months' pregnant, changed her mind and decided to join her husband Archie, a local pastor, and several Sunday school children on a fishing outing. And, she noted in her diary, she even "baked a *chocolate cake* for the occasion." The children, four boys and a girl, were Sherman Shoemaker, age eleven, Jay Gifford, Edward Engen, and Joan Patzke, all thirteen, and Joan's brother Dick Patzke, age fourteen. Archie loaded his 1931 sedan with fishing equipment, picnic lunches, the cake, the five children, and his wife. It would be a wonderful day, he thought. But he had not anticipated the rough and slippery roads they encountered as they motored through the Ponderosa pine forest. At an especially muddy spot, where the road dipped toward a creek, Archie braked the car to a halt. Fifty yards ahead of him was a road crew that had just extricated a grader from the mudhole where it had been stuck. Archie asked the men about both road and fishing conditions; the road was impassable, they replied, and the creek was probably too muddy for fishing. Meanwhile, the rough ride had made Elsye slightly carsick; she was thus happy to get out of the car to take the children on an inspection tour of the creek. It was about 10:20 a.m. Archie had started the car, and as he began to move it to higher ground, he heard Elsye call, "Look what I found, dear." Stopping the car, he shouted back, "Just a minute and I'll come and look at it."[26]

Richard R. Barnhouse, foreman of the road crew, had started the grader and was following Archie up the hill. He stopped too. From the high seat of the grader, he could see Elsye and the children about 100 yards away in the woods staring at something. He could not tell what it was. "As Mr. Mitchell stopped his car," Barnhouse wrote later in his official statement, "there was a terrible explosion. Twigs flew through the air, pine needles began to fall, dead branches and dust and dead logs went up." Stunned, the four unhurt men rushed down the hill just as another bomb, a smaller one, exploded. They were not injured, but the two bombs together seemed to eliminate the possibility that Elsye or any of the children could survive. The mangled bodies of the four boys were scattered around the bomb crater. Joan Patzke lived through the immediate blasts, but died later. Elsye too was still alive, but her clothing was on fire, and she died while Archie was struggling to smother the flames with his bare hands.[27]

A bomb disposal expert, who located and disarmed other unexploded bombs lying nearby, guessed that somebody had kicked the bomb. The tragedy, of course, devastated the townspeople. Grief-stricken, Archie Mitchell accompanied his wife's body to her hometown. A mass funeral for the four boys, attended by 450 people, was held in the Klamath Temple, in the neighboring town of Klamath Falls, because their local church seated only 150. Four caskets lined the front of the sanctuary, surrounded by vast numbers of spring flowers, and the Temple was crowded with students from the Bly elementary and high schools. The Bly Boy Scouts, fellow troopers of the dead boys, served as honorary pallbearers. Three of the children were buried under one headstone, which read: "All Killed by Enemy Balloon Bomb." Another child's grave marker stated: "With Five Companions Killed by Enemy Bomb." Surprisingly little news about the explosion reached the public. To avoid panic, federal government officials prevailed upon the local coroner to conclude, "The cause of death, in my opinion, was from an explosion of undetermined source." The people of Bly were, after all, patriots who would seal their lips, should their government so suggest, as part of their contribution to the war effort.[28]

The homefront children's fears and anxieties were stimulated not only by air raid drills and blackouts, but also by rumors of invading enemy troops, by newsreels, and even by "suspicious-looking people." Some children had nightmares so frightening they were still vivid years later. One mother confessed that her five-year-old daughter had been "terrified ever since war was declared." Not only did she ask "anxiously if every airplane she heard might drop bombs," but she also had "dreams and night terrors that her baby brother has been killed, that the enemy are coming right into the house." This mother forbade all war talk in the house, including news broadcasts, "but it hasn't helped much."[29]

A larger group of children also suffered terrifying nightmares, but they did so alone, not knowing how to share such frightening feelings with their teachers and classmates, or with their parents and siblings. In many of these dreams, the enemy was invading the children's homes. One girl frequently dreamed that Japanese soldiers "would appear while I was hiding in either an old outhouse . . . or in the basement of our house. There would be many soldiers and they would always find our hiding place." In some dreams, the children could not escape; one girl's nightmare was that "Japanese soldiers were coming through the woods with bayonets in attack positions and I couldn't run."[30]

Sometimes it was newsreels from the war already raging in Asia and Europe that shaped the nightmares. Marlene Larson was sitting on the front porch of the family home in Whittier, California, when her mother flung open the door to announce that Japan had attacked the United States. "I remember being so frightened. All the war scenes from the newsreels . . . came across my mind and I felt sick." Her mother also announced that "there would be changes in our lives— Daddy was going to re-enlist the next morning—Everything was so confusing. . . . I just stood there feeling sick." Faced with this upset, Marlene could not believe it when her mother told her to go to the movies. "I don't remember the movies showing—I just remember the newsreels . . . the bombed homes, the sick and dying people—children with no homes and no families. I cried." That night began a stretch of "sleepless nights and terrifying dreams" that lasted for years.[31]

Gay Hollis, born in 1940, lived with her grandparents on a Texas farm. Perhaps provoked by her aunt's certainty that the war was proof of Biblical prophecy, Gay recurrently dreamed of a bomb exploding in a field and "the fire spreading toward the house with the intent of destroying me. For years," she recalled, "I dreaded going to sleep because I knew I was going to have to run from that fire all night— it never caught me." Pat Campbell was ten when the Japanese bombed Pearl Harbor, and while she believed the war had not disrupted her life, "I . . . remember a recurring dream which probably was evidence that I was 'bothered' more by the war than I . . . knew—the dream of standing alone . . . and suddenly realizing that we were about to be 'strafed' or bombed by a squadron of planes (usually Japanese)."[32]

The children also dreamed about the German enemy. Diane Peoples, born in 1936 and living in rural North Carolina, recalled a nightmare that was "so vivid I can still see it—involving a German plane landing on our lawn and our house being on fire." Born in 1933, Barbara Sackman lived in a Polish neighborhood in Toledo, Ohio. Her recurrent dream was of "Nazi tanks coming down our neighborhood main street . . . shooting and running over my friends." Janice Bohnke, born in 1935 and living in Selma, Alabama, dreamed that she and the neighbor children "were walking to school and enemy soldiers blocked off the street in front of us. We ran back and they blocked off the other end of the street and kept converging from each end of the street until we were cornered, scared and helpless." Adolf Hitler himself was a source of nightmares. Charlotte Anderson, a girl in Coffeyville, Kansas, was delirious with pneumonia when "I thought I was having my hands smashed by Hitler.... Instead of ghosts and witches, my generation of kids had nightmares about Hitler and Hirohito."[33]

Some of the children's most frightening dreams were those in which the entire family was imperiled. One girl had "a terrible nightmare about our family being hurt and separated." She, her mother, and her sister had all been wounded and "placed in multiple layer bunks that lined both sides of a bomb shelter.... Daddy was hurt," too, but "all of the bunks near us were full of people upset and crying so he did not have a place to stay near us unless someone died to free up a bunk." In another child's dream, the "family sat huddled together in the kitchen, looking out into the blackness of night as an army of tanks rolled into our front yard." Most children awoke before death struck down them or members of their families. Still, some dreaded nightfall. "Lots of nights," recalled a girl living in Anaheim, California, "I didn't sleep." She would lie there thinking of her family being killed by falling bombs or invading troops, and so she made a plan: "I was sure I'd kill myself if war [threatened] my home."[34]

America's homefront children contracted anxiety from their parents and teachers as well as from each other. "In every situation," observed Dr. Joseph Solomon, "where the parent ... showed evidence of fear or panic the children reacted in a similar manner, usually to an exaggerated degree." Some children suffered from something quite different—that is, parental unconcern with their anxiety. Carol Vealey lived with her family in a two-story house in northern New Jersey. Like other little girls, she had to go to bed early. "When the siren went off I was scared-to-death. I remember standing at the top of the stairs crying very hard and wanting to come downstairs with everyone else, but my mother would say go back to bed, it will be over soon. Then I would lay there waiting for something to happen." Likewise, teachers spread panic. On the third day after Pearl Harbor, reported Dr. Solomon, "a whole group of extremely frightened school children were seen running home from school in the middle of the day without hats or coats." Their teacher, having heard the siren of a passing ambulance, had "sent the children scurrying home in the face of her own personal panic." On the East Coast, too, school guidance counselors recognized that children took cues from their elders. "Misconceptions with regard to children's reactions," wrote a psychiatrist with New York City's board of education, "appear to be due mainly to adults' displacement of their own types of anxiety upon the children." While parents and teachers generated much of the panic that children felt, the fears produced were terrifying nevertheless.[35]

In the wake of Pearl Harbor, a debate erupted involving parents, teachers, and child-care experts. The issue was whether to follow the British example and evacuate children from coastal cities to the countryside. Yet most parents, while concerned for their children's mental and physical well-being, echoed the opinion of a mother in Wilmington, Delaware, a defense-production area. "Whatever the chances are," she told her two young daughters, "we are going to face them, unless we are ordered to go." Her husband was a war worker, and, she announced, "we are not going to desert him."[36]

The most compelling arguments against evacuating American children came from England itself. According to studies done by Anna Freud, Dorothy Burlingham, Susan Isaacs, John Bowlby, and other psychiatrists and psychologists, the uprooted English children were suffering serious emotional upset. Sudden removal from the home and separation from the family produced depression, insecurity, and anger. A. T. Alcock, who studied children taken from London and the East Coast to a reception area in Huntingdonshire, found a sharp increase in bed-wetting, not to mention "anxiety, depression, fecal incontinence, epileptiform fits, [and] sleepwalking."[37]

Freud and Burlingham particularly discouraged the United States from adopting England's policy of evacuation. They recorded sad stories, such as that of a four-year-old girl who sat for several days in the exact spot where her mother had left her. Refusing to speak, eat, or play, the girl had to be moved about like a rag doll. Freud and Burlingham conceded that their "own feelings revolt against the idea of infants living under the conditions of air raid danger...." Still, "for the children themselves, during days and weeks of homesickness, this is a state of bliss to which they all desire to return." Among the girls and boys who remained in the cities during the blitz, they found no similar symptoms of nervous distress. "Love for the parents is so great," observed Anna Freud, "that it is a far greater shock for a child to be suddenly separated from its mother than to have a house collapse on top of him."[38]

These English studies were well known among American medical experts months before the Pearl Harbor attack. Dr. Martha M. Eliot, deputy chief of the United States Children's Bureau, visited England in early 1941 and returned home to warn against any efforts to emulate the British policy. And one could not quibble with experts like Dr. J. Louise Despert, an American psychiatrist, who concluded in 1942: "In summarizing the English literature, the impression is gained that if the British had to do it over again there might not be any evacuation."[39]

In the absence of evacuation, however, American officials had to figure out how to protect urban children remaining with their families. Advice on air raids was not long in coming. Invariably, the first warning was to avoid panic. In early 1942 the American Committee on Maternal Welfare discouraged all talk of air raids in front of children, "no matter how young." Moreover, all advice concerning babies was directed at the mothers. Upon hearing the siren, she should take the baby to the nearest shelter, or to the safest room in the house, to a closet, or under the stairs, a table, or a bed, "so that he may be protected from flying debris. . . ." She should also cover the baby and block his ears with cotton balls to absorb the concussions. Finally, the mother should have with her a bag packed with diapers, a pillow, several baby blankets, a toy, a flashlight, freshly boiled water, baby bottles and formula, and sterilized nipples.[40]

Some people, including even the nation's first family, had unusual notions about how to prevent "fright psychosis" among American children. A week after Pearl Harbor, Eleanor Roosevelt explained that her son John, a naval ensign stationed in San Diego, was teaching his two-year-old son "to say 'booooom' every time he hears a loud explosion from practice firing at the naval base. Now," she explained, "the child thinks he is creating the explosion and is quite delighted every time he hears one. By this method the child will not be frightened if there is a real bombing."[41]

As the weeks lengthened after Pearl Harbor, boys' and girls' fears of air bombings subsided. Afterward, they rose only sporadically, as in June 1944 when the Germans began their V-rocket attacks on England. American newspapers explained how the "infernal machines" flew, advised that one should "duck" whenever one heard the V-rocket's "whizzing" overhead, and stated that these devices were just one more example of German brutality, since the Germans had no military objective, but rather were killing innocent women and children simply for revenge. "Hitler's new device," said a front-page story in the *New York Times*, "resembles an enlarged steel model of a child's fourth of July rocket, equipped with wings and explosives."[42]

Fifty years later, anxieties generated by the war are still with some of the homefront children. One is the fear of airplane sounds. Marian Hickman remembered once loving to watch airplanes. With the outbreak of war, however, "airplanes became a great source of anxiety to me as I began to realize they were a big part of the war. Each time I would hear a plane, I'd wonder if something terrible was going to happen."[43]

Marian Hickman, who was six when Pearl Harbor changed her life, has retained vivid memories of the war years. In fact, most of the homefront children born in the 1930s have flashbulb memories of December 7, 1941, when they heard the news. For the younger girls and boys, however—that is, for those born during the war years—Pearl Harbor was an event to be learned about only after the fact. This is not to say that these little children were unaffected by the Second World War—far from it. But for them it was a later national tragedy that would provide the flashbulb memories: the assassination of President John F. Kennedy on November 22, 1963.

So far, this book has focused on the girls and boys who had reached the age of consciousness by December 7, 1941. Now it is time to introduce the rest of America's homefront children, those who were born during the war years. Moreover, when placed in juxtaposition with the older homefront children, the experiences of those who were babies, infants, and toddlers during the war validate the contention that age is a significant historical variable. Most students of human development agree that age or stage of development—whether called oral, anal, oedipal, or latency, or infancy, early childhood, preschool, or school age—is a crucial factor mediating the effects of one's psychosocial environment. In short, what is important is not only *what* happens in a child's life but also *when* in the child's life—that is, at what age—the event occurs.

Certainly this was the case for the American homefront. With this thesis in mind, it is time to welcome the "war babies" to the homefront.

CHAPTER 2

Depression Children and War Babies

AR WAS ON THE MINDS of the homefront children. "I was born
September 3, 1939," wrote one girl. "Some date, right?" Germany
had just invaded Poland, thus triggering the Second World War.
"For me," she continued, "there was always War until I was six years old." But
the effects of the war were variable, depending upon numerous factors, one of
the most important being chronological age. Clearly, children's perceptions of
what the war was varied from one age to the next.[1]

An important book, based on observations made by children during the war,
was published in 1946. Entitled *The Child from Five to Ten*, the book's authors,
Arnold Gesell, Frances L. Ilg, and their colleagues, asked children to give their
perceptions of "war." Before age five, the children had little comprehension of
the subject, but at five and six they devised plans for ending the war: "Get up
behind Hitler when he isn't looking and shoot him," suggested a six-year-old girl.
It "would be fine," she explained, "if there were millions of Americans on one
side, and only one Jap on the other side. That Jap could be Hitler. Then we would
say [cooing tones], 'Come over here.... We won't hurt you.' Then when he got
here—Bang! We'd blow him all to pieces. No more Hitler."[2]

Beginning at age seven, children talked more openly about their worries; fearing that spies were living in their midst, they had nightmares about the war. One
girl dreamed that a spy asked her: "Do you like your government?" At age eight
and nine, however, these fears subsided, and factual questions began to arise: Why
did the war start? What was the difference between democracy and fascism? Children also became interested in war movies and comic books, and increasingly
they participated in the war effort through scrap-collection drives and by purchasing war bonds. They also played war games. Although psychologists advised
parents to furnish outlets for children's fears, one mother expressed concern that
her eleven-year-old son, far from being upset, "seems delighted at the war.... He
and his friends talk constantly about going out and 'blasting' Hitler. 'Just my

luck,' he says, 'to live where there's no fighting. Boy! Would I love to have an air raid!'"[3]

An age-specific perspective helps to sort out the history of children's wartime experiences. "The imprint of history," writes the sociologist Glen H. Elder, Jr., "is one of the most neglected facts in [human] development." And what constitutes that history? As Elder has explained, "Lives are shaped by the settings in which they are lived and by the timing of encounters with historical forces, whether depression or prosperity, peace or war." These "settings" include all the significant environmental factors, such as the family, neighborhood, and school, not to mention gender, race, socioeconomic class, ethnicity, religion, and region. "Timing" is the second crucial factor. In other words, was the child in infancy when a historical event occurred, or was she or he in early childhood, preschool, or school age? Using this perspective, one can appreciate why children's development during the Great Depression was distinguishable from that experienced during the homefront mobilization of the Second World War, or, later, during the generally prosperous baby boom era, during which television arrived to bring monumental changes.[4]

Throughout the life course, from adolescence to old age, these factors continue to shape individual development. Indeed, Elder has added, the life course itself is an example of "the timing of encounters with historical forces," since the life course refers to "pathways through the age differentiated life span, to social patterns in the timing, duration, spacing, and order of events; the timing of an event may be as consequential for life experience as whether the event occurs and the degree or type of change." As the following chapters argue, children's encounters with wartime changes had variable effects depending upon the setting, the child's age, and the timing of events. And as these chapters show, these events were often monumental.[5]

Finally, what underscores the significance of age in this context is that the homefront children's cohort can be split in two. According to sociologist Norman B. Ryder, "A cohort may be defined as the aggregate of individuals (within some population definition) who experienced the same event within the same time interval." Thus, the boys and girls who were born between 1932 or 1933 and 1945 and were still preadolescents when the war ended form such a group. Though affected in many ways by the war, they were still largely innocents in 1945.[6]

On the other hand, much separated the homefront experiences of the girl or boy born during the baby bust of 1933 from those of the child born during the baby boom of 1943. It is clear, in fact, that—with so much dividing the two halves of this cohort—it truly consists of two distinct "intragroup cohorts." During the 1930s there was a decline in babies born in the United States; between 1941 and 1945, there was a surfeit of newborns. The first group was born into an economy—and mentality—of scarcity; the second emerged during a period of full employment and victory. During the war, the first group of children was in grade school; the second group was in nursery school or still in playpens. Indeed, considering the contrast between childhood in the 1930s and the first half of the 1940s, America's homefront children were split not only into intragroup cohorts, but even between two generations, divided by the Pearl Harbor attack of late 1941.

Two questions frequently asked during the war were: "Should we get married?" and "Should we have a baby?" In other words, the history of the "war babies" is, first of all, a history of decisions. This chapter thus focuses on the women and the men and the decisions they reached, decisions that not only resulted in war-time marriage and baby booms but also reflected the widespread and intense longing Americans at war had for domestic life.

"In 1940, there was the threat of war," explained a young woman named Nancy. "All talk was of war. I had been going with George for nearly four years. We had delayed marriage until he could finish his education, which he did in the spring of 1940." With passage of the Selective Training and Service Act, "men were being drafted rapidly and the war news was alarming." Some of their relatives opposed their getting married, but they decided to go ahead anyway. "All went well for some time but the war pressure became too heavy. Many of our friends had already gone. We could make no future plans at all. . . ." For one thing, Selective Service began to draft married men. Fathers were still exempt, but this would change too. In 1943, expecting to be called at any time, George enlisted in the Army. Just after he had departed for service, Nancy learned she was pregnant. "George and I saw each other twice in the next eight months, when he was sent overseas. Our son was born after he left." As Nancy remembered, there was a sad but not unusual addendum to this couple's wartime history: "George came home after an absence of more than two years; we were more or less like strangers." As for the returning soldier's fatherhood, "he could hardly imagine he had a two-year-old son and didn't know how to respond in any way to him. He would say that I had pampered and spoiled him, that strict discipline was what he needed."[7]

Actually, the marriage boom preceded the Pearl Harbor attack, primarily for economic reasons. The infusion of billions of defense dollars into the economy not only produced full employment but also stimulated decisions to marry. Earlier, during the massive unemployment of the Great Depression, American women and men had reluctantly set aside their dreams of starting families, sadly acknowledging that they would have to wait until economic well-being returned. But sustained prosperity eluded the United States throughout the 1930s. Even low-paying jobs were scarce, unemployment was always high, and with families struggling to make ends meet, levels of consumption remained low. Reflecting economic distress as well as people's uncertainty about the future, the marriage rate dropped from 75.5 per 1000 unmarried females over fourteen years of age in 1929, to 56.0 in 1932. Beginning in 1933, the marriage rate rose slowly, reaching 78.0 in 1937, but with the return of economic depression it slumped to 69.9 in 1938.[8]

Sensitive as always to changes in the economy, the marriage rate rose to 73.0 in 1939, as the early rumblings of rearmament began to be felt. As jobs appeared in large numbers in the early 1940s, millions of women and men readily embraced their long-deferred dreams of marriage. The marriage rate jumped to 82.8 in 1940, to 88.5 in 1941, and to 93.0 in 1942. "With both men and women in war industries," wrote the social scientists Andrew G. Truxall and Francis E. Merrill, "earning unusually high wages, plus an absolute minimum of unemployment, economic conditions were ripe for a rush to the altar."[9]

By the end of 1942, most of the couples who had deferred marriage in the 1930s had finally acted on their dreams. Thus, it was during wartime, not peacetime, that families began to embrace the present. The marriage rate slumped a little after 1942, but it was still higher than for any year in the 1930s; and in 1946 it skyrocketed to 118.1 marriages per 1000 unmarried females over fourteen, thus reflecting the consummation of a second large category of postponed marriages—those legal unions which war-separated couples had delayed entering into for the duration of the conflict.[10]

There was a passion for marriage in wartime America. The Census Bureau, comparing the number of wartime marriages with the number that "would have occurred . . . if the marriage rates had been the same as the average for the years 1920 to 1939," concluded that 1,185,000 marriages took place that would not have occurred during peacetime. For 1942 alone, the Census Bureau estimated the number of such unions to be 387,000. And in early autumn 1943, *Keystone*, a jewelers' trade magazine, announced that the supply of wedding rings was dangerously low, not only because of the number of weddings but also because of the increasingly popular double-ring ceremonies.[11]

Clearly, in stimulating the marriage boom, the upturn in the business cycle had coincided with men and women's wartime-heightened needs for romance and intimacy—for someone to hold on to and return home to. And could anyone doubt that within a year of marriage many of these couples would also be parents, thus creating a baby boom as well? In May 1944 the Census Bureau estimated that there were 37,040,000 families in the United States, compared with 35,124,380 families enumerated in the 1940 census. "World War II," Truxal and Merrill observed in 1947, "saw not only the largest number of families in the nation's history but also the largest percentage in history of men and women who were married. . . . Apparently the family is here to stay," they concluded.[12]

As so often happened during these years of rapid change in the United States, the marriage rate differed from region to region, naturally being highest in the areas to which young women and men of marriageable age had recently migrated. "Spurred by loneliness," one reporter noted, these people "discover that they need each other, and disproportionate gains in marriage are reported from many mush-

rooming defense towns." In the Far West, because of the influx of migrants during the war, the marriage rate remained high months after it had begun to fall in
the East and the Midwest. Likewise, some of the highest birth rates recorded in
the early years of the war came from war-production and military centers in Connecticut, Washington, Nevada, and California.[13]

Full employment, while probably the most significant reason for the wartime
marriage boom, was only one of several factors involved. Another stimulus was
the Selective Training and Service Bill, which Congress debated throughout the
summer and early fall of 1940. Under this legislation, all men between the ages
of 21 and 35 would have to register for the draft; in addition, the act would
authorize the immediate conscription of 1,000,000 men for a year of training.
Paralleling the public debate were countless private discussions over whether marriage—and, in time, a child—would provide an exemption from military service.
In fact, during the summer months of 1940, while Congress deliberated over the
peacetime draft, the marriage rate began to grow until it was 25 percent higher
than for the same months in 1939. The increase was even greater in some places.
A sixteen-state survey reported that the marriage rate had soared almost 50 percent from May to June 1940, the month the Selective Service bill was introduced;
and in the months to follow, it continued its upward trajectory. Some cities experienced long lines at marriage license bureaus. Reports from Boston claimed that
the rush for licenses on July 31, 1940, had been equaled only once before, in the
spring of 1917. In Brooklyn and Staten Island in August, the lines for licenses
began forming as early as 6:30 in the morning. And in Rochelle Park, New Jersey,
officials reported a 150 percent increase in such activity in September, just before
President Roosevelt signed the Selective Service Act. According to the Census
Bureau, the upward spiral in licenses "reflected increased industrial activity and
also the enactment of the Selective Service Act in September of that year which
caused a sharp increase in the number of marriage licenses issued during the late
summer and early fall."[14]

Whether so intended or not, wartime marriages did serve to exempt many
newly married men from military service. Men who were married prior to or during the Second World War were "much less likely" to serve than those who stayed
single. Indeed, in late 1941 and early 1942, Selective Service officials charged that
many recent grooms were merely "trying to escape provisions of the Selective
Service Act through dependency deferments." Marriage to evade the draft would
not be tolerated, warned Brigadier General Lewis B. Hershey, director of Selective
Service, in mid-January 1942: "Those who seek to avoid duty with the armed
forces by marriages of convenience will get little sympathy from the Selective
Service System." Going a step further the next month, Hershey announced that
Selective Service would act on "the presumption that most of the recent marriages, contracted after the summer of 1941, might have been for the purpose of
evading the draft. . . ." In urging draft boards to quit deferring married men who

had no children and whose wives worked, he also warned working women not to give up their jobs just to safeguard their husbands from being conscripted.[15]

As frenetic as this marrying activity appeared, it was, as one observer wrote, "only the beginning. The attack on Pearl Harbor set off a frenzied rush to the altar." Marriage bureaus issued record numbers of licenses, and on December 23, extra police officers were assigned to the Brooklyn bureau to maintain order in the line of 1000 people. In December 1941 the marriage rates for many states were 50 percent greater than for the corresponding month in 1940; and the differential kept growing. Statistics for Seattle in December 1941 showed a 290 percent gain over December 1940; the figure for New York City was 80 percent. The year 1941—with 1,696,000 marriages—witnessed the largest number of marriages in America's history to that time. That record stood for just one year, however; in 1942, America saw another "all-time matrimonial high" with 1,772,000 marriages.[16]

Certainly the business cycle was central to the marriage boom, but wartime emotions also promoted matrimony. William Fielding Ogburn, the sociologist, wrote in 1943 that the period "from December 1941 to April 1942 inclusive indicates the influence of two factors, war and business prosperity, on marriages; but the percent increase in these five months over the similar period a year earlier . . . is an approximation of the influence of the war alone on marriages." Many young women and men, some of them already serving in the armed forces, worried about surviving the worldwide conflict. They "wanted at least a brief period of marriage," Truxall and Merrill observed, "before vanishing—perhaps forever—into the great maelstrom of war."[17]

During these months, newspapers and magazines contained numerous articles debating the issue of "war brides." Some couples—like Nancy and George—had courted each other for years prior to October 29, 1940, the day on which Secretary of War Henry L. Stimson reached blindfolded into a large glass bowl to select the first draft number. In these cases, when the soldier returned home on leave before the final embarkation overseas, the longtime sweethearts decided that there might not be another occasion for them to marry, at least not for months or even years, so they visited the preacher or the justice of the peace and took their vows.[18]

Other wartime couples, however, seemed motivated entirely by infatuation. Intensely—even fiercely—romantic, these lovers cast caution aside in their march to the altar. "With no personal acquaintanceship other than that gained through a few brief visits spread over a few weeks or months," wrote Truxall and Merrill, "a friendship begun at a canteen or USO dance ripened with precocious rapidity into love, with marriage as the final romantic culmination." One woman who feared that her marriage would not survive the war recalled that she and her husband had been "college 'steadies' for six months with no mention of marriage ever made between us. . . . Then he was drafted." They corresponded, and "gradually his letters became more and more ardent. And soon he proposed." Then he was home on furlough. "We weren't alone for an hour before he was pressing the marriage issue with all the high pressure tactics I had heard of plus a few more." Though she doubted she was in love, events transpired that left her "no longer virginal." They were married on the sixth day of the furlough. The couple spent just seven weekends together before he was shipped overseas; soon afterward, she discovered she was pregnant, and when she wrote the news to her husband, "he was very disturbed . . . and couldn't help revealing the fact that the role of father was incomprehensible to him under the circumstances." Also during his absence, she fell in love with another man. "I haven't the slightest idea how it will all turn out but I must confess . . . that I can see many possible outcomes but none that is satisfactory."[19]

Some critics of the marriage boom—the male commentators especially—warned that conniving women were interested not in their soldier and sailor husbands' welfare, but rather in their government insurance policies and allotment checks. Others argued that man-hunting women simply wanted to make sure they got their husbands before the supply ran out. "Strictly insurance against spinsterhood," explained a woman in Chicago. "If this shortage keeps up," added a female student at Northwestern University, "the government will have to start rationing husbands." The pickings did appear slim to many observers, who predicted "a dangerous drop in the birth-rate," accompanied by a breakdown in morality "caused by violent competition for husbands," a rapid spread of polygamy, and a large increase in prostitution, illegitimacy, and mental illness among women of marriageable age. The Metropolitan Life Insurance Company predicted in 1942 that "some six to eight million women will go through life without husbands."[20]

According to several writers, women's response was panic. "Many women," wrote family sociologist James H. S. Bossard in 1944, "are today picking husbands like they shop on bargain days. They do not exactly want the article which is selling so rapidly, but if they do not buy it, someone else will, and besides, one may need it later on." Bossard painted an unflattering picture of "war brides" as vultures preying on vulnerable young men stationed far from home. "The soldier who wants solace, and the frustrated private who needs to regain his self-confidence are obvious types who propose earnestly to Helen today and if rejected, marry Sarah tomorrow."[21]

There were compelling reasons for caution when people's lives were in such flux, especially when the experts predicted that most "war marriages" were doomed to divorce. Eleanor Roosevelt used her Sunday evening radio broadcast to advise college women not to enter a hasty marriage "because your beau is going into the Army." She enjoined coeds to remain at their studies and "gear their present work to real preparation for the future." In various publications, there was an ongoing debate "for and against war marriages." Most experts agreed that an abbreviated courtship of a few days or weeks was an insufficient foundation for building a married life together. The young man might be a lonely soldier or war worker; the woman might be barely more than a teenager. As the family sociologist Ruth Shonle Cavan concluded, such marriages "have a less firm basis than the marriage of two [people] who have progressed through courtship and have weighed the relative advantages of immediate marriage or waiting." Cavan identified still another problem: the "differences in religion or social class, overlooked by the young people in their desire for immediate marriage," which may "become issues of great importance" later on.[22]

Still, most new marriages were not total surprises. "The number of soldiers who married girls they had met at USO dance or overseas," Truxall and Merrill observed, "was small in comparison to the number who married a hometown girl before entering the service, on a brief leave during training, or on an embarkation leave." In fact, as two scholars of marriage later explained, "One of the reasons that nuptiality prospered in wartime was that men entered the army and *then* married. . . ." Many young couples, moreover, held to moral standards that did not permit them to engage in sex before marriage; and since they were impatient for sexual intimacy, they were eager to marry. These speeded-up marriages exemplified the sociological thesis that while wars rapidly accelerate social change, they seldom initiate it.[23]

Finally, some romances and marriages were simply not to be denied, and many of these unions have endured over the years since the war. *Miss You* is a book that prints the war letters of two lovers who had their first date in August 1941 and were secretly married in April 1942. Private Charles Taylor was stationed in Fort Leonard Wood, Missouri; his sweetheart Barbara Wooddall lived in Fairburn, Georgia; and between October and their marriage, they saw each other once. They wrote each other constantly, however, using letters "to develop and nurture their love." In August 1943 a daughter, Sandra Lee, was born. The next year, Charles, then a lieutenant, fought in Europe; landing at Normandy, he was wounded twice before entering Germany with the first wave of American troops. They kept their romance alive in their letters to each other: "I love you, I love you, I love you, Charlie baby—so don't ever forget it. Always—Your Barbie." Charlie apologized that his letters sounded "dry and real dull." And both yearned for the lives they would have together in peacetime. As Charlie wrote: "I know all of the waiting is worth the waiting for what joy and happiness we shall get out of it in the end."[24]

Most couples weighed the pros and cons of marriage. On the one hand, some people said marriage was worth the gamble because, at worst, they would "at least have memories" and possibly also babies. "Take happiness while you can," was a frequent response. One women said she would "rather be a widow than an old maid," another, crassly, that marriage to a soldier would provide financial security no matter what happened to him. But marriage was risky, especially when war separated the newlyweds. "Each may change so much," noted one observer, "as to make future adjustment difficult or impossible." Under these circumstances, some couples did not want children. "It would be unfair to a possible child," said a young woman.[25]

During the war, several professors surveyed students' attitudes toward marriage. One study reported that while these women and men were "realistic" about the hazards, "a considerable majority" was, nevertheless, "favorable to war marriages." Moreover, women were readier than men to accept war marriages. After all, asked a woman in the *Woman's Home Companion* in 1942, "What in the world would be the use of fighting a war if people are not to be allowed to decide whether or not they should marry or have a child?"[26]

In 1942, Paul Popenoe, director of the American Institute of Family Relations, argued that "now is the time" not only to marry but to have children as well. Popenoe, who also taught courses on marriage and family life at the University of Southern California, pointed out that in 1941 more than 2,700,000 babies were born in the United States. "Their parents," Popenoe explained in the *Ladies' Home Journal*, ". . . believe in themselves, believe in America and believe in babies." Buttressing Popenoe's enthusiasm for babies was his commitment to rigidly defined sex roles. Like other "marriage educators," the historian Beth L. Bailey has noted, Popenoe taught that it was not only natural but indeed necessary for society to have "masculine men and feminine women."[27]

Just as people throughout the country debated the issue of "Shall They Marry in Wartime?" so too they asked, "Shall They Have Babies?" Here also the predominant issue was the uncertainty of the future. Dr. Katharine Taylor, for one, grappled with the problem, as a consultant in family life education for the Seattle public schools. "There will be many war babies during this war epoch," she wrote, "and many of them will be left fatherless. A major concern of our nation must be to provide really adequate nurture for these builders of a stronger democracy." Was the nation up to the task? Did Americans even want to shoulder this burden? No one knew for sure. Meanwhile, Americans would ask, "Shall war brides have babies?" As with marriages, the answer more often than not was "Yes."[28]

The Second World War ushered in a baby boom in the United States—and this after a decade in which Americans had delayed having children. Birth rates in the 1930s, demographers Irene B. Taeuber and Conrad Taeuber have written, "were lower than in any previous period in American history." Even as part of a general decline that had been going on for 150 years, the birth rates in the 1930s were particularly low; in fact, in that decade, the number of children relative to the rest of the population was the lowest ever for the United States. In 1940, fewer than 25 percent of Americans were below age fifteen. On the other hand, the population of age sixty-five and over was growing, so that the "age pyramid" was becoming inverted. Most experts agreed that an overall population decline was imminent. The number of children in 1940 was smaller than in 1930, and there would "be still fewer in 1950, if recent trends continue," concluded the 1940 White House Conference on Children in a Democracy.[29]

But then came the Second World War and the baby boom, which began in

The Wartime Marriage and Baby Boom

Year	Marriage rate (per 1000 unmarried women over 15)	Total marriages	Birth rate (women, 15–44)	Total births
1939	73.0	1,404,000	77.6	2,466,000
1940	82.8	1,596,000	79.9	2,559,000
1941	88.5	1,696,000	83.4	2,703,000
1942	93.0	1,772,000	91.5	2,989,000
1943	83.0	1,577,000	94.3	3,104,000
1944	76.5	1,452,000	88.8	2,939,000
1945	83.6	1,613,000	85.9	2,858,000

Source: U.S. Census Bureau, *Historical Statistics of the United States, Bicentennial Edition* (1975), Part 1, 49, 64.

1941 and escalated in 1942 and 1943. According to the Children's Bureau, the birth rate's increase between 1940 and 1942 "was the greatest reported for any period of equal length since the establishment of the birth registration area in 1915." Katharine F. Lenroot, chief of the Children's Bureau, told a congressional committee in 1942 that the war dominated young people's thinking and feelings; and because of anticipated separations, they "are getting married . . . and the number of babies is [also] increasing for that reason and will increase very markedly . . . in the next year or two." Lenroot was correct—more babies were born in September 1942, for example, than in any month since February 1924. And the boom continued. "The peak in the birth rate," explained J. C. Capt, director of the Bureau of the Census, "was reached about one year after Pearl Harbor." Moreover, it stayed at that peak for months; and in January 1944 the Census Bureau announced that "the 1943 birth rate was the greatest in United States history. . . ." Three months later, the *Woman's Home Companion* noted: "We are in the midst of the greatest baby boom in our history."[30]

Keeping in mind the fecund years of the early 1940s, the Taeubers have defined the baby boom as constituting the entire period from 1940 to 1960. The United States' population, far from sliding into the predicted slump, increased 13.7 percent from 1940 to 1950 and 18.8 percent from 1950 to 1960. Renewed immigration after the war accounted for part of the increase, but mostly it resulted from the baby boom. And the arrival of so many babies set off "waves of increase [that have] moved upward in the age structure year by year." Indeed, by the 1980s this baby-boom group constituted nearly one-third of the nation's population.[31]

During the war, peaks in the birth rate followed—by nine or ten months to a year—peaks in the marriage rate (see table). Between April and July 1941, for example, the birth rate jumped. "It is more than a coincidence," observed Philip

M. Hauser, assistant director of the Census Bureau, "that these peaks occurred approximately nine months after the introduction in and passage by Congress, respectively, of the Selective Service Act." And there was another flurry of births ten months after the declaration of war on Japan. During the early years of the war era, the fertility rate for American women 15 to 44 years old rose from 77.6 in 1939, to 79.9 in 1940, 83.4 in 1941, and 91.5 in 1942. Increasingly, other wartime factors—such as people's desire to perpetuate the family name in case the soldier-father did not return—became as important as the availability of jobs in encouraging people to have babies. Based on a survey of marriages, births, and divorces during the early war years, William Fielding Ogburn concluded that in 1942 "the war directly, aside from prosperity, increased the birth rate." Just as 1942 was the top wartime year for marriages, so a year later, in 1943, the fertility rate peaked at 94.3—the highest rate since 1927, during the height of the 1920s economic boom. Also in 1943, the United States, with 3,104,000 births, exceeded the three-million mark for the first time since 1921.[32]

During the war, there was an important institutional reason for the upsurge in births: the federal government established the Emergency Maternity and Infant Care (EMIC) program, under which the wives and children of military personnel in the four lowest enlisted pay grades received wholly free maternity care as well as medical treatment during the infant's first year of life. In 1941 and 1942, large numbers of wives either accompanied their husbands to, or joined them later at, the badly congested areas surrounding the military camps. "To many of these women," a study of the EMIC reported, "children were born under substandard conditions because of inadequate funds to procure the services of the medical personnel and facilities that existed. In a very short time the acute nature of the problem became obvious." Beginning with a few hundred maternity cases in the state of Washington in 1941, public funding spread to 28 states by March 1943, the month in which Congress approved the first federal appropriation for EMIC. From then until November 1946, EMIC, working with the states, authorized payment for 1,163,571 maternity cases. The Children's Bureau, which administered EMIC, estimated that 85 percent of the prospective mothers eligible for EMIC applied for it. At its height in mid-1944, EMIC funded an estimated one out of every six or seven births in the United States. In wartime Nebraska, for example, EMIC paid for 88 percent of all births to servicemen's wives, amounting to 16 percent of total births in the state.[33]

"Are children necessary?" asked *Better Homes & Gardens* in October 1944. The editor's answer was unequivocal: "Yes, happiness lies in conforming to the rules of life, and the first of these rules is that we shall lose our lives in the lives of our children. Perhaps there is not much more needed in a recipe for happiness." The ideology of "family togetherness" was evident in American life not only after the Second World War but certainly during it as well. Indeed, it was the deferred dreams from the Great Depression and the Second World War that, in the postwar years, girded the American "cult of domesticity," the suburban lifestyle, and the desire for additional millions of babies. "The idea," recalled one war bride, "was to live through the war years, get back with our husbands, have kids and raise a family."[34]

"Motherhood's Back in Style," proclaimed a September 1944 article in the *Ladies' Home Journal*. "When young Mrs. Betty Conrad, of Minneapolis, comes out enthusiastically for big families—and proves it by having seven children, with ten as her goal—she may not realize it, but she's the harbinger of a new American trend." Certainly, the *Ladies' Home Journal* insisted, women would have to have their own reasons for choosing to have large numbers of children; "abundant and happier families . . . will not come from any high-pressure campaign or from any argument that motherhood is a duty to the state." No, "if we are to have more children of the kind we want, they must arise from the innermost desires of women, from the heartfelt conviction that motherhood will make their lives richer, more interesting and more secure—that it's a full-time job and one of the most rewarding of professions. This is what women like Mrs. Conrad are trying to tell their sisters."[35]

A poll in 1945 reported that women wanted to have an average of 4.2 children each. "But what of those who *have* four?" asked the *Woman's Home Companion*. "Are they content? Not at all! Mothers of four want still another. Mothers of five want still another. Even the parents of six are tempted." This poll probably exaggerated the truth, but clearly during the war, as well as after it, people longed for the security and fulfillment of family life.[36]

Interestingly, despite the statistical evidence of the wartime baby boom, as well as the qualitative evidence of the domestic ideology supporting it, demographers predicted that the population explosion would be short-lived, and some experts forecast a shrinking population. In 1943 *Fortune* predicted that the postwar generation "will be smaller," since after the war "the birth rate will continue its long and gradual decline." According to *Fortune*, the overall trend of the population would be, first, non-growth from 1965 to 1980, followed by a decline. In 1943, the Scripps Foundation for Research in Population Problems predicted that, in 1980, the American population would stand at 153,000,000; the actual count in 1980 would stand at 226,504,825. Expressing the concerns of politicians who feared that a baby deficit would be harmful to the country, North Dakota's Senator William Langer proposed in April 1945 that the government begin to pay a "baby bonus," with the parents receiving $500 for a first child, $750 for a second, and $1000 for a third.[37]

Demographers and politicians alike failed to recognize that the Second World War was a watershed; their mistake was in looking at the past, instead of the present, to predict the future. But as Landon Jones, the author of a study of the baby boom generation, has asked: "How was anyone to know, peering into the tunnel of the future, that the baby boom was highballing at them from the other end?"[38]

As might be expected during wartime, illegitimate births climbed—though exactly how far, it is impossible to say. This was not a statistic to which people proudly laid claim. Officially, total illegitimate births stood at 103,000 in 1940 and 128,000 in 1945; unofficially, the total was much higher. The presence of illegitimate children was a problem especially for returning war veterans, some of whom had no idea that a baby not of their own making would be awaiting them.[39] Although the rate of illegitimacy grew in all women's age groups, the largest increases by far were for women between the ages of 20 and 29.[40]

During the wartime baby boom, there was naturally an increase in the proportion of babies who were first-born children. This factor, combined with the illegitimacy rate, suggests the youthfulness of many of the parents. Just as important, however, was a second category, which consisted of those couples who, having postponed having children during the jobless 1930s, took advantage of wartime prosperity to have their first babies.[41]

Indeed, so many people wanted children that a black market in babies made its appearance in late 1944. "Baby brokers" sold illegitimate infants to eager adoptive parents, reportedly for as much as $2000 each. In Houston, the top price for healthy infants was $500. Margaret Markle of Houston's probation office told the story of a couple who had advertised their unborn child for sale. A woman offered, instead, to buy their fifteen-month-old daughter. The couple assented, whereupon the father pocketed the money, lost $150 in a dice game, and used $200 for the down payment on a car. Another Houston couple turned a profit by buying a baby for $112 and then selling it for $218. Fewer than half the states had laws to prosecute unscrupulous baby dealers who offered to pay the expectant mothers' hospital expenses in return for the babies, whom they then sold for big profits. Mothers who later fought for possession of their offspring had no legal recourse. One city in which the black market flourished was San Francisco, where the percentage of young mothers was growing. "It is the old, old story all over again," explained Charles Wollenberg, director of the California Division of Social Welfare. "The sweethearts of service men will come to San Francisco for a 'last goodby.' Then they will remain here to have their babies and leave them before they return to their home towns." Other expectant women were married, but not to the fathers of their babies. A frequent situation, Wollenberg said, was that of the wife of a serviceman who had been overseas. She would have a child and then try to place it with another family while her husband was still away. Meanwhile, baby brokers openly advertised in the personal columns, offering to pay pregnant women's expenses. At the same time, the demand continued to outstrip even this growing supply, providing perhaps a final gauge of the strength of the domestic ideology underlying the wartime baby boom.[42]

While it was during the Second World War that American couples embraced their dream of having babies, what were the outcomes for the "war babies" themselves? Many of these babies faced the prospect of living their first few months or even years in the absence of their fathers. The question for numerous homefront families was the same: "Will Daddy go to war?" And when fathers did leave, separation and absence, not to mention possible death, saddened and worried families in all parts of the country. Psychologists have shown that of the children who suffer separation and absence, none are more affected than the little children. This was certainly true for America's homefront boys and girls born during the war.

CHAPTER 3

"Daddy's Gone to War"

"BEING BORN in 1937," wrote Ruby Anglea in 1990, "and my father serving in World War II, I thought I could be of some help to you in gathering information for your book. But the strangest thing happened shortly after I began to write down my memories—I couldn't. I was recalling my father being separated from us, then my mother leaving to join him in a stateside camp, my living with my grandparents—and suddenly as if I was paralyzed I could go no further." Ruby was not alone. To these homefront girls and boys, nothing was more unsettling than the father's departure for military service. In other families, the wartime absence of an older brother was equally upsetting. The safety of uncles, cousins, and neighborhood fathers and brothers also concerned boys and girls. These were anxious and painful years for the homefront children.[1]

For some, recalling wartime fears and sadness forty-five years later has been cathartic. "I felt a cleansing of my mind and soul," wrote a homefront girl; "many tears have fallen, while putting my memories on paper." Others agreed. Leona L. Gustafson's father was drafted in 1945. "Mom began to cry as she and Dad packed his things; she didn't stop crying for what seemed like forever. I asked Dad when he would be home and for the first time in my life I saw tears in his eyes as he answered that he didn't know. . . ." The effort to remember her sadness and fears, Leona confided, had served as a catharsis: "These tears have remained inside for forty-five years; it's past time they were shed."[2]

Seven-year-old Kay Branstone recalled lying in bed at night "crying myself to sleep, because I was so scared my Daddy would go to war. I didn't ever mention this to anyone," being "too scared to talk about it." She fervently hoped there would be "something wrong with him" so that he would be IV-F, physically disqualified. While Kay's father was not drafted, her two uncles were, and Kay, who often stayed with her grandparents during the war, lived with their anxieties.[3]

When Maureen Dwyer's father returned from his draft physical, he told her that he had been certified I-A: "I can still feel my stomach drop when he said that." Actually, few fathers were drafted during the first three years of the Selective Service System. From its establishment in October 1940, Selective Service Class III-A had protected men with dependents from induction. A year later, 10,102,000 of the 16,100,000 selective service registrants were classified III-A. Rather than draft fathers, Selective Service first lowered the draft age for the unmarried, then began drafting "less-well-qualified men" both physically and educationally. Not until late 1942 did there commence a full-fledged debate on the conscription of pre–Pearl Harbor fathers. While some politicians contended that it was essential to draft all able-bodied men, including these fathers, others responded that there were still plenty of single men available and that Selective Service could fill its quotas by conscripting "those unmarried men who shun work and are found in pool rooms, barrel houses, and on the highways and byways. . . ." During the first half of 1943 Congress debated legislation to exempt fathers altogether, but adjourned before passing a bill. In the interim, Selective Service, which predicted that there would be a shortage of almost 450,000 in filling calls if fathers were not considered, issued a decree abolishing Class III-A as of October 1, 1943, thus removing the ban on inducting fathers.[4]

In October 1943 the number of fathers drafted was 13,300, or 6.8 percent of the total. In November the figure doubled to 25,700 (13.4 percent), and in December it doubled again to 51,400 (26.5 percent). It continued to rise, and in April 1944 more than half of the inductees (52.8 percent, or 114,600 men) were fathers. The figure declined after April only to rise again between February and June 1945. Indeed, between October 1943 and December 1945 the number of fathers drafted was 944,426, or 30.3 percent of the total. By V-J Day, of the 6,200,000 classified fathers aged 18 to 37, one-fifth were in the service; of the youngest fathers, aged 18 to 25, more than half (58.2 percent) were on active duty.[5]

Statistics do not exist to tell precisely how many wives and children suffered from the absence of a husband and father during the war, but it is possible to estimate their numbers. In June 1945 a total of 2,818,000 Army wives (and about 1,825,000 children) received family allowances from the Office of Dependency Benefits; in line with this, the Bureau of Labor Statistics estimated that there were 3,000,000 Army wives. At that time there also were 3,855,497 persons in the Navy and Marines, an estimated 35 percent of whom, or about 1,350,000, were married; so the total of armed forces wives exceeded four million. Finally, there is a most helpful statistic that places the subject of male wartime absence— whether of fathers, sons, or brothers—in a broader context. During the war, nearly one family of every five—18.1 percent—contributed one or more family members to the armed forces.[6]

While some fathers were drafted, others volunteered. Some fathers argued that it was their patriotic duty to join up; sometimes, their wives disagreed. This seemed to be especially true in doctors' families. Doctors were badly needed, but most of those who volunteered were past draft age and had several children. Carol

Helfond was five when her father, a physician, joined the Army; her mother had argued strenuously that he was too old to go and was very upset when he decided to join anyway. But as William MacTavish Whortle, a surgeon, wrote his wife in 1943: "Naturally, one could have, as other [doctors] have, avoided service. However, I never would have allowed such to happen. . . . The sooner all realize that it is part of their job to win, the sooner it will be over." Recalled another doctor's wife, "My heart sank in silent anguish" when he "spoke of his plan to enlist. . . . Yet, I was not surprised, knowing him well."[7]

Whether draftee or volunteer, father's departure was usually traumatic. Many of these farewells took place in crowded railroad stations. Carol Helfond's family escorted her father to the station. "My mother held up well until the train carrying my father pulled away. Then she ran into a phone booth and sobbed her heart out." Ruth Ann Grinstead was eight when her father joined the Navy, leaving her mother with the sole responsibility for ten children. Seeing the human interest angle in this sailor's departure, a newsreel company staged the event and filmed it. "We were all lined up on the sidewalk in front of our home, with mom holding the baby and my Dad walking away from us, and we all waved goodby." Her father actually left the next day, and, later, the family went to the local theater "to see the newsreel of us waving my Dad goodby, and we all cried. . . ."[8]

Some children resented their fathers' decision to leave home; they felt abandoned. The infants, however, were more confused than anything else. Sandy Newton's father, a fighter pilot, was shipped overseas in 1943, when she was only three weeks old. To keep Sandy's father's memory alive, her family would point to the sky every time an airplane flew over and say, "See, Sandy, airplane. Your Daddy flies an airplane." In time, she came to think that her father *was* an airplane, "a fantastic flying machine." A little girl born in 1944 had a different problem; she thought every man in a sailor suit was her father.[9]

Children missed their fathers deeply. "How I miss you Daddy," wrote a seven-year-old girl in 1943. "Gee when in the dickens are you ever going to get to come home. I'm not putting the Blame up to you but I do wish you would get a furlough. . . . Please won't you tell me if you . . . get word from the Major? Will you call up when you get orders [?] . . ."[10]

Some children were taunted about their absent fathers. Ruth Larson was born the week after Pearl Harbor; when her father was shipped overseas, she and her mother returned to the family farm near Arapahoe, Nebraska. One day an older cousin told the little girl that she had no father. "I do," she protested, but she remembered this episode of cruelty. Ethel Geary, a second-grader, was the daughter of a sailor. She too was taunted, in this case by a boy at school. "'You don't have a father,' he began to chant for all to hear." Ethel waited for him after school; she handed her eyeglasses to another child and then "proceeded to beat the kid up with my lunch pail."[11]

What the children and their mothers feared most, of course, was that Dad would not return. A homefront girl in Newark, Ohio, wrote that she cried herself "to sleep many nights thinking about my father being killed by the Japs." These fears persisted even after the war. Gayle Kramer was seven when the war ended, but her father, a naval officer, had to remain in the service until his ship was decommissioned. Other fathers were returning, but hers had not yet done so. One day the woman living on the corner asked where her father was, and she replied that he was "out fighting the war." At this, the woman laughed and said, "Honey, the war's been over a long time; maybe he's not coming home."[12]

America's girls and boys who were separated from their fathers or other important relatives probably endured more heartache than any other children on the home-front. Clearly, the war's consequences were great for those boys and girls who experienced the sudden departure of loved ones for faraway battlefields. In line with these observations, this book assumes that the most important institution in the lives of children, especially young children, is the family. When the family's circumstances change, so the child's life invariably changes as well. The wartime absence of the father was a prime example; other examples were the new and expanded roles for women, with mothers filling voids both at home and in the factories; the mass migration of more than 30,000,000 men, women, and children to Army posts and war-boom communities; and booms in both marriages and births.

From this perspective, then, this book about children is also about their families. A pioneer in the historical study of families, Tamara K. Hareven, has written that family historians begin with the assumption that "the key to an understanding of the interaction between personal development and social change lies in the family." Or, as James H. S. Bossard, a sociologist of the family, explained in 1944: "So comprehensive and fundamental are the changes wrought by war, and so closely is the family interrelated with the larger society, that there is perhaps no aspect of family life unaffected by war." This was particularly true during the Second World War, which brought such immense economic, social, and cultural change to the United States that historians have called it a watershed in American history.[13]

Still, it is difficult to generalize about the effects of a father's absence on his family and children. "Reactions to their absence," wrote one therapist during the war, "are as varied and numerous as were the reactions to their presence." Families adapted in different ways, depending upon the family's prior history and the ages of the family's members. Reuben Hill, sociologist and author of a book about "war separation and reunion" on the homefront, observed that men who went to war had "played widely different roles in their respective families"—from companion, handy man, lover, and disciplinarian to bread winner—and, Hill concluded, "each family missed its man in terms of his special role in the home."[14]

Long before the outbreak of the war, for example, many of America's fathers had relinquished their paternal responsibilities to other family members, such as older sons. There were absent fathers during the 1930s too, but those who left did so either to search for jobs or to escape the shame of being unemployed. Suffering from joblessness during the Great Depression, many demoralized fathers also turned to heavy drinking. These fathers became less important to a family's functioning than older brothers who had assumed the duty of directing younger siblings' lives. Two doctors from the Judge Baker Guidance Center in Boston told of a ten-year-old boy, "the youngest of 7 children, of poor family background in that both parents drink, the mother to excess. . .." The boy "was referred for bunking out and picking up food and clothing wherever he could find them." His older brother, who had become the surrogate father, had been the only family member the boy had admired, and they got along well. But then this brother, who had enlisted, was killed overseas. The boy received a portion of his life insurance, which he gave to his mother so that she could enter a hospital for the treatment of alcoholism. Even though he was the youngest person in the family, the boy was assuming some of the responsibility that his brother had formerly shouldered.[15]

Although psychologists have assessed the impact of a father's absence on a child's identity, there is a void in the literature—and this void is particularly galling because the issue of identity is central to human development. In the father-absence research done on America's homefront children, there is precious little on girls. During the war years the published cases focused almost exclusively on how boys dealt with father absence. These articles were entitled "A Boy Needs a Man," "What Shall I Tell Him?," and "Sons of Victory." No article on father absence asked, "What Shall I Tell Her?"[16] Likewise, postwar research on the life-span results of father absence has generally examined only boys.[17] Thus, to locate the evidence and assess its importance, the historian needs to look not only at what is there but also for what is not there.

There is a retrospective solution, however—the recollections of the homefront girls as well as the boys, as expressed in letters. Even if not put down on paper until years later, these letters provide valuable information, including deep personal insights into tightly held feelings.[18]

To begin, one must recognize that father absence is more complicated than it might appear. As one psychologist has written:

> Children growing up in a single-parent home headed by the mother may be affected by any of the following: the altered family structure and consequent differences in maternal role behavior; . . . the presence of surrogate caregivers associated with the mother's employment; or qualitatively different maternal behavior vis-a-vis the child because of the emotional meaning the father's absence has to her. There are many other factors which also may operate either singly or in concert with each other, allowing absolutely no possibility for delineating the "true" causal agents in the child's development.[19]

The children's first-person testimony about the war years, garnered from their letters, validates the contention that for numerous families the operative issue was not really the absence of a father, but rather the manner in which the mother responded to that absence.[20] And here, as one might expect, the variety was wide. Eleven percent of the wives in a wartime study of Iowa farm families, for example, "welcomed the separation as a release from an intolerable marital situation or as an opportunity to think through an unsatisfactory relationship." Indeed, for some families the most traumatic wartime event—though ostensibly the happiest one— was the return home of the soldier or sailor father.[21]

Other mothers, however, were devastated by the separation and absence. Leona Gustafson, who was four when her father reported for induction, recalled that "beginning on the first night Dad left . . . my mother had awakened me with her crying. I can remember going into her room and stroking her forehead while telling her every thing would be all right, that Daddy would be home as soon as he could. I became what I was to remain for the rest of my mother's life—her daughter, her best friend and, in a sense, her mother."[22]

The memories of Lois B. Heyde, born in 1937, are the opposite. She pictured her mother not as a woman needing comfort and support, but as an heroic part of the war effort. Her mother, who worked on the flight line at an aircraft factory, had skin that was "burned deep tan from the reflection of the summer sun and the aluminum." When Lois's father joined the Army in 1942, her mother took a job at the Boeing plant near Atlanta. Eventually Lois and her younger brother went to live with their grandparents on a farm north of Nashville. Every few weeks, their mother would ride the bus to visit. Since the bus stopped at a nearby town, her mother would have to walk five or six miles to the farm, sometimes at night. And since she was quite pretty, she carried a revolver "to discourage persistent types who wanted to give her a ride."[23]

Because of absent fathers, numerous children lived in families of women. "While my Father was overseas," stated Rachel Love, born in 1938, "my Mother and I lived with Grandmother in her house." Two aunts with husbands in the service and another girl, a cousin, also resided there. "It was a house made up entirely of women. The sisters all worked and my grandmother looked after the children." Anita McCune, whose father was a Marine fighting in the South Pacific, lived with her mother Eva, her Aunt Goldie, her thirteen-year-old sister Lahoma, and Aunt Goldie's daughter Margaret Ann, who, like Rachel, was seven. "Most of the time it was just us five women," she remembered, and they not only survived, they prospered. Her mother and aunt raised chickens and slaughtered them in the backyard, and they painted and fixed up their old house.[24]

When Jane Coad's father was killed in Germany in November 1944, she was living with her mother and two sisters in Omaha. Asked about her father's death, she replied that "the long-term effects on my sisters and myself were that we grew up in a family of women. Strong women, not brought up to show affection." Her mother idealized her eight-year marriage, "so we really had no idea how a marriage truly worked. We didn't realize that couples argued and yet still loved one another. We laugh now, but for years we were really terrified by a male voice raised in anger."[25]

Wartime separations resulted in the rearrangement of family roles and thus wrought great changes in the lives of American families. But one thing was clear: It was the mother whose response to wartime change most affected the home-front children. Secondarily, it was grandparents who filled the voids caused by changing family circumstances. In order to understand the effects of father absence, it is necessary to inquire into the contributions, as well as the tribulations, of mothers and grandparents.

Mothers suffered no shortage of advice during the war. During a father's absence, they learned, one of their primary tasks was to keep "the memory of Daddy crystal clear through long periods of separation." As a step in this direction, boasted one mother, "We even celebrated Daddy's birthday in absentia with a party, cake, candles and all." A letter from Daddy, wrote a child-guidance expert, could become "a piece of happiness to share" with the child. After reading it aloud, the mother and child could work together to prepare a reply and then walk to the corner for the child to place it in the mail box. Similarly, mothers read that they could use photographs, gifts, foods, and sports to tell children about their father. For example, a mother could show the child pictures of Daddy from the time he was a little boy until he donned a military uniform. As for food, explain that this "is the way Daddy likes [his oatmeal] best." If Daddy is a good tennis player, tell the child that "when Daddy comes home, he will play tennis" with you. "I was amazed," wrote the mother of a two-year-old, "at how many opportunities present themselves for talking about Daddy, for making him a part of their lives."[26]

Mothers were told to give frank answers to their children's questions. "'No, we don't know how long Daddy will be gone. . . . Yes, of course we hope Daddy will come back safe, and we believe he will. Yes, some men will be killed on our side too, but that is what war means. . . . No, you will not be alone. If Daddy doesn't come back I will still take care of you.'"[27]

Dorothy Humphrey took such advice seriously. She and her doctor husband had two children, ages three and one, when he joined the Army in 1942. During her husband's long absence overseas, she wrote him daily, and from reading her loving, detailed letters, it is evident that family conversations revolved around "Daddy." Likewise, she filled her letters to him with stories about the children's accomplishments, their "quaint sayings," and compliments they had received, not to mention their physical ailments and emotional struggles. Dorothy wrote about "her unsatisfied physical longing" for her husband, and she told him that "the terms of single-parenting are demanding." Yet she persevered; Dorothy Humphrey and other mothers of father-absent children were homefront heroes.[28]

Homefront mothers served as powerful examples to sons as well as daughters. Clearly, these wartime "Moms" were inspirations to their own children. Jean

Beydler, who at age seven had moved with her family from Ashland, Kansas, to southern California, recalled that her mother "knew that history was in the making and had an adventurous spirit. She hauled us kids—on the streetcars and on foot—everywhere." Mother and children frequented the Los Angeles Coliseum for bond drives and patriotic rallies featuring Generals Jimmy Doolittle and George Patton. They took in the Hollywood Christmas Parade, visited museums, and got tickets to such radio shows as Fibber McGee and Molly, Spike Jones, and Al Jolson. They also went to dances at the Venice Pier Amusement Park, seeing Bob Wills and the Texas Playboys so many times that Wills knew the children by name. "We were a busy little troop," Jean remembered. "We saw everything ... that was happening in Los Angeles during those war years." As Jean happily recalled this period, she said, "My mother was the real hero in our lives."[29]

James Shields, born in 1939, also viewed his mother as a wartime hero. One day, James was riding a Chicago bus with his mother and sister when they overheard a woman say, "Boy, I hope this war goes on forever [because] my husband is making more money than he ever could in peacetime." With that, James's mother got up, stood by the woman, and proclaimed, "Lady, maybe you don't have any friends or family in the war, but I do." And then she punched her in the face. "A scuffle ensued and we were thrown off the bus."[30]

Likewise, Carol Helfond was proud of her mother's assertiveness and courage. While Carol's father was serving in Europe in the Army medical corps, she and her family returned to their house in Queens, Long Island. One day, a neighbor paid a visit to complain about how Carol's brother was behaving. Her mother thanked him for coming and assured him that she would take care of the problem. As the man left, he muttered under his breath, "Dirty Jew." Carol's mother literally pounced on him, hitting him several times and ordering him "never to come near our house again." Carol remembered "feeling horrified and yet, at the same time, very proud of my mother, my father's patriotism, and my Jewishness."[31]

Pat Adler was four when her father joined the Navy in 1943. Like many families, Mrs. Adler and her two young children moved in with one of the grandparents. Despite her father's two-year absence, Pat said that those years hold some of her "fondest memories." She could picture her mother "in a green and white dress, with her hair in a pompadour, putting on her fur coat," to join other service wives in the neighborhood for an evening out. "Looking back," Pat wrote, "I realize my mother was only twenty-five or so, but we thought she could do everything."[32]

The homefront children recalled their mothers' strength, but embedded in their wartime memories were also images of great sadness. Wives and mothers were anxious about their husbands' safety and welfare. Many were extremely lonely. A wide variety of events could trigger tears. Some children had never seen their mothers cry until the day their fathers left for the service. Another occasion for mothers' tears was the arrival of V-mail. Lucretia Spence's mother had been only twenty when Lucretia was born in 1940; her father but eighteen. He enlisted in the Army Air Corps soon after the Pearl Harbor attack, and while he served overseas, she worked at the shipyard in South Portland, Maine. She was pregnant when her husband left, but stayed on the job. She returned to work shortly after giving birth to a second child; her sister and parents cared for the children during the working day. Lucretia's mother also had a brother fighting in the infantry in North Africa; she worried about him too. Worry and stress consumed her life. She spent her days working and "her evenings writing letters to my dad and," Lucretia remembered, "reading letters from him. I recall her crying often because of loneliness. Her praying and tears are much a part of my early memories. . . ."[33]

Song lyrics also triggered memories and tears. "I'll Be Home for Christmas," a popular wartime song, was so highly evocative that it reduced one girl's mother to tears each time she heard it. "It got so I was forbidden to play the radio when she was within earshot." Sally Applebury was five in 1942 when her father, a small-town doctor in Missouri, left for the service. He was gone until she was nine. Sally recalled the "incredible sadness at holidays. I can still see my mother standing on the back porch crying on Thanksgiving and Christmas. I think the war changed their relationship," she continued, "because once having been submissive . . . she had to be the one in charge for 4 years all alone; she was not the same person when he returned. Needless to say, neither was he."[34]

Often for girls and boys born during the 1930s, it was not fathers but older brothers who were gone. Clare Haddock was eight when she heard President Franklin D. Roosevelt give his "day of infamy" speech, and she felt that "my life would never be the same. As the youngest of eight children I had always had the security of all those older siblings . . . ," but now her brothers and sisters went to war. Two brothers enlisted in the Navy, one sister left to work in a munitions plant, and another sister, a nurse, joined the Army. Clare recalled "the nights I would hear my mother crying in her bedroom. This would happen when my brother Frank (a sailor on a submarine) would be out on a mission in the Pacific. . . ."[35]

Ann Laramie, who was the same age as Clare, had eight brothers in the service. Despite her mother's prayers, Ann "just assumed" that her oldest brother, who was drafted shortly after Pearl Harbor, "would die." But he returned; indeed, Ann believed, "It was my mother's prayers that brought them all safely back." Another believer in the power of prayer was Concepcion Amaro, born in 1938. Her three oldest brothers all joined the service in 1942. The final blow was when her sister Carmen entered a convent at the same time; her mother became despondent. "She would cry a lot, sometimes she would go for days not speaking, she would just sit there." Her father too was wracked with worry not only about his sons and his wife, but also about his inability to earn more than a paltry income. But the brothers all returned, and as family lore had it, V-J Day began just as her sister Carmen had "finished a Novena for the War to end. So her prayers," like those of Ann Laramie's mother, "came to be answered."[36]

Incessant tears and prayers, however, were depressing to children and maddening to other family members. On December 7, 1941, Barbara Abbett, eight years old, went with her parents to visit her brother, 165 miles away, to celebrate his purchase of a house. But once the report of the Japanese attack came over the noon news, her mother began to cry. The parents worried that Barbara's two older brothers would both have to serve. Her mother cried all day, and she began to pray, "Oh Lord, please bring our boys home safe from the war." Her brothers did go to war, and, Barbara recalled, "The prayer that I was to hear for the next 3¾ years started that day," and she heard it not once, but "at least a dozen times a day." In June 1944, when the news of the Normandy invasion reached the United States, with its stories of dead men on the beaches, Barbara's mother began to cry uncontrollably and say her prayer. Her father could stand it no longer. Taking a swipe at the table, he knocked everything on the floor. Then he "jumped up and screamed at Mom, 'Stop that praying. If there is a God, why doesn't he stop that? How could He let them mow our boys down like wheat rows? Our boys are probably dead.'" But in the summer of 1945, both sons returned home, and before they were shipped to the Pacific, Japan surrendered.[37]

For some children, it was not the older brother but the older sister who was absent. Kathy Scanlan's sister, who was sixteen years older, volunteered for the Red Cross and served in Scotland, England, France, and Germany. "By having a sister overseas," she wrote, "rather than a brother or father . . . , I enjoyed a unique position among my friends. . . . Knowing that women could & did serve in dangerous areas of the war gave us all a different perspective."[38]

In their sorrow and anxiety at the absence of sons and daughters, some parents neglected the younger siblings, particularly their emotional needs. Barbara Ceaser, whose two older brothers were in the service, recalled that both she and her little brother, Rick, were "often lost in the shuffle. No one paid us too much attention." Rick remembered that their mother cried when no mail came, "and my father would cry with her after he came home from work"; both were "sick with fear" that their sons would be killed. To escape their tears, Rick frequently retreated for hours into the flat's walk-in closet. Contrarily, other younger children benefited from increased parental attention in the absence of older brothers. One homefront girl wrote that she became "essentially an only child. As the youngest, and only girl, I was something of my family's pet, and since I had all my parents' attention I had gotten a little spoiled. When both my brothers came home . . . things changed."[39]

At most, father absence involved one father per family, but certain families witnessed the departure of six, seven, and even eight sons and daughters. Especially difficult for some children was the absence of an older brother who had become a surrogate father. Howard Dolloff was eleven at the time of Pearl Harbor; his older brother Bob was twenty-two. Bob was like a "'little father'" because Howard's father was "more like a grandfather, in age," and because Bob was "so

kind" to him, taking him to concerts and introducing him to "all kinds of music."
Thus Howard suffered a double loss when his big brother went to war.[40]

Regardless of the relationship, the younger siblings were always afraid, as one
remembered, "that the dreaded telegram would come to our house." In some
neighborhoods, the bearer of ill tidings was an Army officer who would drive up
to the house and knock on the door. "If I looked out," recalled one homefront
girl, "and saw a man in uniform, I felt dread. . . ."[41]

Clearly, one of the most unheralded homefront stories was that of the grandparents who took in daughters, daughters-in-law, and a multitude of grandchildren.
Some children moved in with their grandparents for periods of a few months or
a year, while others became part of the household for the war's duration. One
grandmother offered the proposition that "If anyone should ask for a name for
this war, it's 'Grandmother's War.'" For their part, the homefront children who
lived with grandparents generally remembered these as happy years—years of
affection, if not adoration. When Ruth Larson's mother was working, her grandmother cared for her, from the time she was a baby until she was almost four. "I
really think I was bonded to Grandma!" Ruth remembered. "I'm sure she instilled
a lot of moral values & genuine love in me."[42]

Perhaps it is not surprising that, in the absence of father, it was grandfather
whose love and care were remembered by many of the homefront children. When
Carol Strachan's father left for sea duty, she and her mother moved in with her
maternal grandparents. Looking back, she wrote, "I realize that my Grandaddy
took over for my Daddy in many ways. He read the funny papers to me, . . . took
me to the park and movies, and bought me a cherry smash . . . after church. . . .
[And] he would sing me to sleep at night with 'You Are My Sunshine.'" Tutti
Cantrell, born in September 1942, boasted that she was her "Grandpa's" darling.
Her father had been inducted into the Army when she was just ten days old, but
Grandpa had promised his son-in-law that "he would take care of his baby while
he was gone. And that he did!"—becoming not only her protector, but also her
buddy, bundling her up on Saturday afternoons and, she recalled, taking "me to
town . . . to show me off and buy me ice cream." When Tutti's father returned
two and a half years later, "he was sure that I was spoiled and that I needed him
to straighten me out."[43]

Grandparents lavished affection on their homefront grandchildren. "My
grandparents loved and spoiled me," recalled a homefront girl, Kathy Lynn Betti.
Kathy's father was home on leave when she was born on Christmas 1942, but she
did not see him again until the war was over. Meanwhile, "my mother and I lived
with my Italian maternal grandparents, and I was pampered to the hilt. When my
father did come home I wanted nothing to do with him. I remember hiding. . . ."[44]

Grandparents were also protective of these children. Phil Wright, who was
born in 1938, moved to his mother's hometown of Protection, Kansas, when his
father was shipped overseas. Phil noted that while "I loved my father very much,

... I believe that my mother and grandmother must have been very skillful at shielding my younger brother and me. I cannot remember ... ever contemplating that he would not return."[45]

Surrogate fathers were also important, giving needed attention to the home-front children. "If you want to do something worth-while try this," advised *Parents' Magazine*, "let your child invite Jimmy or Jane whose father is overseas to dinner one night. . . . Competent as any mother may be to direct her child's energies and activities, there is simply no escaping the fact that when a man takes a hand in the youngsters' games a sort of rough and ready masculinity adds to the fun." It was the "'strong man stuff,'" noted Dr. Milton I. Levine, a pediatrician with New York City's Bureau of Child Hygiene, "that little boys need and that mothers can't supply. . . ." One beneficiary of such advice was Phil Wright, who explained that in addition to loving relatives, a neighboring rancher "'took up the slack' and acted as a surrogate father." Many mornings before dawn, the man drove over in his pickup truck and took Phil to his ranch for fishing. For an urban boy, Billy Whortle in Detroit, there were two men who filled in for his father. One was the father of Billy's best friend, whose family included him on trips to their cottage in northern Michigan. The other was a middle-aged man who had two grown daughters, but no son, and who took Billy to Detroit Tigers baseball games.[46]

But neither grandparents nor surrogates could entirely fill the void caused by absent fathers and mothers. Barbara J. Carter's parents took two of their children and relocated in San Francisco, but left six-year-old Barbara to be reared by her grandmother and aunt. "I did not get to see my parents from the time I was 6 until I was 11 years old. By then they were complete strangers to me," as were her brother and sister, who were four and seven when they left, but nine and twelve the next time she saw them.[47]

Some homefront children bear scars of abandonment; they were expendable. David Childers's father "disappeared into the Army Air Corps" in 1942; at the same time, his mother and aunt began to talk about job opportunities in the defense plants. "Little did I know that at 8 years of age I too was about to be affected by the war. The plan was this: Mother and her sister will go to New York, get jobs, and split the expenses. Son, me, gets to live with my maternal Grandmother, a widow who ran a Boarding House. So that's the way it was." David's parents divorced at the end of the war.[48]

Moreover, not all grandparents were warm and loving. When Lois Heyde's father joined the Army and her mother took a war job, she and her brother were left with their great-grandmother. "This was a very unhappy time for me. Separated from my parents and living with someone who was not a caring person and always seemed angry with me. I missed my mother very much and recall . . . crying and begging her not to leave when she came to see us."[49]

Naturally, grandparents who worried mightily about their sons at war transmitted these anxieties to the grandchildren. A homefront girl wrote that of the seven uncles on her father's side, all but one served. "At one point," she added, "birth certificates had been 'doctored' so that" two of her underage uncles could enlist. Barbara Balch's father was "the only one of 9 sons *not* to be in the military." When Barbara's Uncle Jack was killed in France, her grandmother "screamed and cried for three days, pulling on her hair so hard she pulled tufts of it out."[50]

Thus, the homefront children—seeing grandparents cry and sharing their grief—had to worry about their uncles, in addition to fathers and older brothers. Virginia Reading, born in 1934, had an uncle in the Marines. He had made several island landings in the Pacific when the news reports began to arrive about Japan's inhumane treatment of prisoners of war. Upon hearing one such report, Virginia's grandmother burst out crying, "I would rather see him dead than a prisoner."[51]

It rarely fell to children to notify their grandparents of an uncle's wounding or death, but it did happen. Ann Marie Diamond, age ten, was home the day the telegram from the War Department arrived for her grandparents; her uncle had been wounded in Europe. Her grandparents were Italian immigrants, so "in my most comforting, loving, voice, without cracking, without shedding a tear, I proceeded to tell them. . . ." Only later, when she had gone to her bedroom, did Ann Marie break down and cry. An even more difficult challenge confronted Edward Kaucher, a fourteen-year-old living in Cincinnati. On December 27, 1944, the telegram arrived notifying the family of his uncle's death. Edward's grandmother lived across town and had no telephone, so his mother handed him and a cousin bus fare and told them to tell her. "It was a clear night . . . ," he recalled. "My grandmother's first reaction was a scream that brought neighbors on a dead run." "'My baby, my baby,'" she cried.[52]

Most of the men who served in the war were either fathers, brothers, uncles, or all three. Their lives were at risk, and tens of millions of relatives worried about them. And it was their letters home that maintained the essential connections, no matter how imperfectly, between the soldiers and sailors overseas and their homefront families. When war separated a family, wrote James H. S. Bossard, "a face-to-face relationship gives way to a letter-to-letter relationship." Letters told war stories, and they offered advice to the boys and girls. In May 1943, a Marine lieutenant stationed in the Solomon Islands advised his infant daughter to "set a goal and strive . . . to attain it." More than that, he wrote, "Your Mother and my Wife is a wonderful woman. Be like her." Letters often concerned family responsibilities. It "looks like it might be a while before your daddy sees you again," Will Whortle, an Army surgeon, wrote his eight-year-old daughter Susan. "You must in the meantime be a good girl. Be a help to your Mother and always do the things she tells you to because she knows best and is older and wiser. Study hard in school. Then when the war is over we will have a lot of fun together. Your part . . . is to be good & work hard & help Mother with the boys."[53]

Children got these messages whether they could read or not, since families congregated in living rooms and read the letters aloud. Letters home even became neighborhood events. When Gayle Kramer's father shipped out to Okinawa, she and her mother moved into her grandmother's house in New Orleans, which was bulging with aunts, uncles, and other relatives. Mail was supremely important to this household. Their postman would "shout through the screendoor, 'Carrie, you got a letter from Malcolm.'" He also would tell people up and down the block, so that later in the day, "neighbors would walk by and say, 'I heard you got a letter from Malcolm. How is he?'" For some families, however, letters were an impossibility. In one Tennessee farm family, the soldier father was illiterate. "My father could not read or write," his daughter recalled, so her mother "never received a letter from him in the four years he was gone."[54]

Letters certainly buoyed the spirits of wives and mothers, and their happiness, however momentary, heartened the children. Years later, one of the homefront children expressed the pleasure and surprise she felt when she read the letters that had passed between her parents. "It seemed so strange," she wrote, "that these romantic letters were to and from my parents. They became . . . real people with real feelings."[55]

"I do love you so much and always will," Will Whortle wrote his wife Geneva on August 2, 1943. "You are the whole world to me and I will only be happy when it is all over and I am back with you and our babies forever." Stationed at Camp Carson, Colorado, Major Whortle expected to receive orders soon, and his letters became increasingly affectionate and philosophical. On August 6, he told her: "You are so wonderful and my darling I love you more each day. The years past have been short and those when I come back cannot be long enough." At Camp Carson, rumors were rife that Whortle's unit was due to be shipped out. On August 11, he wrote his last stateside letter to his wife, telling her that "when all is over & you come to meet me, we will have a hell of a time and it will be a great day." Then he was gone. His next letter bore an APO address in care of the postmaster in New York City; Whortle was in North Africa. "Keep everything together by hook or crook," he implored. She did.[56]

Fathers corresponded with their families not only by letter, but also through photographs, recordings, and drawings. "Daddy sent us a record that he had made in a recording booth in Hawaii . . . ," recalled a homefront girl. "He spoke with a very clear, slow voice. . . . He sounded so far away and so alone. It was a mixed blessing." Not all the recordings evoked sadness, however. One uncle, an accordion player, took his instrument with him when he was inducted into the Army. He sent his family a small yellow disk on which he had recorded "The Sheik of Araby." Families not only played such recordings "over and over," they also responded in kind. Ten-year-old Nancy Hart and her younger sister Barbara made a recording for their father, singing "Coming in on a Wing and a Prayer" and telling him about recent events in their lives.[57]

Men overseas also drew pictures for the children. Barbara Hooper's father was drafted in 1943, when she was seven. While on a ship in the South Pacific, another serviceman taught him to draw a dog in profile. "We looked forward to his sending us a drawing of that dog. We copied it so much I believe I can still draw it."[58] More ambitious were the cartoon drawings done by Joe Wally, a talented artist, for his daughter Diane, whom he adored. Joe dedicated his 1945 war diary to "my daughter and first born child Diane. Like my life it is dedicated to her...." Stationed on Guam, Joe created a cartoon character named "lil Lulu," who became Diane's friend and role model. Lulu, who wore long braids like Diane, practiced the Golden Rule, went to Sunday school, said her prayers nightly, shared her toys with the little girl next door, held "Mommie's hand" when she crossed the street, took "a nap each day without complaining," and was "always kind and never selfish." Joe Wally mailed several dozen Lulu drawings, often to the dismay of censors. In the corner of one cartoon, he explained: "*Censor*: To My 3 Year-Old Daughter." Joe even gave Diane drawing lessons, showing her in one letter how to draw a stick figure in motion and add clothes to it. Diane, who would become a commercial artist, stated that Joe was "the predecessor artist in my genetic make up, I know." More than that, however, it was the "lil Lulu" drawings that made Diane feel very loved and important and taught her many worthwhile lessons.[59]

Naturally, the most wonderful letters for children were those written to them only, especially by their fathers. "Hello Putsie," wrote Maureen Dwyer's father, a gunner on a PT boat in the South Pacific. "Don't feel bad about me at Christmas because we will have everything but home out here ... so you just be a good girl." In their loneliness, fathers told their children how much they missed them and loved them. Terms of endearment abounded. "Such a ... lovely picture," wrote Cheryl Kolb's father from England. "I wonder who the little girl is on the swing? Do you suppose that could be my Cherry-Pie—my little wrinkle nose— my little button nose or my little blond bomber?" While on duty in England and France, Cheryl's father wrote her every few weeks. "And how is Daddy's little angel? ... You're my little bunch of sweetstuff with the great big grin." He composed poems for her on her sixth and seventh birthdays, and she in turn sent him her drawings and told him, "I love you sooooooooooo much!"[60]

Major Charles R. Kolb did return from the war. Other fathers, big brothers, and uncles did not, however. These absences became permanent, and the home-front girls and boys struggled with their grief.

Blue and gold star service flags hung in the front windows of homes across the United States. Hanging from a gold cord with a field of white bordered in red, the flag had a star in the middle. A blue star designated that a family member was in the service; a gold star signified that a father, husband, brother, or son had been killed in the war. The American death toll from the Second World War was 405,399. In addition, 670,846 Americans were wounded. But how many "war orphans" were there? During the Second World War, about 183,000 homefront children lost their fathers.[61]

The children were very aware of these flags and what they meant. Mary Malo-ney, a grammar-school girl in Davenport, Iowa, walked sixteen blocks to school. One morning, she decided to count all of the flags and stars on her side of the street. She counted 86 flags and 183 stars, showing that many of these families had two or more members serving. Four of the stars were gold, which her father explained meant making "the supreme sacrifice." After that, Mary began watch-ing for new gold stars, "and sadly they increased in number." One day, she saw a woman sitting on the front porch of a house with a gold star hanging in the win-dow. "I stopped and looked up at her. I wanted to say I was sorry.... I stood there for what seemed like a long time and finally gave a little wave. ..." Because of her sadness, Mary stopped counting the gold stars after that. The enormity of what was transpiring did not escape the children.[62]

Moreover, homefront children found death to be a fascinating subject. They had countless conversations about it, but comprehension of it varied develop-mentally with age. According to Arnold Gesell and Frances L. Ilg, children in the first three or four years of life have a "very limited concept of death." At age five they develop an awareness of peripheral aspects of the subject, such as coffins, burial rites, and graves. More important, for the first five years of life, children share their parents' anxiety and worry about who will take care of them if their parents die. At six and seven, there is a "new awareness of death" and the "begin-ning of an emotional response to the idea of death." Developing too is the "idea of death" as the "result of aggression or killing." From five or six to ten, many children are preoccupied with the war, especially boys, for whom the war could be "a vehicle through which they can express their own hostilities and aggres-siveness." At eight and nine, the child begins to accept "quite realistically ... that when he is older he will one day die," and to ponder "what happens after death." And at age ten, important changes are evident that continue into adolescence. Although Gesell and Ilg found some children who still believe that the dead "stay in a 'long sleep,'" they also interviewed girls and boys who asked "theoretical questions" about reincarnation and heaven. At ten or eleven, children understand that death is irreversible. Recent studies have concluded that it is children between ten and fifteen who "are the ones most frightened by the threat of war" and who "get more sleep disturbances ... because of the fear" that "'I'll go to sleep and not wake up.'"[63]

Invariably, the advice to war widows was the same: "If you can take it, your children can." No one said that the adjustment would be anything but excruci-atingly difficult for everyone, but as a widow wrote a year after her husband's death, "we are finding out how to live as a family even without Daddy. We don't intend to take our man along as an unhappy ghost but as a happy memory to laugh with and remember with love." Numerous families, however, could not report the same measure of progress in grieving; for them, the ghost was as handy as a framed photograph on the wall or mantle.[64]

Whatever the age, the homefront children's curiosity about death was excited by the war. From little children, there were innocent questions. Marilyn Silverstein remembered sitting on her grandfather's lap while he listened to the war news on the radio, and asking him what death was like. It is "like going to sleep," he told her. But some of the older children had to see for themselves. Joan Padgett and other members of her Girl Scout troop used to walk past a mortuary on the way to their meetings. "One afternoon," she remembered, "we got up the courage and peeked in the windows. Lo and behold, . . . there was the corpse of a soldier, laid out, in full uniform, ready for the funeral."[65]

The homefront children encountered death in different ways. Some saw the 1944 film *The Fighting Sullivans*, a true story in which all five of the Sullivan brothers enlisted in the Navy, and all were killed when their ship went down. Emily Roman, the youngest of nine children, six of whom were in the service, saw the movie with her mother. In the film when the Western Union delivery person knocks on the front door, Mr. Sullivan asks which of his five sons is dead. Emily remembered the sadness that swept the theater when the answer was that all five had been killed. Other children confronted death in school and at church. Nancy Murray was in the third grade when the principal and two men in uniform entered her classroom and told her teacher to come to the office. The teacher's husband of less than a year had been killed in battle. During the war Jerome Burns, an altar boy at his local Catholic parish in western Minnesota, attended the priest at burial masses and internments at the cemetery. The funerals of soldiers and sailors, he recalled, were "sadder than the normal deaths because the servicemen were all so young." Betty Hall's experience was similar. Her father, a member of the American Legion post in Jenkins, Kentucky, was in charge of providing the firing squad for funerals. "Sundays," Betty said, "when my friends were going to movies, the park and other places, we were generally at a funeral. . . ." She attended twenty-nine funerals in one year, beginning with her brother's in 1943.[66]

For boys and girls with family members at the war front, death seemed to inch closer as the war went on. Children as well as adults realized that the odds of one's returning diminished over time. First, it was young men in the neighborhood who were killed in action. In Tallulah, Louisiana, Alta Thomann saw the mother next door "screaming & screaming" after learning that her son had been killed in Europe. Jon Presnell's family lived next door to the Webbers, in Enid, Oklahoma. George was the oldest of four sons, Donny the youngest, and they all used to play together in the barn behind the Webber house. "Early in the war George went into the Army—throughout the war one by one all four brothers joined up." Only Donny returned. "The father passed away soon after . . . of a broken heart," Jon Presnell recalled.[67]

"We all sat and cried," Barbara Ceaser, a thirteen-year-old living in Detroit, wrote in her diary. "It's hard to believe." Her cousin Bobby Rollo, a recent graduate of the University of Colorado, a fine athlete, and the family's pride, had been killed in action. As soon as he received the telegram, Bobby's father, Barbara's

Uncle Nick, a widower who had raised Bobby, went to the home of his sister-in-law, Angeline Ceaser. Recalled Rick Ceaser, who was eight at the time, "my mother screamed and cried with him in the kitchen." To the Ceasers' flat came other relatives as well as many young people, who were Bobby's friends as well as friends of the four Ceaser children, two of whom were also in the armed services: Michael, who had joined the Marines at nineteen, and Harold, who had enlisted in the Navy at seventeen. "My mother stayed in the kitchen with Uncle Nick," Rick remembered, "and the living room soon filled with people talking quietly." Indeed, the only person absent was Gaspare Ceaser, a barber who, after closing his shop late on Saturday afternoons, enjoyed doing the grocery shopping. But then everyone heard his steps coming up the stairs. "I will never forget the look on his face when he opened the front door," Rick wrote. When he saw all the young people, friends of his sons, he knew something was wrong. He stopped very briefly and said, "'Hello boys, hello girls,'" but then "in sort of a half run" he hurried into the kitchen. "I knew he thought one of my brothers was dead...." For a moment, there was silence, but then Rick "heard the most agonized wail [he] had ever heard.... I didn't realize then, but... it was a cry of both relief and pain. The sound carried through the house like there were no walls to stop it."[68]

In assessing the impact of the Second World War on America's homefront children, it is clear that the most devastating event was the death of a loved one. Without exception, homefront girls and boys who experienced such losses have stated that the war fundamentally altered their lives. The war "totally changed my life," wrote Kay Britto, "—it took away two brothers I never got to know, brought *great* grief to my parents and changed our whole family structure." Bill Moore was twelve years older than his sister Erlyn, but she adored and idolized him. Bill was her mother's favorite too. In 1941 he joined the Army Air Corps and became a bomber pilot. When he was shipped overseas, his young wife and baby girl moved in with the Moores. On September 27, 1944, his bomber was lost over Germany, and the life went out of his family. "As the years went by Bill stood higher and higher on the pedestal mom had put him on. He was the last one in her thoughts when she died at 81 in 1980." As for Erlyn, she named her first son after Bill. She often talked with her children about him. "I want them to know him.... Not a day goes by that I don't think of him."[69]

During the war Henrietta Bingham lived in a small town in western Montana; she was nine at the time of the Pearl Harbor attack. "The war changed my childhood & ultimately my life. In Dec. 1941 all the war meant to me ... was excitement.... By 1945 that excitement had turned our family to tragedy & sorrow...." In 1942, Gerald, her sixteen-year-old brother, falsified his age, enlisted in the Army Air Corps, and became a top turret gunner on a B-25. He was killed the next year on a mission in the Pacific. Henrietta's parents asked their next son, Gene, not to enlist until he had graduated from high school. The day after graduation, in May 1943, he joined the Air Corps and became a nose gunner on a B-

24; the next year, he was killed on a training mission. Gene's body, which was shipped to Montana, was escorted by his best friend, Corporal Marvel Best. Marvel became part of the family. That fall he was killed in Italy. "My parents felt they had lost another son." In 1945, when Gerald's class was graduated from high school, "they left an empty chair for him.... I was then 13." Her parents had aged greatly during this period; her "young pretty mother had turned almost white," her father "appeared as an old man." "I remember sitting at that graduation ceremony," Henrietta wrote, "& remembering how excited I was at age 9 when the war began in 1941 & how I now felt just a few short years later. I had lost my childhood."[70]

Nothing did more to destroy a homefront child's sense of security than the death of a father or brother. But sudden departures and long absences also resulted in fears of abandonment. Because of the war, many homefront girls and boys suffered a premature loss of invulnerability. All else paled beside this fact.

CHAPTER 4

Homefront Families on the Move

"FORD WILL Build Plane Plant Here," boomed the eight-column headline on February 20, 1941, in the *Ypsilanti* (Michigan)*Press*. The plant, to be named Willow Run, would be located thirty miles west of Detroit in southeastern Michigan. The next month, bulldozers began clearing the flat fields. Soon the factory took shape, "a huge sprawling upthrust of red I-beams and endless windows." Within weeks, access highways and a railroad spur from the main line of the New York Central appeared, along with acres of parking lots and a mile-long airfield. By Christmas time—even though the I-beams at the east end of Willow Run were still open to the wind—production workers began filling the plant, ready to build B-24 Liberator bombers. Willow Run became the largest plant in the world: the main building covered 67 acres, while the airport spread over an additional 1,434 acres and had six runways. The first B-24 rolled out of the plant on September 10, 1942, and by war's end Willow Run had produced 8,685 airplanes. At its height the plant turned out one bomber every 63 minutes.[1]

Between 1940 and April 1944, some 200,000 people moved into the four counties of the Detroit–Willow Run production region. Almost half of these migrants—97,477, to be precise—were children under age ten. Migrants came from the southern Appalachians and from the Deep South. Many also came from farms and small towns across Michigan and from both peninsulas of the state. The daily work force, zero in 1941, peaked at 42,331 in June 1943.[2]

One family of newcomers, the Castles—John, Evelyn, and their seven-year-old son Tommy—drove to Willow Run in the fall of 1942 from a town in central Michigan. Unable to find rental housing within forty miles of the plant, they instead bought an 18-foot house trailer. "Arrived here about eight o'clock this morning in the rain!" Evelyn Castle wrote in her diary. "Drove miles, it seemed, to find a park. . . . Finally found this place with a vacancy. It is easy to understand why there was a vacancy as we are parked in a water-hole with mud all around. 'Muddy Lane,' we call it." While John found work as an inspector at the bomber plant, Evelyn signed a contract to teach fifty-eight second-graders at the local elementary school.[3]

Most of Evelyn's diary entries concern not her job, but her family's "life in the trailers." Trying to find an available washing machine in the camp's small laundry room was always a challenge, since there were but two machines for sixty families. Sewer lines laid too close to the surface froze; overburdened electrical circuits blew out, causing lightbulbs, percolators, toasters, and radios to go dead; "the small garbage pail which must serve six trailers is always overflowing." And everywhere there was mud. "Rain, rain, and more rain!" Evelyn groaned. "I don't believe there is a more depressing sight than a trailer camp on a rainy day." Snow replaced the rain in late November, but "yesterday's snow was today's puddles. Mud and water all around us. It's like being marooned on an island, except we're not alone!"[4]

The Castles' miserable housing conditions were luxurious compared with the quarters of other migrant families in Willow Run who—due either to their poverty upon arrival or the severe housing shortage—had to concede defeat in their quest for decent housing. They moved into tents, open lean-tos, shanties, chicken coops, and abandoned garages.[5]

Evelyn Castle worried as much about her son Tommy's life in the trailer camp as she did about the mud and the long lines for toilets and showers. Indeed, virtually all the mothers in the Willow Run area confided that their children were their biggest worry. "A trailer camp is no place for a child!" Evelyn wrote in her diary in November 1942. For one thing, Tommy had started "playing with boys who had not been taught the right and wrong way of doing things." That night at the dinner table, Tommy announced, "I don't want to play with Billy. He *swears* too much!" Three days later, Evelyn noted that the trailer camp lacked playground equipment, having "only one swing which becomes a source of trouble when about 35 children all want it at the same time." And, of course, "the little tots just love to wallow in the mud. . . ." For the second time in three days, Evelyn wrote in exasperation: "A trailer camp is no place for a child."[6]

The schools in Willow Run were also badly overcrowded. Tommy attended the two-room Spencer School, near the trailer camp. In June 1940 the school's enrollment stood at 78; a year later, there were 142 on the rolls; by June 1942 the number had climbed to 188. Construction to add four classrooms and a basement began that summer; the expanded school would accommodate 210 pupils. But when Spencer School opened in September, there were 300 girls and boys waiting to enroll. In December the school adopted half-day sessions, for by that time the student body had swelled to 410. In addition, all pupils above the sixth grade were bussed to the Ypsilanti schools. By then, school officials realized that some of their third and fourth graders had special educational needs, particularly the fourteen-, fifteen-, and sixteen-year-old boys and girls, fresh from southern farms and mountains, who were trying to overcome deep academic deficiencies.

Not until September 1943 did the local school district relieve the pressure on the Spencer School by opening three new elementary schools.[7]

For Evelyn Castle the trailer itself was a major concern. One problem was space. "The trailer isn't large enough to play in," Evelyn noted, "—not the way *he* likes to play—with model airplanes being his hobby!" On rainy days, she would let Tommy take his shoes off and play on his bed with his toy soldiers and airplanes. But "it is a problem in the evening," she continued, "when John tries to study and Tommy gets his planes warmed up and drops bombs on his targets." There was a more serious problem, however. One evening three weeks after the Castles had first parked in the camp, a trailer just 100 feet from theirs burst into flames. "It was a blazing inferno coming from the oil heater." Tommy watched as the fire destroyed the trailer and all its contents. He "is sleeping with us tonight," wrote Evelyn. "He was thoroughly frightened by that fire and begged us to turn our stove off." Evelyn had a judgment to render about her family's experience: "Life is not what you make it but how you can take it—in a trailer camp!"[8]

Well over 30,000,000 Americans migrated from one part of the country during the Second World War. In fact, the migration's importance begins with its sheer size and diversity. There were so many migratory streams running in so many different directions that a master collagist would be better qualified than a historian to depict this scene. In fact, what follows is somewhat of a collage, in mixed colors and hues.

Between 1941 and 1945 the Census Bureau tracked America's population; a sample enumeration in March 1945 showed that about 15,000,000 civilians were living in different counties from the ones in which they had lived on December 7, 1941. Added to the civilian migration was the military migration; by the end of the war, 16,354,000 men and women had served in the armed forces. The wartime migration between 1941 and 1945 thus easily exceeded 30,000,000, or one-fifth of the nation's population. Over all, the war stimulated the largest mass movement of the American population up to that time: 25,000,000 Americans moved and did not return home after the war. Thirteen million had changed counties within the same state and twelve million had changed their state of residence.[9]

Even these figures are low, however, since numerous Americans moved more than once during the war. Migrating from one Army or Navy base to another, or from one construction site to another, were tens of thousands of men, women, and children. An example was Rachel Love, whose father, an Air Corps instructor, was reassigned every two months, and as "he went from base to base all over the United States—we went with him." She attended five different schools in one year. But another "military brat" surpassed that total, attending 36 schools in six years. Another homefront girl, Barbara Reiniker, was the daughter of a construction worker at Army camps in Missouri, Louisiana, Utah, and Washington. In between visits to grandparents in Neosho, Missouri, Barbara and her family followed him from place to place. Trips were difficult. Barbara recalled driving back to Missouri from Yakima, Washington, in a 1934 Plymouth "that had a door on the driver's side which would not stay closed. The door had to be tied so the window could not be closed . . . resulting in a very chilly trip."[10]

The major destination of the migrants was the West. In the first two months of 1942 alone, a million people migrated to California, Oregon, and Washington. Texas, Arizona, Utah, and Nevada also received large numbers of soldiers, sailors, defense workers, and their families, as did states along the East Coast, such as Florida, Virginia, Maryland, and Connecticut. The District of Columbia, with its large influx of government workers, also boomed in population. Likewise, the Great Lakes region, with large migrations into Michigan, Indiana, and Ohio, experienced sizable growth during the war.[11]

The largest wartime population shift was from farms to cities. The rural population of Oklahoma, Kansas, and other plains states, for example, decreased by about one-fourth between 1941 and 1945. The major loser, however, was the South, which—with 1,500,000 people moving to other regions—contributed five times as many people between 1941 and 1945 as it had between 1935 and 1940. In November 1943, midway in the war, the Census Bureau counted the number of ration book registrations and concluded that the metropolitan population had risen precisely at the expense of rural America. During the war years, one of eight rural Americans left their farms forever.[12]

Not only was the rural population declining rapidly, but it was the younger families who were leaving. The Census Bureau classified 21 percent of the children born between 1940 and 1947 as "migrants," noting that the proportion of children actually eclipsed that of adult migrants. The greater mobility of the children "may be partially explained by the younger average age of the parents involved . . . younger adults always being the most mobile. . . ."[13]

For one thing, young couples were at an age to make lifetime decisions. As Conrad Taeuber, the demographer, pointed out, "Two-thirds of the young [farm] men who had been between 20 and 25 years of age in 1940 had migrated or entered the armed forces by 1945." The exodus from the farm naturally included many children. From April 1940 to April 1944, the number of farm children under age fourteen decreased by 1,200,000, evidence that families were on the move. In August 1942 an expert on defense housing estimated that of the additional 1,600,000 war workers who would relocate by July 1943, only one-third would move as individuals. Families "were fairly mobile," Taeuber explained, "when children were young, less so when the children had reached adolescence."[14]

During the Second World War, farm families migrated to the booming bomber factory at Willow Run, to the massive shipyards in San Diego, and to numerous Army bases and Navy posts scattered across the United States. Joining the farm exodus was the small-town migration from the South, the Great Plains, and other regions. For all these migrants, life was radically different in the "war-boom" areas, such as Los Angeles County, which between April 1940 and October 1943 experienced an influx of 568,143 migrants. And the migrants brought their children; from 1940 to 1943, the number of children under five increased by 136 percent in Wilmington, North Carolina, 128 percent in Pascagoula, Mississippi, 80 percent in Wichita, Kansas, 60 percent in Savannah, Georgia, and by large percentages in Galveston, Fort Worth, and Beaumont–Port Arthur, Texas.[15]

An example of spectacular growth was Richmond, California, which in April 1940 had a population of 23,642. Located diagonally north across the bay from San Francisco, Richmond was a port city. According to Zelma Parker, the city's supervisor of child welfare in the public schools, Richmond was "a nice little American town." Within three years, however, the United States Maritime Commission had constructed four shipyards there, employing 90,000 people. In addition, various industries had also built factories and refineries there, including Standard Oil of California, the Pullman Company, and the Atchison, Topeka & Santa Fe Railroad. By April 1943, Richmond's population had soared to 135,000. Children were everywhere. "Teachers are so busy registering new pupils," Zelma Parker wrote, "that regular classes cannot be held. School rooms are filled with unfamiliar looking children. Shock-headed, dressed differently, dirty, poor teeth, rickets, queer talk." The influx of so many people put severe strains on the city's educational, health, and welfare facilities. Richmond exemplified other problems common to the war-boom areas: racism, nativism, classism. As Parker explained, the influx not only strained but frequently exceeded the "natives'" tolerance, thus activating hostility toward the newcomers. "'Peace on earth, good will toward men.' But not good will," Parker stated, "to the Okies, the Arkies, and the Texans."[16]

The movement of war-production workers and their families was but one of the streams of the wartime migration. Another consisted of "camp followers," those families hopping from one military installation to another. During the Second World War the wives and children in America's service families accompanied "Daddy" on his several reassignments until, finally, he was shipped overseas, and they could follow him no more.

Camp following was a happy time for some homefront children, a disaster for others. Elizabeth Schluemer was nine in September 1942, when her father, a Lutheran minister in Sioux City, Iowa, joined the Army as a chaplain. She recalled sitting on the family's front porch "watching our furniture and toys and all our everyday things being loaded on a moving van to go into storage. It would be three years and three months before we would see our belongings again in a home of our own." During the 1942–43 school year, Elizabeth and her younger sister attended five different schools and, while on the road between duty stations, missed eight weeks of classes. Elizabeth's family spent a few weeks here and a few weeks there, staying with relatives in Ann Arbor, Michigan, Pittsford and Seneca Falls, New York, and North Platte, Nebraska, before moving to duty assignments at Fort Lewis, Fort Worden, and Fort Flager, all in Washington State. In March 1943, when her father received orders to report to an infantry division on Kodiak Island, Alaska, Elizabeth and her family moved back to Sioux City. "I know this all was harder on my parents than it was on us as children." In fact, "to us it was exciting. . . ."[17]

For most homefront children, being a camp follower was preferable to the alternative of living without father. One girl's family moved in with her grandfather in Des Moines, Iowa, when her father joined the Marines. Eventually, however, he received orders to Camp Pendelton, California, "and joy of joys, we could go be with him. . . . We moved into a quonset hut and . . . we loved it. . . . Then some brand new Navy apartments opened up and somehow we were one of the lucky few families who got to move there. It seemed beautiful and luxurious to us."[18]

Elaine Dunn was seven in 1940 when her father, a member of the naval reserve, was called to active duty. Only twice during the next five years, including a period when her father's ship was torpedoed and he was missing in action, did Elaine see him for extended periods. Once, her mother took Elaine and her two younger brothers by train to Philadelphia, where her father's crew was on temporary duty. Another time, they joined him in Tacoma, Washington. "These few months," Elaine recalled, "were the only times I remember that we lived as a 'real' family. . . . I marvel that my mom, herself only 24 years old, and with the complete responsibility for three children, would set out for the east and west coasts with nowhere to land when we got there!"[19]

A service wife, writing in *Parents' Magazine*, tried to buoy up other women by showing that camp following could be "a rewarding and heart-warming experience for all," parents and children alike. "This is our second home since being in the Army," she wrote, "and the adjustment to two neighborhoods in the past few months has not been easy for [my] young [son] John. But we are now convinced that, far from being harmful, it has been very good for him. He is much less babyish than he was. . . ." Best of all, she explained, the family was together. Facing the inevitability of an overseas assignment, "we treasure as never before every chance to be together. . . . We are all storing up memories of family good times that will help us through the days ahead."[20]

The child's perspective, however, was not always this rosy. Homefront girls and boys remember lonely days in hotel rooms; others recall their cars breaking down on the highways, with blown tires that could not be replaced because of rationing. Social ostracism was another problem for service children. A girl recalled that she, her older sister, younger brother, and mother joined her father when he began military training in Virginia; they lived in a small trailer outside of town, where "we were always treated like poor white trash." Another problem was loneliness, caused by attending several different schools in one year. "I was *always* the new kid in the classroom," remembered one girl, "the one who didn't stay long enough to make any friends. As soon as I would just about have a friend, we would move on."[21]

A difficult emotional burden was borne by those little camp followers whose fathers, and sometimes mothers as well, did not want to bring them along. Called "Bureau-Drawer-Crib Babies" by the clinical psychologist Arthur L. Rautman, these children slept in bureau drawers in a series of tiny rented rooms until one day the father was shipped overseas. In such families, Rautman explained, "husband and wife have not lived together long enough to establish a psychological family unit"; "their only life together has been under unstable and abnormal conditions, perhaps limited to a short honeymoon or to interrupted periods of casual housekeeping in tourist cabins, transient hotels, or furnished rooms." Resented because they cried and demanded attention at a time when the serviceman and his wife wanted to focus on each other, some of these babies never did bond with their fathers.[22]

For some service families, the last wartime opportunity to live with Daddy came during his final furlough before being shipped overseas. The fathers, of course, had their own anxieties to contend with. "While most men express a great desire to see family members again before leaving for overseas," wrote an Army counselor at Fort Leavenworth, "many worry about the final good-bye and whether or not the emotional anxiety of the last few moments together is worth the happiness of the leave." The experts consistently gave families one piece of advice: Say goodbye at home, rather than in a crowded train station. *Parents' Magazine* told mothers not to take their children out of school in order to have a final farewell ceremony. One concession, however, was to let Daddy call for the children at school: "To have their classmates see their Daddy in uniform will set the children up tremendously. . . . And don't," the magazine insisted, "end Daddy's leave with tears and dramatic farewells."[23]

Often the last furlough did not follow this script. Some children had premonitions that their fathers would not return. Jane Coad was seven in July 1944 when her father said goodbye. "When he left us at the end of his furlough, I *knew* I'd never see him again. I was a very out-going, happy child . . . who wasn't given to crying, but I became hysterical when I got on the troop train with him (to leave his bags near his seat) and he had to have my mother leave with me." Her father was killed four months later in heavy fighting against the Germans.[24]

In his January 1939 State of the Union address, President Franklin D. Roosevelt submitted a $9 billion budget containing $1.32 billion for national defense. Just a week later, he asked Congress for additional appropriations totaling $525 million for building a "two-ocean" navy, strengthening the nation's seacoast defenses, and manufacturing military aircraft. And in March and April, Roosevelt requested further appropriations. Throughout 1940, additional millions flowed into rearmament programs. Actual outlays for the Departments of War and Navy rose from $1,368,000,000 in 1939 to $1,799,000,000 the next year, an increase of 31 percent. In 1941 the total soared to $6,252,000,000; in 1942 to $22,905,000,000; in 1943 to $63,414,000,000; in 1944 to $75,976,000,000; and in 1945 to $80,537,000,000.[25]

These federal billions for defense practically eliminated unemployment in the United States; indeed, many areas reported severe labor shortages. But there was another factor involved in the elimination of joblessness: the entry of millions of people into the armed forces. In 1939 the number of military personnel on active duty was only 334,473. Two years later, the figure had spiraled to 1,801,101. In 1942 it doubled again to 3,858,791, and in 1943 it more than doubled to 9,044,745. In 1944 the total rose again, and in 1945 it peaked at 12,123,455.[26] And as men and women flocked to the nation's war factories and induction centers, unemployment plummeted, falling from 8,120,000 (14.6 percent) in 1940 to its wartime low in 1944 of 670,000 (1.2 percent).[27]

Clearly, by 1940 the homefront war was already under way. "In a social and economic sense," wrote the sociologist Philip M. Hauser in 1942, "American participation in the war antedates December 7, 1941. The defense, lend-lease, and victory programs had already been initiated when Pearl Harbor was attacked." Interstate migration was such a part of these changes that in March 1941 the Tolan Committee, which had been established the year before as the Select Committee to Investigate Interstate Migration of Destitute Americans, renamed itself the Select Committee Investigating National Defense Migration. "As the defense program accelerated," committee member Sparkman observed, "a new problem emerged: The interstate migration of workers drawn to defense industries."[28]

With rising defense expenditures, internal migration in the United States not only grew, it changed in composition as new streams of migrants took to the highways. By early 1941, for example, a million servicemen had moved to duty stations then being constructed in out-of-the-way places where land was cheap, but rental housing nonexistent. Paul V. McNutt, administrator of the Federal Security Agency, testified that "some of the men of the armed forces will move their families with them, and this swells the size of the population of nearby communities." Also swelling these populations were thousands of construction workers, who had been unemployed or underemployed since the collapse of the housing industry in the 1930s. Nearly 700,000 workers would soon be employed on construction projects, McNutt explained, "building these new camps and new factories." Still other migrants were moving to major defense-production areas. "The idle machines of Detroit, Chicago, and Pittsburgh," McNutt stated, "have begun to hum again and are tended partially by the local unemployed and partially by the newcomers from surrounding areas."[29]

While soldiers and sailors' families constituted a large part of the wartime migration, this stream was secondary in size to the flood of men, women, and children flowing into production centers. In big places and small, the migration, as Detroit's superintendent of public welfare explained, was "principally of young families and of young workers who were particularly adaptable for the streamlined type of industrial employment developing in Detroit. . . . One had only to walk along the streets . . . to notice how young was this city's population." This was a migration of hope after years of economic desperation.[30]

The defense migration was not without its ironies, however. For just as defense funding created jobs, so too it destroyed them. For example, workers in industries judged by the government to be nonessential to the national defense found themselves jobless. Some companies had to reduce production or shut down entirely, not enjoying high enough priorities to obtain valuable production materials such as copper and aluminum. The "displacement of labor resulting from material and equipment shortages," stated an economist with the Work Projects Administration, "will occur in all parts of the country." For a while, labor displacement plagued the automobile workers; production fell from 3,779,600 cars in 1941 to 220,800 the next year, and companies laid off more and more workers. The Michigan Unemployment Compensation Commission estimated in January 1942 that 175,000 automobile workers had lost their jobs. Likewise, manufacturers of stoves and refrigerators in Wisconsin, Ohio, Indiana, Michigan, and Pennsylvania laid off tens of thousands of workers. In January 1942, Eleanor Roosevelt—in her capacity as assistant director of the Office of Civil Defense—told the Tolan Committee that she had received numerous letters from Detroit's unemployed. One wrote, "We have been laid off, we don't know what is going to happen to us." Rumors abounded in the city. "How long will it take to convert plants?" asked another person. "The cost of living is rising. Our unemployment compensation is inadequate. Our whole situation is insecure."[31]

This would soon change, as industry rebounded with the conversion of automobile, refrigerator, and stove factories into bomber, tank, and ordnance plants. Construction jobs also became available, and thousands of carpenters, masons, painters, and plumbers joined the migration. Their job was not to build houses, however, but to build the war-boom towns and mammoth defense installations arising in Virginia, Georgia, Texas, California, and elsewhere. One such place was Childersburg, Alabama, which, in December 1940, was selected as the site of an $80 million powder plant. But Paul McNutt, for one, bemoaned Childersburg's future. "A southern town of 500," he said, "is expected to attain a population of at least 5,000 as a result of a nearby military establishment with an aggregate military strength of 13,000 men." The town lacked a bank, a hotel, and a movie theater; vacant houses were rare, apartments nonexistent; and Childersburg's water supply was even "inadequate for its normal population. . . . There is no sewer system whatever. . . . There is no pasteurized milk available in the area. . . . There are no hospital facilities within a distance of 40 miles."[32]

Some people who came to Childersburg were unwilling migrants who had been forced off their lands by the federal government. As the American military and industry turned pastures and cotton fields into air bases and ammunition factories, the people who had worked these lands for generations had to move. As early as January 1942, some 14,500 families had abandoned their farms because of the military's purchase or lease of more than 4,000,000 acres of rural land. By the end of the year, the number of displaced farm families reached 30,000, mostly in the South and Midwest. At Camp Stewart, near Hinesville, Georgia, 1500 farm families had to vacate 360,000 acres of piney woodlands. For the first time in years, land values in the South rose; but this resulted in another problem, as the white landowners, rushing to sell, evicted the black tenants who had spent their lives on these farms. Again, Childersburg is an example. Initially, the project to build the Alabama Ordnance Works was delayed due to confusion in surveying the plant's location. Original plans called for 27,000 acres, but living on the land—four miles north of the town—were 210 African-American farm families. Thus, when the white landowners sold these lands to the Army, these families had to move. Some migrated to nearby counties, others went to cities in search of industrial jobs. Some found employment at the ordnance works. But only later did the Army realize that it had made a mistake in Childersburg, and needed only 13,500 acres for the powder mill.[33]

On the Great Plains, too, landowners sold large parcels of land and evicted their tenants, or raised rents so high that only the newcomers—construction workers, for example, their wallets flush with overtime money—could afford to pay. In November 1941 in Saunders County, Nebraska, the army purchased 17,000 acres for a bomb-loading plant, thereby dispossessing 112 families. Both landowners and tenants were angry, arguing that the government was cheating them out of their land and crops. Milo Stall, a tenant forced off his farm in Iowa because of an ordnance plant, complained that the government had given him "kind of a fake deal.... They came around and told my wife we had to move in 10 days." The announcement "scared a lot of people around here and they started to hold sales of their stock right away." Stall, age thirty-five, his wife, and four children then "moved over here to this hog-house that we cleaned out."[34]

For millions of rural Americans, however, the war provided the opportunities for which they had yearned for over a decade. Because of falling prices, then drought and dust storms, farmers in the Dakotas, Nebraska, and Kansas had suffered hard times throughout the 1930s. But then, explained Fred Seaton, a Nebraska newspaper publisher, "all of a sudden, the Government began taking our young men and women into the civil service, and the mechanics into the defense plants. And, of course, the draft has taken some of our boys, too." In fact, Nebraska was losing its brightest, most ambitious young people—high school graduates "accustomed to handling machinery and accustomed to making decisions." And for many families, a way of life was changing. One Nebraska farmer—asked by the Tolan Committee whether he believed his son, who was working in a Denver defense plant, would ever return home to farming—replied sadly: "I doubt it very much, sir."[35]

Migrants from the northern tier of the Great Plains, including the Dakotas,

Nebraska, and Kansas, tended to move to the Pacific Northwest, whereas those from Oklahoma and Texas went primarily to California and Arizona. Some migrants stopped in Arizona to pick cotton on the way to California; others picked crops in California before seeking factory jobs. Once in California, however, the migrants quickly realized that practically everyone applying for war production work was a migrant. In mid-November 1941, only 11.4 percent of the employees of San Diego's Consolidated Aircraft Corporation were native-born Californians. Almost as large a percentage (11.1) was from Texas, while Missouri contributed 7.3 percent, Oklahoma 6.7 percent, Kansas 5.7 percent, and Illinois 5.4 percent. At this time, too, 35 percent of Consolidated Aircraft's work force had been in California less than six months.[36]

But what about the children's welfare? Did their parents believe they had made the right decision in migrating? These are difficult questions for the historian to answer, but many such parents provided their own, generally affirmative answers, at the end of the war. Some remained where they had moved, others went someplace else; but relatively few returned home. Studies in July 1944 of wartime migrants in Portland, Oregon, and in San Diego showed that three-fourths wanted to stay there after the war. By the end of 1945, the Census Bureau reported that there had been no "appreciable return movement of civilian migrants" back to the farm.[37]

For some of America's wartime migrants, there was no real need to decide whether or not to uproot the family—for these units were already on the road. Among America's migratory laborers were Mexican-American families who toiled in fields miles from home. In 1941, however, new opportunities became available for Hispanic-Americans. As Gerald D. Nash, the historian, has written, the Second World War "created hitherto undreamed of possibilities for jobs, education, and training. Invariably, these drew Spanish-speaking Americans to towns and cities, thus accelerating their urbanization." Leaving the farm fields behind, these migrants headed to the war-boom centers of the West and Southwest. In Los Angeles, for example, 17,000 people of Mexican descent found shipyard jobs where before the war there had been none available to them. And to replace these farm workers as well as the Japanese-American farm workers who had been interned, the United States in 1942 signed the "Mexican farm labor agreement," better known as the *braceros* program (for *brazos*, meaning "one who works with his arms"), in which Mexico, a wartime ally, agreed to supply seasonal workers for agriculture and the railroads. During the war years, a total of 168,000 *braceros* labored in the United States.[38]

Another stream of migratory farm labor that got diverted during the war was one that began each spring and flowed northward from the Florida Everglades. African-American families loaded into jalopies, ancient buses, and rickety pickup trucks. The children over age seven picked potatoes alongside their parents in North Carolina; strawberries, tomatoes, and more potatoes in Virginia and Maryland; and finally arrived on New York's Long Island in September for two final months of potato-picking. In November, with only a few dollars in their pockets, the families turned their vehicles around and headed back to Florida, riding on threadbare tires.

Once back in Florida, the families spent the winter months picking beans, peas, and celery. As tough as life on the road was for migratory workers, in some ways Florida was even worse, for in that state African-Americans had few rights which whites felt bound to respect. Just weeks after Pearl Harbor, growers in Belle Glade and Pahokee tacked up posters in the quarters, juke joints, and other places where black men congregated: "Work or Fight: This means you." The growers, who sat on the local draft board, also threatened to place any man working only two to three days a week on the preferred list to be inducted into the army. But in early 1942, after years of exploitation, growing numbers of African-Americans saw a chance not only to escape oppression but also to share in the nation's wealth and perhaps even in its democracy. Foremost in the minds of many were the children who, having their entire lives ahead of them, would be the chief beneficiaries. Not everybody left the area, of course, but many did. "Negro vegetable pickers," the *Belle Glade Herald* reported in June 1942, "have . . . either gotten other and better paying jobs or have gone into the army. . . ."[39]

For many Americans, in Florida and elsewhere, a way of life was coming apart. Among the wartime migrants were Dust Bowl refugees, displaced black tenant farmers, Chicanos from the Southwest, and jobless women and men from the urban East and Midwest. Joining them were millions of other Americans who crowded the nation's train stations and highways. Some were national defense workers, others were service families. And since many of the migrant families were headed to the same destinations, it is important to appreciate what happened once they converged in America's war-boom communities.

Fueling this mass migration was a combination of optimism and patriotism. In remaking their lives, the migrants demonstrated not only their ambition but also their willingness to take risks. These qualities were admirable and even heroic. Indeed, the migrants were America's new pioneers.

Sometimes, however, the resolve of these pioneers flagged in the face of the prejudice and hostility that awaited them, for the established settlers saw them as anything but heroic. Willow Run is a prime example; when the residents decried "the invasion" taking place, the newcomers' readjustment difficulties became glaringly apparent. Of the migrants to Willow Run, those most disdained, if not despised, were whites from Appalachia and African Americans from the deep South.[40]

Most whites migrating to Willow Run came from Kentucky, Tennessee, Alabama, and Mississippi. According to historian Alan Clive, the migrants who faced "the greatest difficulties in adjustment" were those from the southern Appalachians. So many seemed to be arriving that a wartime joke began with the question, "How many states are there in the Union?" "Forty-six," was the answer.

"Tennessee and Kentucky are now in Michigan." The punch line of a more hostile version was: "There are now only forty-five states in the U.S.—Tennessee and Kentucky moved to Michigan and Michigan went to hell!"[41]

"Southern riffraff," the natives called the newcomers—"white trash," "Kaintucks," "briarhoppers," and, of course, "hillbillies." One sociologist concluded that "a certain class of white southerners was conceptualized as an identifiable group, one disliked even more than Negroes...." Often the natives' initial contact was with the migrant children. "You can't be sure of these people," explained Mrs. Willard Cuff, a widow who had lived near Willow Run for more than thirty years. "A little boy from one of those families came into our back yard and picked some pie plant. Of course, the yard is not marked off. One of my neighbors told him to go home. You know what he told her? 'I'm not afraid of the state police or any of you big shots!' That's what he said," Mrs. Cuff continued. "I guess he meant *because we live in a house.* His parents must have told him that."[42]

The established settlers were openly critical of the "trailerites." Trailers not only crowded the trailer camps, but because parking spaces were at a premium, they also dotted the vacant lots in townships around the plant. Numerous trailer families had outdoor privies, and some buried their garbage in the backyard. Increasingly, the natives warned that the newcomers were contaminating the water and risking outbreaks of typhoid fever. Still worse, they complained, the "trailerites" showed no interest in restraining their rude and belligerent children. Indeed, the children's behavior became an important measure of the dangers posed by the migrants. "We're law-abiding people and mind our own business," explained Mrs. Wenborn, an eighteen-year resident of a home near the bomber plant. "I wish our neighbors would do the same." She reserved special censure for "one family on Harris Road that has no control over its children—three girls and two boys from 7 to 14. Those children threw rocks into our yard last summer. . . . Then they threw green apples into our garden, broke the tomato plants, and ruined the crop. They are regular little vandals, always destroying something."[43]

What the established settlers chose not to see was that these children were also victims—of inadequate food and nutrition; of unheated trailers; of perpetual sniffles and runny noses, not to mention whooping cough, meningitis, and even polio. Some of the children were the victims of fatigued parents, their patience and energy stretched to the limits by the dual demands of the workplace and the home. All the natives could see, however, was the squalor. Paranoia and revulsion were the lenses through which established residents peered at the newcomers.[44]

Rumors and snap judgments magnified fears and apprehensions on both sides. "There has been less recognition," sociologist Thomas R. Ford has written, "that the rural provincialism of the mountain migrants is matched by the urban provincialism of city dwellers, and that this latter also figures in the maladjustment." A woman whose home was next to a trailer camp complained not only that the school was overcrowded with migrant children, but also that their parents "look as though they have tuberculosis, and many of the children are very dirty. I'm afraid of skin diseases. And the stories we hear about the kind of men in the [company's] dormitories, and the possible danger of sex maniacs attacking children—well, I've been taking my own two children and their cousin to and from the ... school."[45]

But what did the war-boom experience look like from the other side—from the perspective of the uprooted southerners who lived along the muddy lanes in the trailer camps? One consolation was that the newcomers from Appalachia did not feel entirely isolated. Families and friends had migrated in chain networks, with people from the same hollows and villages helping to populate the same trailer camps and schools. The state of Kentucky supplied the largest group of families; of the 1900 pupils registered in the schools in Willow Run Village, 437 were Kentuckians. While the child folkways transplanted from the Appalachians proved shocking to the old settlers, the children clearly brought commendable traits with them as well. Malcolm Rogers, the local school superintendent, made this assessment: "Parents from the South allow their children to use tobacco at a very early age," and profanity was more prevalent among the new children, as was "some petty thieving because so much property of interest and attraction has not been properly protected." But, added Rogers, "these children ... do a minimum of fighting. They are unsophisticated and surprisingly honest. They are inclined to be kindly and friendly."[46]

Like many European immigrants who had preceded them to America's urban-industrial centers, whites from Appalachia had led rural, traditional lives. When they moved to the North, this changed. No longer regulated by the rhythms of the seasons, nor even by the sunrise and the sunset, the new settlers responded to the factory whistle, which signaled the beginning of the work shift, and to the industrial discipline governing their behavior once inside the plant. The factory timetable, as one student has noted, was "foreign to much of the older mountain culture, where time was marked by seasonal change, important events, and the length of shadows."[47]

For Appalachian people, the economic need to move was pressing. "There's no work in Kentucky," explained a migrant to Detroit. Recognizing this, the Ford Motor Company sponsored mass meetings in the South to recruit workers to Willow Run. Local radio stations and newspapers also carried patriotic appeals to stimulate the migration. Finally, relatives and friends who had earlier made the trek to Detroit wrote letters and sent money back home, encouraging others to join them. "By bus, train, and car they poured into Michigan," Clive has written. "Thousands of people passed through Detroit's two railroad stations and the Greyhound bus terminal each day, sometimes carrying all their possessions in a single battered suitcase or canvas bag."[48]

The migrants arrived somewhat dazed at their destinations. They were generally unfamiliar with the amenities of urban life, ranging from electricity and running water to public schools and buses. A young man from Tennessee had difficulty mastering Detroit's traffic signals and street signs, so he improvised. Riding the bus, he counted the number of trees between his home and the plant; he knew that ninety-four trees down the road he would arrive at his job.[49]

Still, the southern Appalachian migrants were better educated than the people they left behind. They were also younger, numbering many young families in their ranks; indeed, it was the couples of child-bearing age who were leaving in the largest numbers. And they were ambitious for a better life.[50]

Nevertheless, the old settlers had little but scorn for the "hillbillies." Their accent was the primary identifying mark; once northerners heard it, they cringed and tended to close their hearts and minds. As a Protestant minister in Willow Run Village, a government project offering housing to 2500 families, explained: "We of the North have a greater reserve and think that the southerners are forward; the southerners feel that the northerners are cold," and "the difference . . . accounts for a great deal of the enmity that exists between the two groups." The southerners agreed. One, David Crockett Lee, wrote a letter to the *Detroit News* in May 1942 explaining why he had decided to return home. "We found Detroit a cold city, a city without a heart or a soul. . . . So we are going back to Tennessee . . . where men and women are neighborly, and where even the stranger is welcome." But while some left, the great majority stayed and adapted.[51]

During the Second World War, hundreds of thousands of migrants moved to such "congested production areas" as Willow Run. Between 1940 and 1944, for example, San Diego grew from 289,000 to 609,000, while Hampton Roads, Virginia, the site of naval bases and shipyards, jumped from 343,000 to 656,000. San Francisco's Bay Area climbed by 583,000 people, while Los Angeles' increase was 518,000.[52]

A second category of war-boom areas received the governmental designation of "new manufacturing areas," since most of these areas had been farmland with no manufacturing activity prior to the war. Located near remote towns—such as Choteau, Oklahoma; Listerhill, Alabama; Milan, Tennessee; and Parco, Wyoming—these new plants in the countryside produced explosives, synthetic rubber, ammonium nitrate, anhydrous ammonia, and aviation gasoline. Workers at these plants also assembled bombs and loaded shells, and children living nearby feared explosions. Pat Jury, whose parents worked at the Sunflower ammunition plant in Kansas, recalled that when the air raid sirens sounded while she was in the plant's child-care center, she felt "an overwhelming sense of dread."[53]

In 1943 a subcommittee of the House Naval Affairs Committee investigated living conditions in "congested areas." Beginning in Hampton Roads in March, the subcommittee also listened to testimony in San Diego, San Francisco, Los Angeles, Newport, Rhode Island, and in the Puget Sound and Columbia River areas. What the members heard was not surprising: the migrants encountered substandard housing and exorbitant rents; health hazards ranging from scabies, ringworm, impetigo, and other skin conditions to life-threatening epidemics such as meningitis, rheumatic fever, and polio; open sewers, leaky septic tanks, loose refuse and garbage, and other sanitary deficiencies; shortages of fresh fruits, vegetables, meat, and milk; inadequate child-care facilities; crowded schoolrooms, teacher shortages, and upsurges in juvenile delinquency.[54]

Migrant housing was a disgrace. The homefront children recalled overcrowding and dilapidation. Corinne Brown's family lived in a "dump" in Omaha. "Housing was impossible to obtain," and she judged that "that house today would be condemned!" Corroborating the children's memories are the House Naval Affairs Subcommittee's hearings. The committee learned about crowding in housing when it came to San Francisco. Dr. Herbert R. Stolz, a medical doctor and assistant superintendent of schools in Oakland, told about two Oklahoma families consisting of four adults and eleven children who lived in a single room; they slept on the floor and cooked on a one-burner hotplate. At another address, two families with a total of twelve people slept in the same room, but as Dr. Stolz explained, "They don't all sleep at the same time—it's a 'hot bed' up there." In another single room lived a family of seven, including a child who had tuberculosis; "The mother is trying to isolate it in one corner of a room and there are these two adults and five children and no toilet in the place at all." Yet, as bad as the overcrowding was for whites, for black families it was even worse. Clearly, explained a housing official, "There is no money in housing the poorest people well." But there were handsome profits to be made in housing poor people poorly.[55]

Middle-class families also faced deplorable housing conditions. They too lived in cramped quarters and paid exorbitant rents, in spite of the Office of Price Administration's efforts to control rents in defense areas. Regardless of race or class, families with children suffered the most. Landlords evicted families and replaced them with single war workers who collectively paid much higher rents. Landlords also specified in newspaper advertisements that they were interested in renting to "married couples only" or to "adults only." And couples with children complained that their landlords failed to make repairs or to exterminate bugs, in the hope of encouraging them to go elsewhere. Bugs—especially cockroaches, silverfish, and bedbugs—were such a frequent complaint that their size and speed became legendary in the retelling. Because of an infestation of cockroaches, Bill Ratell, a welder at the Philadelphia naval yard, wanted to move his wife and two infants to another apartment. But, as he told the Tolan Committee, "When I mention children they don't want to talk to me any more." Responding that this practice was spreading all over the country, Chairman Tolan sadly observed: "It used to be an honor to have a large family, but it seems to be a handicap now. . . ."[56]

In December 1943 another congressional committee, this one under the chairmanship of Senator Claude Pepper of Florida, traveled to Pascagoula, Mississippi, a port on the Gulf of Mexico between New Orleans and Mobile. For three days in the library of the town's high school, the committee members met not only local officials but also workers and their families, nursery school and grade school teachers, and welfare workers. With Pascagoula, the story comes full circle, for Pascagoula was Willow Run in miniature.

In the late 1930s the Ingalls Shipbuilding Corporation had moved to Pascagoula, a sleepy town of 4000 people. Within months workers at the new shipyard began welding the hulls of steel ships, and by the autumn of 1943 the population had grown to 30,000. As in Willow Run, there were shortages of milk, meat, and ice, not to mention of doctors and child-care centers; similarly, the elementary school day was split into two shifts to accommodate the migrant children. And even substandard housing was at a premium. E. L. Mancil, president of the local metal trades council, told the Pepper committee that one of the federal housing projects for war workers currently "had 300 applications on record for houses and no houses to put the people in." As a result, housing in the vicinity of the shipyard was so overcrowded that it endangered people's health. In November 1943 the Public Health Service inspected thirty of the trailer, tent, and cabin camps sprinkled in and around Pascagoula: "Several of these camps consisted of tents and wooden shacks. . . . Many lacked the basic sanitation facilities," such as running water and garbage cans. The worst health hazard was "the effluent from septic tanks and cesspools . . . as the majority of places were drained directly into a . . . ditch adjacent to the camp grounds."[57]

Health problems in the war-boom towns included overworked doctors, who sometimes made critical mistakes. Mrs. Harvey McMinn, who had moved to Pascagoula with her family in January 1943, lost her two-and-a-half-year-old son to medical malpractice. Sickly for months before his death, the boy had suffered from asthma, acidosis, frequent vomiting, and anemia. One night, he went into convulsions. Mrs. McMinn telephoned the doctor and, as she recalled, he "said he was tired and had to go to bed and get some sleep, and he didn't come." They did not try to get another doctor that night; instead, at 5:30 the next morning, the McMinns began the five-hour drive to their home in Jackson, Mississippi, where the boy could be treated by the family doctor. But he died enroute.[58]

Perhaps the worst hazard to these homefront children was one common to poor people across the country: fire. Testifying before the Pepper committee, the representative of a tenants' association in Pascagoula told of a recent trailer fire. "I believe the people went on a party and left the children with another lady to take care of them. This lady went out and while she was gone the children got hold of some matches and set fire to the gasoline used in the cookstove. . . . [Next,] the gas heater overturned, the trailer caught on fire, and burned. The man who lived next door saved the lives of three of the kids, but one child died."[59]

The federal government belatedly recognized the wartime housing problem. Some of its measures were stopgaps; for example, the National Housing Agency launched a "Share Your Home" campaign, urging families to "move over" and make room for the migrants. And families responded. An estimated 1,500,000 families shared houses, flats, and apartments with relatives, friends, and strangers. In fact, among the homefront children's fondest memories are those of the lodgers with whom they shared their homes during the war. In 1943, when Mary Crownover was eight, her family moved to Fort Worth, Texas, where they shared an old house with another family of war workers. "Both men worked nights," she explained, "so it was great to have two mothers." A homefront girl who lived in Hawaii recalled that her "extended military family" included a Jewish naval officer and a Native American chief petty officer, both of whom "had been rudely rejected by their Hawaii contacts." Joalyn Hopper lived several thousand miles away in Childress, Texas, which became the site of a school for bombardiers; but her recollections were similar. Joalyn, who was born in 1940, slept in her parents' bedroom, while the dining room became the living quarters of a succession of lodgers. Vera Schock's mother was working as a secretary in her rabbi's office in Camden, New Jersey, when a man from New York City walked in to request help in finding housing. His name was Eli, and he was an engineer at the local war plant. Vera's mother said that he could stay with her family. "My sister moved into my room and Eli . . . became our boarder. This was my fondest memory of the war years." Eli, who played chess by mail with a friend, was not only intelligent, but he also took the time to talk with and listen to Vera.[60]

Homefront children expressed pride that by taking in lodgers, their families were helping to win the war. Delores Naydean lived in Hastings, Nebraska, which had a naval ammunition depot. The government urged local families to open their homes to the war workers. Her mother moved Delores and her sister out of one bedroom and into the dining room, emptied another bedroom by putting a rollaway bed in the living room for her brother, and then rented the bedrooms to two couples. "We made do," Delores remembers, "& my mother felt she was fulfilling 'her duty' in the war effort." Her mother was also pleased to have the extra income.[61]

While commendable, voluntary efforts were inadequate. Leaders in industry and labor urged the government to build emergency housing. And during the war, the government did produce about two million dwelling units—by building low-cost housing projects and by converting other buildings into apartments. War workers' villages sprang up across the country. They had quaint names, like Hilltop Manor and Plainview, both in Wichita. And they were massive. For some homefront children, government housing represented a major improvement in their standard of living. Marsha Spencer was ten years old in 1942 when her family moved into Redbank Village, which housed shipyard workers in South Portland, Maine. "How exciting this was to move into a brand new house!" Children did suffer some mishaps, however. Esther Williams had moved with her family from a tenant farmer's shack in the cotton fields near Fort Worth to "a mammoth housing project near Wichita, where all the duplex houses were alike." She recalled one day getting "lost coming home from school" and walking "into the wrong house to find the man of the house sitting in his underwear drinking a beer. The house number was right, but the street was wrong."[62]

But defense housing was in short supply. Beyond the nation's "critical war needs," explained the National Housing Agency, "we were not able to spare materials and manpower to take care of the normal new housing needs of the nation, to accommodate the increased number of families, or to replace substandard housing or slums." In short, about two million migrating war workers and their families—or half the total—had to find quarters "in existing dwellings," which were terribly overcrowded and consigned many homefront children to squalor.[63]

All things considered, the suffering that many children experienced in the war-boom communities was not only physical but emotional. Katharine Guice, a welfare worker in Pascagoula, told of the sadness felt by children who "have come in here and have been uprooted. . . ." Some "are really sick emotionally," she explained, "from the strain they have been put under due to moving, parents who are discontented and who work long hours, and are irritable." Children's upset expressed itself in bed-wetting, temper tantrums, anger and violence, stealing, and fire starting. They "could be helped," Guice believed, "if we could get them to a psychologist. . . ." The principal of Pascagoula's South Elementary School agreed that "a great many of our children are emotionally upset. There is so much hubbub going on. They go home and everybody is working and talking about the war." Most negatively affected of any age, she added, was the "older child around 10 to 14."[64]

Parents and children alike were lonely in their new surroundings. One mother living in a housing project explained that her family could face shortages, squalor, and overcrowding "if there were only friends near-by with whom we could laugh about troubles and worries." As it was, she knew only one woman, and she lived ten miles away; "She might as well be in Africa." Yet it was then "more than ever" that her family needed "reassuring friendliness." "I guess I'm just a small-town girl," added another woman, "because I miss having neighbors drop in during the afternoon. . . ." And as another mother explained, "I think that because we are a defense family, and obviously here only temporarily, nobody takes us seriously as neighbors."[65]

Some workers commuted long distances because their families could not find housing in the vicinity of the war factories. Others did so because they did not want to expose their children to the hazards of war-boom community living. Commuters to Pascagoula drove up to 140 miles a day. Explained Robert Bateman, secretary of the local metal trades council: "Many employees would rather go to the trouble of riding that distance if by doing so they can leave their children in a school, get a doctor once in a while, get their groceries, get their checks cashed, and have a few chickens to help them out in this high cost of living that you have to put up with in a boom town." But commuting too added to family distress. One commuter's wife said her two-year-old daughter had not seen her father for six weeks. "He leaves so early and gets back so late that the child is always asleep when her father is at home." Most families, however, decided that, despite the drawbacks of the war-boom communities, they would rather live together as a family, even if the best housing they could find was a chicken coop.[66]

As much as the war battered the mothers, fathers, and children who lived in the war-boom communities, it also deepened Americans' appreciation of family togetherness. Although the families in Willow Run, the San Francisco Bay Area, and Pascagoula dealt with adversity, their desire for marriage, babies, and togetherness was as strong as ever, and perhaps even stronger. The social forces afflicting American families were powerful, and a persistent wartime question was, "Can Families Take It?" "The ability to 'take it,'" one writer noted, "grows with time and practice. . . ." Through it all—the migrations and the war-boom towns, prejudice against newcomers, housing shortages, medical problems, loneliness—the family's challenge was to persevere. And judging by the record, most families succeeded against the odds.[67]

CHAPTER 5

Working Mothers and Latchkey Children

G AIL HAGAR was not yet two in 1942 when her father entered the United States Army and then her mother left her "with a lady" and "got a job at the camp so she could be near daddy." For most of her life, Gail recalled, she was unaware that this had had "any effect on me at all. . . ." But she became a grandmother, and one day her daughter was "phoning around . . . looking for a babysitter so she could go to work. I became very angry with her & told her she should be ashamed [for] just trying to unload Amy on anyone and going off & leaving her that way." Gail's anger surprised and concerned her, for she realized that "my feelings were not about my daughter and Amy . . . [but] about my mother leaving me." She was astonished that she harbored such "great resentments from even that early age." Yet there was another reason for her upset. Like her childhood feelings of abandonment, this too was a life-course change related to her mother, who was eighty and ill. Gail, at fifty, was now struggling "with the fear all over again that she is going to leave me & what's going to happen to me when she's gone, and who will take care of me?"[1]

The American latchkey child was one of the most pitied homefront figures of the Second World War, and his or her working mother was not only criticized but even reviled. Such pity was usually misplaced, however, and such calumny was undeserved. Critics viewed the working mother as a homefront culprit who indulged her own selfish desires while neglecting her children and housework. But in denouncing mothers for abandoning the home in favor of the war plant, these critics failed to comprehend not only that the nation's factories were crying out for workers, but also that millions of American working mothers were their families' breadwinners.

"Rosie the Riveter" was a wartime reality—her muscles bulging, her hair in a kerchief, holding a large pneumatic gun in her hands. Clearly too, as one writer observed: "The hand that holds the pneumatic riveter cannot rock the cradle—at the same time." Still, it is worth emphasizing that many of the forlorn stories about neglected infants and toddlers were exaggerations, the ulterior purpose of which was to discredit the practice of mothers working. Clearly, America's working mothers appreciated their maternal obligations.[2]

"A mother's first duty," intoned Father Edward J. Flanagan, the founder of Boys Town, "is to her husband and children, and the proper maintenance of a home is definitely a full-time job. Consequently, a mother employed in industry must necessarily neglect one or the other of her occupations.... Even if arrangements have been made for the physical care of her children during her absence from the home, harmful effects may and do result in innumerable instances." Judge Michael J. Scott of St. Louis concurred, contending that in America there was "a new form of broken-home condition ... where both parents are out of the home working." One thing was certain, the judge explained: "There is no way you can replace the mother in a home ..."; break "down the home life and you're breaking down the very foundation of your country itself." Mrs. Alfred J. Mathebat, president of the American Legion Auxiliary, went the judge one better in predicting that "if we are careless, we will have children here that will make the Russian wolfpacks look like kittens."[3]

Self-appointed experts, lay people, and politicians alike feared that wartime social changes were demoralizing the family, thus diminishing children's prospects for happiness and individual fulfillment. Political conservatives and liberals, including governmental officials as disparate as J. Edgar Hoover, director of the Federal Bureau of Investigation, and Katharine F. Lenroot, chief of the Children's Bureau, shared a deep disquiet over the presumed relationship between working mothers and neglected children. "Millions of Americans are now fighting or moving toward the battle fronts," wrote Hoover in a 1943 article entitled "Wild Children." "Other millions labor on day, night, and midnight shifts in war factories," often leaving the girls and boys unattended at home. According to Hoover, never married or a father, the latchkey child had a dismal future, including the possibility of "stumbling into the dreaded maze of delinquency and disease, of reformatory and prison, or, if they are not apprehended, of maiming and plundering." The nation's leaders "must realize," he concluded, "as should every American, that the boys and girls are our most priceless national asset, that their preservation is as important as any objective in this war." Hoover was correct in cherishing America's children, but his view was myopic.[4]

At issue for Hoover and others was male power; many saw the entry of millions of women into the paid labor force as threatening to the patriarchal goal of relegating women to the lifetime performance of unpaid, largely domestic, tasks. The indictment of working mothers was especially cruel in the case of servicemen's wives, many of whom were the impoverished mothers of young children. With passage of the Servicemen's Dependents Allowance Act in 1942, the government began to pay monthly allowances to the families of service personnel, but the allowances were insufficient to run a household, particularly in cities such as Detroit and New York.[5] By early 1944, a total of 1,360,000 women with husbands in the service were working outside the home, of whom 280,000 had children under age ten. More servicemen's wives would have taken jobs if they could have, but in numerous locales mothers had extreme difficulties in arranging for child care. In New York, for example, the head of volunteers for the Army Emergency Relief announced that 1000 Army wives, who had no income except for their allowances, were destitute in the city: "From ten to twenty young mothers with babies under 2 years old come in every day asking for help." These women were caught in the middle, for until 1944 no public care was available for children under the age of two.[6]

In assessing the harsh criticism rendered by Hoover and others, it is important to keep in mind that even at the height of the war in 1944, while 2,690,000 mothers worked in vital defense industries, a much larger number stayed at home. In 1941, about 30,000,000 women were homemakers, with no paid employment; in 1944, seven out of eight of these women were still engaged entirely in homemaking. During this same period, 3,700,000 homemakers joined the labor force. Yet more than half of this increase was offset by the 2,000,000 women who had been employed in December 1941, but had returned to their homes by 1944. As John Modell, the historian, has explained, "whatever the pressures, whatever the opportunities for attractive or rewarding gainful employment during the war, marriage and particularly parenthood still militated heavily against employment."[7]

It is also notable that the major changes in the proportion of the female population who were at work came in two age groups, 14 to 19 and 35 to 44. The change was least for women in the prime child-bearing years of 20 to 34. During the war, as the historian D'Ann Campbell has noted, the shift in labor force participation was "least for mothers of children under six . . . and greatest for older wives without children to care for. . . ." Indeed, the labor force participation rate for mothers of children under six years of age increased only from 9 to 12 percent during the war. (In 1950, the figure still stood at 12 percent. By contrast, forty years later in 1990, over 50 percent of married women with children under one year of age worked outside the home.) Finally, as Campbell also observed, "while the war certainly caused an increase in the average number of women employed, it did not mark a drastic break with traditional working patterns or sex roles. . . . The housewife, not the Wac or the riveter, was the model woman." Thus did domestic ideology reign at this time, and it would continue to do so throughout the 1940s and 1950s.[8]

For America's homefront mothers, work outside the home was not an assault on patriarchal dominance, but usually an economic necessity. Women headed between 17 and 18 percent of all families in the United States, or almost one in five; they had to work to support themselves and their children. Other mothers worked to supplement low family incomes, such as servicemen's wives who could not feed their families on allotment checks. Still others took jobs to boost their families' standard of living, which had taken a beating during the Great Depression. Moreover, when women worked in manufacturing during the war, they took home only 65 percent of what men in the same industries received. Had women earned their fair share, they might have been satisfied to work fewer hours. Finally, patriotism motivated many American women, just as it did many American men. "The motives for a mother's working are usually complex," wrote Hazel A. Fredericksen of the Children's Bureau. Fredericksen then asked a pertinent question— one that was not often heard: "Who shall say at a time when labor is so necessary to a nation's winning the war that a mother's right to work should be questioned?"[9]

Nevertheless, working mothers continued to bear a stigma in the eyes of numerous people. The disdain emanated from Americans' conviction of the 1940s that mother's place was in the home; in their eyes, gender was the appropriate determinant of familial, economic, and social roles. But Irene E. Murphy, a social worker who coordinated the Detroit area's child-care committee, came to the mothers' defense. "There is an unfortunate tendency," she declared in 1943, "to blame the mothers who have gone into war industry for their eagerness to leave their children so that they may earn money. We should remember, first, that these women are desperately needed on the assembly lines, and second, that this may well be their opportunity to save for the next 'rainy day.' Those women may still remember the depression, and the public censure directed at those who had not saved enough to tide them over the bad years."[10]

The story of America's working mothers and latchkey children is generally one not of failure and neglect but of perseverance. Working mothers managed their families' lives in order to minimize disruption and damage. "The charge that because a mother goes to work she loses interest in her children is too absurd to comment [on]," asserted Anne L. Gould of the War Production Board. How widespread, in fact, was the incidence of children left in locked cars or at all-night movies? The truth is that journalists distorted reality by focusing solely on the problems without acknowledging the successes. Observers then rightly indicted the federal government for its slow response to the need for child care, but they have made the mistake of using this indictment as the central argument for concluding that many, perhaps even most, latchkey children were victims of neglect on the homefront. There were exceptions, but this was not usually the case, as the evidence makes clear.[11]

The problem is one of perspective. The assumption, for example, that because the federal government never gave its highest priority to child-care programs, the children of working mothers necessarily suffered neglect, is an unfortunate one. In fact, working mothers, fathers, and others largely filled the void caused by governmental shortcomings. Historians and other observers need to adopt the perspective not of the journalists, social conservatives, or earlier historians, but of the parents—especially the working mothers—who made the decision to take war jobs, who made arrangements for child care, and who clearly wanted their children to be healthy and happy.

There is no denying there were latchkey girls and boys who suffered neglect, particularly in war-boom towns where all of life's necessities were in short supply.

Pascagoula, Mississippi, was one such town. In late 1941, Nan Dawkins left her home in Fayette, Mississippi, to move to Pascagoula where she found a job not as a welder at the Ingalls shipyard but as a social worker in the public schools. She also volunteered to do research for the mayor's recreational survey committee. And by December 1943, when Senator Claude Pepper's committee took its probe of "Wartime Health and Education" to Pascagoula, she had finished her study of the town's child-care needs. In her appearance before the committee, Dawkins identified the problem: "The present nursery school was equipped to care for 30 children. The enrollment is now 89." At the conclusion of her testimony, she announced: "I also have some notes I have made on typical cases of working mothers with small children."[12]

"Mr. and Mrs. W. V. S——— work on the night shift and leave an 11-year-old girl and a 9-year-old boy at home to look after a 3-year-old brother and a 1-year-old sister. The 9-year-old boy is at home with the two younger children from the time the parents leave around 2:30 p.m. until the 11-year-old girl gets in from the afternoon shift at school about 5:30." The parents said that since "they both work on the same shift because of transportation,"they believed it was "a better plan for them to be away from the children at night than most of the day." In another family with both parents working the night shift, there were four children under twelve at home alone at night. The children's mother had requested the day shift, but her request had been denied. "Sylvia P———" had also asked for the day shift, with the same negative results. Separated from her husband, Sylvia had total responsibility for "the care and support" of their three children, ages nine, twelve, and fourteen. But to do this, she "works as a welder at Ingalls on the night shift," leaving her children at home alone until 2:30 or 3:00 in the morning.[13]

As Dawkins reviewed her notes with the Pepper committee, it became clear that there were two categories of neglected children based on age. First, there were the preschoolers up to five and six years old. Second, there were the girls and boys between the ages of about six and thirteen or fourteen, many of whom were not only themselves latchkey children, but also caretakers, having the primary responsibility for their younger siblings—some less than a year old—for shifts of up to nine and ten hours. This burden was obviously stressful to the older children, who missed school frequently for reasons of fatigue and illness. Moreover, some of these young caretakers were truant because their parents directed them to stay home from school. But this was not unusual; indeed, in the cultures from which many of the migrant families came, children had traditionally contributed to the household economy. In one family, two boys just six and eight years old "were kept out of school about six weeks to take care of two younger brothers, ages 3 and 15 months." In another, a nine-year-old boy missed school "for several weeks to look after the three pre-school children while the mother worked."[14]

Scheduling for these families produced no-win situations for the children. "Sometimes I don't know which is worse," explained Katharine Guice, a child welfare worker in Pascagoula, "—to have both parents working during the day, and let the older children keep the children, or have one parent work days and another nights, so that somebody is always asleep at home and the children have to be quiet all the time." Fatigued parents collapsed after their shifts. "We've had one particular case where the mother works on the night shift, and she sleeps like somebody dead in the morning." Meanwhile, "the four little boys carry on like bedlam. I've been there to shake her and wake her up, and everything is in disorder."[15]

Although the senators studied all aspects of war workers' lives in Pascagoula, they found the incidence of child neglect to be especially upsetting. The senators were not alone. Also in 1943, Agnes E. Meyer, a journalist and the wife of the publisher of the *Washington Post*, toured the country for the newspaper to survey America's war-boom communities. She called her trip a "journey through chaos." What proved most memorable in Meyer's reportage were her heartrending descriptions of the latchkey children. "From Buffalo to Wichita," she wrote, "it is the children who are suffering most from mass migration, easy money, unaccustomed hours of work, and the fact that mama has become a welder on the graveyard shift." Meyer's descriptions of child neglect were widely quoted by politicians, child welfare officials, teachers, and others who argued for large increases in federal funding for child care. Also quoting Meyer, however, were those people who believed that mother's place was in the kitchen and the nursery and were totally opposed to the practice of mothers working outside the home. (Meyer also represented the Child Welfare League in Washington, which lobbied against group care for the children of working mothers.) In one account, Meyer wrote: "In Los Angeles a social worker counted 45 infants locked in cars at a single parking lot while their mothers were at work in war plants. Older children in many cities sit in the movies, seeing the same film over and over again until mother comes off the evening 'swing' shift and picks them up. Some children of working parents," she concluded, "are locked in their homes, others locked out."[16]

"Who's going to take care of me, Mother, if you take a war-plant job?" So asked the curly blond-haired boy of three or four pictured in the May 1943 issue of *Better Homes and Gardens*. The boy looked sad but resigned to his fate.[17] Actually, the answer to the boy's question was a complicated one. In truth, working parents—mothers especially—devised a diversity of solutions to meet their particular needs.

One irony of the homefront was that while some locales reportedly suffered from desperate shortages in child-care facilities, in others the parents preferred not to deposit their children in public institutions and, instead, made other arrangements. In November 1942 New York City's health commissioner reported that "actually one-third of the child day care units are operating under capacity. . . ." Surprising too was a 1943 Gallup poll in which a cross-section of women responded to the question of whether they would take a war job if their children were provided free care. Only 29 percent said they would, while nearly twice as many (56 percent) said no. Mothers in Detroit explained why they did not send their children to the local day-care center. "I wouldn't have a stranger," said one. "No one could be better than my mother." Health concerns were important to others, who worried, as one said, that "the baby might catch a disease in a nursery."[18]

In September 1943 the War Manpower Commission (WMC) reported that one-third of the defense areas had an "acute" need for child-care centers. In the remaining areas, however, less than 25 percent of total capacity was in use. Rhea Radin of the WMC attributed low usage to several factors. Location topped the list. Many of the Lanham Act centers, Radin said, "used to be WPA nurseries, which were serving low-income groups in locations not necessarily near those sections where factory workers live. . . ." Second, Radin explained, "the stigma of WPA charity clung to these nurseries, too," prejudicing some women against the Lanham Act centers. Other experts reported that many women believed that making private arrangements for their children's care kept up appearances better than using public facilities. And when one of the homefront girls asked her mother why she could not attend a nursery school for the children of working mothers, her mother told her that the school was "only for 'backward' kids." High fees were a third reason cited by WMC field workers as well as by other government officials. Weekly costs in Detroit per child ranged from $6 to $16, "no small sums," according to historian Alan Clive, "for that majority of working women who made less than the average of $47 per week earned by Detroit female employees in 1943." The situation was somewhat worse in Minneapolis, where three out of five mothers using child-care centers were earning less than $24 per week and could afford to pay only 50 cents per week for child care.[19]

Reports from the West Coast were similar. Child-care fees in parts of California and Washington were one dollar per child a day, double the prevailing rate elsewhere. Likewise, transportation was a major problem. In Oakland a doctor who was responsible for the operation of sixteen child-care centers stated that "unless a center is placed about every five blocks the transportation problem will be too difficult to attract many mothers who otherwise might take advantage of the centers." The situation was the same on the East Coast. In New Jersey, for example, experts cited transportation problems "to and from the nurseries" as the major deterrent to attendance. Gradually, however, and despite the many obstacles, attendance in the government-funded centers began to increase. In the fall of 1943, as the number of Federal Works Agency centers rose from 1,620 to 1,951, the average enrollment rose from 24 to 30 percent of capacity.[20]

Lack of information continued to hurt attendance, as did the cultural resistance of people who, having traditionally relied on their own resources, including grandparents, sisters, and nieces, saw no need to place their children in centers. Much of the war-boom migration originated in rural America—whether it began in the black belt of the Deep South, among the ridges and valleys of the southern

Appalachians, or on the parched farmlands of the Great Plains. Thomas R. Wells, school superintendent in Pascagoula, explained that his nursery-school program began with only eighteen children, "and the people in our community had to be educated to use these facilities. Most of them came from rural areas and were unaccustomed to paying for that type of service. The need for [organized] child care was not so great in the communities where they came from. The dangers that exist in a community situated as we are did not exist there. We had to sell the program almost child by child...."[21]

More than a sales pitch was needed, however, because most mothers were apparently satisfied with the arrangements they had made for their children. A 1943 survey done in Milwaukee concluded that only 11 percent of the working mothers were dissatisfied with the child care they had arranged, even though only 1 percent reported that they had placed their children in day nurseries. In contrast, 46 percent relied on relatives; 11 percent employed a housekeeper or left the children with a friend or neighbor or the landlady; and in 3 percent of the cases, the father cared for the children while the mother worked a complementary schedule.[22]

National surveys corroborated the Milwaukee finding that most working mothers were satisfied with their child-care arrangements. According to a U.S. Women's Bureau investigation of ten major production areas, a large proportion of working mothers, ranging from 30 to 45 percent, left the care of their children to relatives in the household other than the husband or older children. In another Women's Bureau study, this one of thirteen war plants, the author concluded: "Care when the mother was away at work was given ... most frequently by adult relatives living in the household. The husband on a different shift from the wife served as a second source for child care. Little use was made of child-care nurseries and only a few families used paid service to care for children."[23]

Further evidence of the variety of arrangements comes from the memories of the homefront children, few of whom complained about their care. Most frequently, they recalled, their parents worked alternate shifts so that one of them was available for the children. Linda Byrd's father worked the day shift at a Kaiser shipyard in Portland, Oregon. "When he came home, Mother went to work and stayed until midnight. Therefore, we were never alone or with a babysitter." The major complaint was that parents were sometimes too tired to play with, or even pay much attention to, their children. Marge Sheridan's mother worked the day shift at a defense plant in Detroit while her father worked the evening shift at the Detroit Edison Company. "My father tried to sleep days while looking after me and I remember him giving me ... coloring books and pencil and paper to play with while he napped. I learned early," she wrote, "to be a loner and play by myself."[24]

In factory towns, extended families had traditionally worked alternate shifts in order to provide care for the children. Doris Poisson, who grew up in a working-class family in Lowell, Massachusetts, recalled that her mother "went to work [in the 1930s] because some of us were old enough to take care of the house after school." During the war, she left her job at the Lawrence Mills to take one at a parachute factory. "She worked the night shift . . . so that my sister-in-law, whose husband had been drafted, could go to work days. There was a baby, and they each took care of it."[25]

When fathers were absent in the armed services and there was no extended family available, some mothers also chose to work the night shift. Larry Paul Bauer recalled that his mother did so in order to return home "to get me up for school and make my lunch." Nan Gardiner's mother was a nurse who "tried to work night hours so she could be home when my sister was awake and I was home from school." In addition, grandparents frequently stepped in to provide care. "I have had my house full of grandchildren for a month," reported one grandmother, "and so have all of my friends whose children are off for war work of one kind or another. Well," she sighed, "I'm glad we are worth something in this time of stress."[26]

The homefront children also took care of themselves. While numerous mothers entered the factories, others took over businesses which their husbands had run. One mother operated the family's tavern, another the family's flour mill, but both tried to be available to their children as they had been before. Moreover, older children took pride in helping their mothers. Jean Richardson was a grade school student in East Nashville, Tennessee, when her mother, a widow making military boots in a shoe factory, saw her salary triple "because of her longer hours and the fact that she was doing a man's job." Jean and her sister did the family shopping at a nearby grocery store. "We could stop on the way home from school and buy things for our dinner because our mother worked late. I soon became very good with our stamp book and knew how many stamps I needed to save to buy three slices [of meat]. It was the same with canned goods," and Jean and her sister shared their stamps "to buy our favorite canned soup."[27]

Finally, neighbors and other private families also filled the void when mothers worked. Kathy Manney's father was already in the service when she was born in 1942. She lived with her mother, who was a painter at a Kaiser shipyard in Portland, and with her grandmother and aunt, who were welders at the yard. In need of help with her eighteen-month-old daughter, her mother placed an advertisement in a local newspaper, and, Kathy recalled, "she was very lucky in the couple that answered her ad. . . . Their entire family adored and indulged me." Bob Mangrum was three when his father died in 1938. During the war, his mother worked at the Dixie Manufacturing Company in Columbia, Tennessee, making military uniforms. His mother arranged "for an elderly lady down the street to keep me. . . . She was a really nice old lady, but she made me obey her, and I did." His mother dropped him off at 6:30 a.m., and picked him up "when she got off from work, whether it was 4:30 p.m. or 9:00–10:00 p.m. each night." Recalling that he never missed a meal and always had clean clothes to wear, Bob wrote: "I don't have any negative effects from this stage in my life. . . . It was what had to be done and we all did the best we could."[28]

At this point, it is necessary to point out that the wartime concern about latch-key children was essentially a concern about white children. With little fanfare, African-American mothers had long before faced the problems caused by leading dual lives. Although black mothers had worked throughout the nation's history, few people in the decades before the Second World War—whether scholars or activists—had lamented the fate of black latchkey children. And during the war, in the voluminous periodical literature published on the topic of latchkey chil-dren, white children were really the subject of concern. Black children's unique circumstances were seldom explored in the magazines; segregation's effects on child development, for example, were never seriously discussed. An African-American child's face seldom appeared in the photographs and drawings illus-trating the articles, even though there were proportionately more black latchkey children than white. During the war, an additional 600,000 black women entered paid employment. Leaving domestic service were another 400,000 black women, and still others left farm work to take more lucrative factory jobs. There was also an increase in black women working in such personal services as beautician, wait-ress, and cook. Although black women were always at the bottom of "the hier-archy of preference," many took advantage of wartime opportunities for better pay and higher status.[29]

"The growing proportion of married women in the labor force during World War II," Jacqueline Jones, the historian, has written, "signaled a trend toward con-vergence of black and white female employment rates." Still, racism meant that the problems that befell the white family in wartime usually hit the black family first and with greater harshness. Moreover, interracial solidarity among women was lacking; in fact, many whites—seemingly oblivious to the fact that black mothers had children too—tended to see black nursemaids as the solution to the problem of white latchkey children.[30]

Regardless of race, the worst possibility for the daughter or son of a working mother was not that of being a latchkey child, but rather being sent away to live with relatives or, worse, being placed in an orphanage. Children could cope with the demands of having working parents, but many developed lifetime resent-ments as a consequence of being sent away. One girl did not see her war-working parents for six years; and a boy recalled that he rarely saw his parents once his father had joined the Army Air Corps and his mother left him behind in West Virginia while she sought work in New York City. "At the end of the war," the boy recalled, "my parents divorced, and I became a pioneer among the 'Latchkey Kids.'" Sometimes the children went to live with their grandparents, and usually this was a happy arrangement; but some grandmothers and grandfathers were angry, cold, or uncaring toward their grandchildren.[31]

The worst option was the orphanage. Judith Mackenzie's father and grandfather had been construction workers at Wake Island and were captured early in the war by the Japanese. When her mother contracted tuberculosis and entered a sanitarium in Colorado, Judith and her sister and brother were placed in a Denver orphanage, but a year and a half later, her siblings were sent elsewhere. "I saw my sister often," she recalled, "but I never saw my brother again." Elizabeth Choate and her sister spent the war years at the Army-Navy Children's Home in Bath, Maine. Their grandparents lived twenty miles away but "were *told NOT* to visit us more than every few months—because we were 'teary' when they left. . . . All families were told their children would 'adapt' better . . . if they were not visited except on holidays and vacations! No book would be complete," she wrote, "without stories of homes like these." As sad as these were, there were even worse stories; practically all of the homefront children who wrote of their lives in orphanages recounted that they had suffered sexual abuse in these institutions.[32]

Most of the working mothers and service wives who sent their children away felt guilty about what they had done, even if there seemed to be no alternative. Barbara De Nike's husband was in the Navy. She could not find housing in the Bronx for herself and her two children, largely because of prejudice against renting to families with children. "It was a very lonely time," she recalled. "I felt so buffeted about." Near the end of the war, she placed her children in an orphanage while she looked for a place to stay. Finally, a nun at the orphanage told her, "'Don't come again until you can take them.'" Lyn Childs, one of the five working women whose lives are the subject of the documentary film *The Life and Times of Rosie the Riveter* (1980), still felt the pain of leaving her daughter behind when she went to San Francisco to work in the shipyards. "And I left her with my mother, and that was one of the most awful things that I think I ever did to anybody, because even today, that child hurts because . . . I promised her that, give me one year away, and I'll come back and get you. And that one year stretched into two years, and finally into three years, and ended up almost five years before I could go back and get her."[33]

There was still another issue for working women. Although the topic of abortion was taboo in much of America, women on the shop floor discussed it often. In a Buffalo, New York, aircraft plant, a woman worker remarked: "There are only three subjects we discuss in the women's rest room—'my operation'; how to keep from getting pregnant if you aren't; how to get rid of the baby if you are." The crux of the problem was the unavailability of maternity leaves for working women. The Women's Bureau agitated unsuccessfully for the establishment of national standards for maternity leaves, mandating that leaves should be granted of not less than six weeks before and two months after delivery and that on request leaves could be extended up to a year. At the war's end, however, no such federal mandate existed; and in forty-two of the forty-eight states, there were no laws at all to protect expectant mothers who worked.[34]

"Although you have heard almost nothing about it," Gretta Palmer wrote in *Woman's Home Companion* in 1943, "pregnancy is America's Number One industrial health problem today." One of the few communities that dealt with the issue was Flint, Michigan, which had a public community-wide program of maternal and infant care. Flint's program authorized not only maternity leaves but also counseling and funds for health care. Most companies and most communities, however, had no such benefits. And without consistent leave policies in the war factories, "women have naturally tried to solve the problem themselves—often behind their employers' backs." They dressed in loose clothing to hide their pregnancies. Sometimes, recalled one woman, "The matrons are good sports—they pretend not to see. We all help cover up the pregnant woman. We know she needs the money." Other women—despite abortion's illegality, its moral stigma, and the risk of death through botched surgery—decided they had to continue working and saw abortion as the only alternative to quitting. Dr. Morris Fishbein estimated that during the first two years of the war, abortions increased by 20 to 40 percent. In one war-boom community in the Midwest, a midwife who performed abortions said, "There's an abortion boom. . . . I had forty-five patients on Saturday. The girls like Saturday because that gives them the week end to rest. They come here straight from the factory, in slacks and overalls." Some did not survive; in 1942, 17 percent of deaths of women during pregnancy and childbirth resulted from abortion.[35]

In 1944 the demand for slots in the government's child-care centers burgeoned on the West Coast, because the new women coming into the factories were younger and had young children. Many were on their own, and having fewer traditional family resources to fall back on, they had a greater need for organized child care. "This gives the child-care program added momentum," explained the director of the Los Angeles child-care coordinating council. "Moreover, mothers are becoming interested who for a long time were skeptical. They are finding that leaving their children with the neighbors was not a satisfactory solution." Finally, the fees had dropped as public subsidization had risen. By late 1944, a mother could leave her child of two to five years for 50 cents a day, which also paid for lunch and morning and afternoon snacks.[36]

Throughout 1944 both the number of and enrollment in federal child-care centers grew. Up to that time, a host of disputes had bedeviled the program. On the one hand, the Federal Works Agency, which controlled the Lanham Act funds for the community needs of defense areas, urged the expansion of group child-care facilities. On the other hand, the Federal Security Agency, the Children's Bureau, and the U.S. Office of Education lobbied against group care, criticizing this approach as a danger to parental authority generally and, in particular, to the mother-child relationship. Dr. Arnold Gesell warned that the young child needed his or her mother full-time; to give the child "a push here and a lift there . . . there is a need for day to day adjustment. . . . If the mother works[,] this must be regarded as insoluble—his care should not be sacrificed to anything." Joining in this effort were the Child Welfare League of America and the National Educational Association, not to mention state's rightists, conservative Protestants and Catholics, and political conservatives who denounced group care as both socialistic and threatening to the family's hierarchical structure. A final division, as the historian Susan Hartmann has shown, was between "elite" women and working-class women. "Unlike elite women," Hartmann has written, "who often addressed the issue in terms of child neglect, union women consistently emphasized working mothers' anxiety over the security of their children and defended child-care programs as essential for both 'the welfare of the child and the peace of mind of the mother.'"[37]

In 1943 the dispute over child care reached a showdown. The Federal Security Agency and its allies prevailed upon Senator Elbert Thomas, Democrat of Utah, to introduce a bill that would end the FWA program. Known as the War-Area Child Care Bill of 1943, the Thomas bill would terminate the Lanham Act funding of group child-care facilities in favor of foster-home and individual child care. Financing would be on a dollar-for-dollar state-matching basis. (Lanham Act funding also was on a 50-50 basis, but the match was not with the states, but with the cities and towns.[38]) Under the Thomas bill, the states, not the local communities, would be responsible for initiating child-care funding requests to the federal government. Additionally, the state offices of education and child welfare, assisted by the U.S. Children's Bureau and Office of Education, would operate the programs. Finally, as Ohio's conservative Senator Robert A. Taft, one of the bill's staunchest supporters, conceded, the new legislation would make it far easier to terminate federal involvement in child care at the war's end, "if we avoid the situation of having one of the Federal Departments attempt to deal directly with the people of the various states." Thus, the program envisioned in the Thomas bill would not only be cheaper than the FWA program, but also easier to dismantle. Although the bill won passage in the Senate, it died in committee in the House, the victim of an incongruous lobby composed of the FWA and its allies, which urged a dramatic expansion in the number of FWA centers, joined by Catholic organizations that opposed federal child care of any kind.[39]

With the House's pigeonholing of the Thomas bill, the FWA's immediate future in child care seemed secure. In July 1943, Congress appropriated additional funds for the FWA to implement the community-facilities provisions of the Lanham Act. Child-care needs were soaring, and experts predicted that by September more than one million children would require some sort of supervision while their parents worked. Shortages were so acute in one-third of the nation's war-production areas, the War Manpower Commission reported, "as to be a serious hindrance to the recruiting and retention of women in industry." And yet by the end of 1943, there were in operation only 2,065 Lanham Act centers; altogether, just 58,682 children were enrolled. Shortages of both physical facilities and trained, experienced staff slowed the program's expansion. In February 1944 a total of

65,772 children were enrolled in 2,243 centers, but Lanham Act funds were nearly exhausted. Unprocessed funding applications from local communities piled up on the federal bureaucrats' desks.[40]

Gradually, however, the numbers grew. For one thing, the proponents of group child care organized a potent lobbying drive to keep the funding flowing. The women's auxiliaries of certain industrial unions joined with community leaders and Federal Works Agency officials in the effort. Also influential were the six women members of the House of Representatives, led by Mary T. Norton, Democrat of New Jersey. In late February 1944, Norton presented to the house "a joint appeal" for immediate funds to expand the wartime child-care program under the FWA. What was at stake, the statement said, was not only "the health and safety of our children" but also "the achievement of our war-production goals." "Therefore, we women members of Congress assume the responsibility of speaking for the millions of working mothers . . . and of impressing upon you the need for action." In the floor debate, the women members took their views to the male representatives. "I cannot believe that there is a man on the floor of this House, and I know there is not a woman," began Norton, "who does not consider that this is an absolutely necessary program." Frances P. Bolton, Republican of Ohio, added that the country had asked "our women to go into the production lines in addition to doing their first and foremost role of homemaking. . . . Heaven knows I am against Federal encroachment on States' rights or local rights. I do not like it. I do not approve it. But it is the Federal Government that is asking women to go into production. Are we going to be penny-wise and pound foolish? Are we going to fail to keep first things first?" Congress responded to these arguments, along with thousands of positive letters from union members, with increased funding for child care.[41]

In the spring of 1944, child-care enrollments began to rise rapidly. One reason was the FWA's decision in late April to make funds available for the care of children under two years of age. The FWA explained that while it "frowned upon" the employment of women with children under two, "we know that, realistically, many women with young children have been forced to take war jobs." By mid-May, a total of 87,406 girls and boys were enrolled in 2,512 war nurseries and child-care centers. Enrollments reached their peak in July 1944. In that month, 3,102 centers served 129,357 children, with an average daily attendance of 109,202. Enrollments had doubled in just five months. Although the FWA centers never served the predicted clientele of 160,000 children, they did help to meet many families' wartime needs. Lanham Act child-care centers ultimately received federal funds totaling $52,000,000, with matching sums from states and local communities. By the war's end, a large number of children—estimated at between 550,000 and 600,000 for the Lanham Act programs—had received some care.[42] Finally, the federal government had made an important declaration, even if belatedly: in this national emergency, the public would provide subsidized care for the children of wage-earning mothers.[43]

Another encouraging development was that groups in various communities established child-care programs of their own. Some began at the neighborhood level, as in Malden, Massachusetts, where in early 1942, twenty mothers scoured their neighborhood for tiny furniture and playthings for the children in their "cooperative nursery school." Seven mothers in a New York town—all with children between two and six years of age—surveyed their neighborhood's needs, conducted fund-raisers, then rented four rooms and filled them with donated furniture, books, and play equipment. Their final triumphant action was to nail up a large sign, patriotically painted red, white, and blue, which bore the legend "Inwood Community Day Nursery." Its opening-day enrollment was sixteen, but by the end of the week, the number had risen to thirty-five. In other cities, girls ranging in age from eight to thirteen took instruction in child care and homemaking. In New York City, for example, girls from large families used this training to take care of their younger brothers and sisters. On St. Patrick's Day, fourteen of these girls prepared a dinner for their working mothers and served it to them at the YWCA, complete with menus in shamrock green and flowers for the tables. And in Gary, Indiana, boys as well as girls took classes in planning, preparing, and serving meals, laundering and mending clothes, and helping younger children to dress and undress.[44]

By mid-1944, happy stories of child-care successes had begun to supplant the sad tales of latchkey children highlighted in numerous magazines just the year before. Both local efforts and federally funded programs were beginning to prosper. So too were a handful of initiatives by private industry. In the fall of 1942, the Douglas Aircraft plant in Santa Monica announced plans to open a nursery within four miles of the plant, but "out of range as a target for the enemy." In efforts to recruit women workers to its airplane factory in Buffalo, the Curtiss-Wright Corporation announced not only that it was offering prizes of $50 war bonds for recruitment of new workers, but also that it would double the size of the nursery school in operation on the plant's grounds. "Each morning," noted a reporter in Buffalo, "company guards pluck children from mothers' sides as they pass [through] the plant gates. Eight and a half hours daily the moppets play, snooze, ingest assorted vitamins, [and] watch test planes zoom by."[45]

Probably the most innovative child-care program was the product of an industrialist's fertile mind. Edgar F. Kaiser, son of Henry J. Kaiser, was general manager of the two massive Kaiser shipyards in Portland, Oregon. Among the large laboring force in the shipyards were 25,000 women workers. To house workers' families, the government erected its largest wartime civilian housing project, Vanport City, Oregon. The town was situated on lowlands just outside the Portland city limits. Recognizing that child care would be an immediate problem, Kaiser consulted child development experts as well as architects for advice on constructing two large centers. Lois Meek Stolz, formerly the head of the Institute of Child Development at Columbia University, became the centers' director.[46]

While stabilizing women's employment in the shipyards, the centers promised to be unique innovations in their own right. For one thing, the buildings were to be located "not out in the community but right at the entrance to the shipyards, convenient to mothers on their way to and from work. ..." Second, the centers were to be large, accommodating 1,125 children each, between the ages of 18 months and six years, during a twenty-four-hour, three-shift working day. And third, they were to be run not by the federal government, nor by the local schools or any other public body, but essentially by Edgar Kaiser and his staff.[47]

Although Kaiser's staff would manage the centers, the company would not pay for their operation. Edgar Kaiser wrote these costs—minus the nominal sums paid by the mothers—into the company's cost-plus-fixed fee contracts with the government, so that the public footed the bill. In addition, he convinced the U.S. Maritime Commission to bear the construction costs for the two centers. The centers' wheelspoke plan placed a large grassy play area—replete with four wading pools—at the hub of the building, accessible to each of the fifteen spacious and cheerfully decorated classrooms along the spokes of the wheel. Forming the wheel's outer rim were the large classroom windows, each of which faced onto the shipyards; benches located by the windows enabled the girls and boys to see where their mothers worked. Each center had an infirmary, with a trained nurse and a social worker on duty, as well as a large and fully staffed kitchen, which prepared not only the children's meals but also take-home meals for the families of mothers coming off their shifts. Called "Home Service Food," this program had been suggested by Eleanor Roosevelt. The take-home meals were nutritionally balanced, neatly packaged, and contained full directions for reheating and for "supplementary salads and vegetables to make a full dinner." Available at 50 cents a portion, each meal was sufficient for a mother and a child. The meal for March 15, 1944, for instance, included fresh salmon loaf and avocado salad.[48]

Attendance at the Kaiser Child Service Centers reached its peak of 1,005 children in September 1944; this was less than half the stated goal of the prior fall. Despite low enrollments, the Kaiser centers still stand as an example of what private industry might accomplish to alleviate the pressures on working mothers while providing real care for the children. As a student of the Kaiser program has written, "their existence serves to prove that quality, center-based child care services can be made available, given the necessary ingredients of priority, leadership, and professionalism."[49]

During the Second World War, government at all levels cooperated in an unheralded but unique and highly successful before-and-after-school program called Extended School Services (ESS). ESS was a simple concept. As Idella Purnell Stone, a mother in Berkeley, California, explained it in a letter to Eleanor Roosevelt in March 1942: "Get the school teachers to care for our children after school until we, the mamas, can call for them. It might even become necessary to give the children their suppers there." And since the school buildings were already in place, the program had only to be implemented. "This would free us to get into war work NOW." ESS became one of the biggest success stories of the war.[50]

In August 1942, in a little-noticed move that proved fateful beyond mere dollars, President Franklin D. Roosevelt made available $400,000 to the U.S. Office of Education (part of the Federal Security Agency) and the Children's Bureau (part of the Labor Department) "for the promotion of and coordination of [Extended School Services] programs for the care of children of working mothers."[51] During the Second World War, many of the homefront horror stories focused on the plight of very young children up to four or five years of age. But since relatively few women with children under the age of six took jobs, the much larger group of children needing care were those between six and thirteen or fourteen. Charles P. Taft of the Federal Security Agency estimated that probably 80 percent of the girls and boys needing care were school age. For these children, the greatest need was for a place to go before and after school—a supervised environment in which to wait until their mothers ended work and could pick them up.[52]

In the fall of 1942 the U.S. Office of Education announced the ESS program for the school-age children of working mothers, and by early 1943 the agency had granted funds to seven states to promote Extended School Service programs. Actual federal financial assistance was minimal. Funds were available only for organizational and administrative purposes; not a penny was to be spent for personnel costs for the operation of nursery schools or child-care centers or for the maintenance of children in these programs. But the $400,000 grant funded 222 positions nationwide and stimulated the creation of 450 more. New Jersey used its initial funds to hire two staff members to work for the State Department of Public Instruction, one serving primarily as a field worker, the other as the state coordinator for the program. Other states followed this pattern. Funding was the key issue, and in New Jersey, as elsewhere, it originated at the local level, with the public schools assuming primary responsibility for school-age children.[53]

The ESS programs were self-initiated: that is, the local school system, welfare department, or office of civil defense first perceived a critical need in the community and then attempted to deal with it. The federal government's function was to provide advice. By February 1943, the Office of Education had granted funds to seventeen state and one territorial (Hawaii) departments of education to hire field workers to help the local communities organize the Extended School Services, and for administrators to coordinate statewide child-care committees. Throughout the spring, the number of state programs grew. In addition, the Office of Education distributed leaflets in its "School Children and the War" series. Leaflets 1 and 2—*School Services for Children of Working Mothers* and *All-Day School Programs for Children of Working Mothers*—endeavored to answer the community's questions: "What is the all-day school program?" "How can a community interested in such a program get one started?" "What general principles underlie the program before and after regular school hours?" "What equipment is necessary for the expanded program?" "What are the possible sources of financial support?"[54]

By June 1943, local communities in thirty-three states had benefited from federal guidance in establishing ESS programs. But with summer vacation approaching, vexing problems loomed for working mothers. "Schools are closing," Bess Goodykoontz, assistant commissioner of the Office of Education, explained. "Children for whom after-school care was needed now need all-day and all-week care. The same school facilities can be used. . . ."[55]

Throughout the 1942–43 school year, cities and towns relied upon publicity to attract parents' attention to Extended School Services. Civil defense block leaders spread the word. Announcements of ESS went home in pay envelopes or were carried home from school by the children themselves. Bell Aircraft in Georgia gave to its working mothers a flyer entitled "Your Child's Life While You Build Bombers," which told them, "If your daughter or son is 6–14 years of age[,] she or he will enjoy before and after school care because it provides: Expert guidance, Active life with youth of own age, Experience in arts, crafts, drama, nature study, music, Supervised games and play." Moreover, the flyer said, the cost was reasonable and the location convenient. Other publicity included signs on streetcars and buses, "announcement flashes" in movie theaters, and posters in employment offices, union halls, and in schools offering vocational training classes for women. The Kansas City, Missouri, school district issued a three-page folder entitled "Children's War-Service Program," in anticipation of working mothers' needs during the summer vacation months. School boards and child-care committees in Oakland and Vallejo, California, and in Toledo and Hamilton, Ohio, distributed similar leaflets. Industrial publications and news flyers also told of children's services; an issue of the *Airview News*, published by plants in Santa Monica and Long Beach, California, used illustrations to depict one boy's day "beginning with the mother's good-bye kiss for her small son [at the child-care center] at 6:30 in the morning and completed with her return at 4 o'clock to take him home." Also active in publicizing these services were the local radio stations and newspapers.[56]

The publicity paid off. Drawing upon various community resources, cities and towns across the country boasted of their ESS programs. To the ESS programs located at the Hutchins Intermediate School in Detroit, for example, children from the neighboring schools were transported daily by students from Wayne University and the Merrill-Palmer School. With guidance from their professors, the students led boys and girls in games, dancing and singing, dramatics, handicrafts and hobbies, gardening, marketing for food, setting up for meals, and cleaning up afterward. In Youngstown, Ohio, the public library assisted the ESS activities by supplying "suitable books" to children, as well as by instructing teacher's aides in storytelling, rhymes, and finger painting. "We are teaching children as young as 2 years how to use books," explained the program's director. Moreover, the aides themselves—senior high school girls taking courses in home nursing and child care—were also benefiting. The director explained that, as the mothers of tomorrow, "they will be able to work more wisely with their own children. . . ."[57]

In the ESS programs in Vallejo, California, enrollment was 1,000: 350 in nurseries located in the public schools and 650 school-age children. As in Youngstown, the modus operandi was to integrate the child-care services with the regular school program by placing the centers in the public school buildings and by utilizing the high school's home economic students as child-care aides. The first children arrived at 6 a.m. "Many of the early arrivals," Bess Goodykoontz explained, "have their breakfasts at school." At about 9 o'clock, following indoor play with blocks, clay, dolls, picture books, and paper and crayons, the little children went outdoors to play with blocks, wagons, tricycles, boxes, a teeter-totter, slide, and sandbox. After a mid-morning snack of orange juice and cod liver oil, they came indoors for a nap. At noon they ate lunch followed by another nap. After playing outside again in the afternoon, "around 4:30 they come in for music, stories, and free play. If the mothers' working hours demand it, the children are given a simple supper before they depart for home around 6:30."[58]

On June 30, 1943, the ESS program was officially discontinued. Seed money for ESS had come from President Roosevelt's emergency fund, but on that day the federal funding expired. By that time, however, the programs in many communities were self-sustaining. Moreover, during the last two years of the war, many communities applied for and received Lanham Act funding for their ESS programs. Cleveland's program, which provided in-school care at lunchtime, after school on school days, all day on Saturday, and during vacations, received funds from a variety of sources. Financing came at first from parents' fees and contributions from civic foundations, but the program gained a more substantial footing when the Cleveland Board of Education assumed the primary responsibility, with the aid of Lanham Act money.[59]

Likewise, in Kansas City, Missouri, enrollment in ESS programs was booming in 1943. Funding came from the school board as well as parents' fees and contributions from the Community and War Chest. As Kansas Citians looked to the approach of the 1943–44 school year, there was no thought of discontinuing the program. "You can't realize how much the program has meant to mothers at North American Aviation," asserted Marian Davis, the company's personnel director. "Extended School Service for the coming year is absolutely essential." Opening at 7:00 a.m. and closing at 6 p.m., Kansas City's schools provided that children arriving early in the morning "should be given a warm cereal with a generous serving of milk." Mothers as well as children ate meals in the schools, though they were spared the doses of cod liver oil that accompanied the children's meals.[60]

In 1945 Will S. Denham, the Kansas City area director of the War Manpower Commission, applauded the school system for rendering "a magnificent service in caring for thousands of these children of workers taking war jobs." It is important to emphasize, however, that mothers employed in all lines of work—and not just those engaged in defense jobs—were eligible to place their children in Kansas City's before-and-after-school programs. Thus, mothers employed as domestic servants, janitors, garment workers, department store clerks, waitresses, and office workers also made use of these services.[61]

Because of the multiplicity of Extended School Service programs and the variety of sources funding these programs, it is impossible to tabulate the total numbers of children served. In mid-1943, the U.S. Office of Education did announce that ESS was caring for about 320,000 children, including 60,000 preschoolers and 260,000 of school age. A year and a half later, however, the Women's Bureau stated that 2,828 ESS units were in operation, with an enrollment of 105,263 children. Since it was not known to what extent the states and local governmental units either assumed fiscal responsibility for Extended School Services or initiated their own programs, these latter figures are surely short of the actual numbers involved. ESS represented a homefront triumph of the first order.[62]

Certainly, many of America's latchkey children experienced suffering during the war years, ranging from loneliness and fear to the actual neglect of their physical and emotional needs. But historical perspective demands that fresh questions be asked regarding the subject of working mothers and their children. There is no doubt that child neglect occurred in the war-boom areas, but even in these areas it was the exception, not the rule. Extended families including grandparents and aunts and uncles shared the child-care responsibilities with the working mothers. If relatives were unavailable, parents—especially mothers—tried to make the best possible arrangements for their sons and daughters, given the cost and location, and also considering that many such parents, newly arrived from farms and small towns, were themselves making major adjustments to the factory system and urban living.

There is no sure way of knowing how many latchkey children suffered, or in what ways. Nor can we know how many shared the feelings of six-year-old Janet, whose mother had taken a job and left her in the care of the landlady: "I'll tell you a secret. When my mother is away and I don't know where she is, I cry. She says she's at work, but I don't know where she is and I get scared." At the same time, as observers at the time noted, there was no way of knowing to what extent children's upset was the result of factors not directly related to child care, such as not being loved at home.[63]

During the war years, there were recurring predictions that working mothers were destroying American society by turning children into misfits. In 1944 Henry L. Zucker, secretary of Cleveland's Emergency Child Care Committee, wrote that the "nomenclature of social work literature has been enriched during the war by such terms as 'latchkey' or 'doorkey' children and 'eight-hour orphans.' Unfortunately ... these new terms foretell a war-bred generation of problem adolescents-to-be in the 1950s and of maladjusted parents-to-be in the 1960s."[64]

But postwar investigations, particularly by developmental psychologists, have indicated otherwise. Fifteen years after the war, Lois Meek Stolz, then a psychology professor at Stanford University, surveyed the results of forty studies done on "the effects of mothers' employment on children's behavior." In her earlier investigation of war-born children published in 1954, *The Father Relations of War-Born Children*, Stolz had found evidence of damage to "the first-born child in the war-separated family," including predictably stressful father-child relationships when the returning father "usurped the child's 'mommy,' assumed personal intimacies, made unexpected requirements for behavior, demanded obedience, and used methods not heretofore experienced by the child."[65] In Stolz's assessment of the studies on working mothers, however, she found "no evidence that maternal employment affected the personality development of children in unfortunate ways."[66] These studies undermined the conventional wisdom of the 1940s and 1950s, as shaped by the case studies of "separation and attachment" written up by John Bowlby, Anna Freud, and other psychiatrists, who had worked with the British child evacuees of the Second World War.[67]

And this was just Stolz's point; the United States children had not been evacuated, and those who were left behind or sent away probably suffered far more than did the latchkey children. Stolz also issued a challenge. Since mother's working was "not such an important factor in determining the behavior of the child as we have been led to think," it was time to look elsewhere "to know why some children of employed mothers develop acceptable behavior and others do not."[68]

During the Second World War and in the decades afterward, the American working mother has been a convenient scapegoat for a variety of societal ills. Forty years after the war, for example, Ann Landers explained in her advice column: "The nuclear family began to fall apart when Rosie the Riveter went to work in the defense plant to replace the men who had gone to war. She liked the money and the independence," Landers wrote. "She also found it stimulating and chose to keep on working." Perhaps historians should take the lead in rejecting Landers's interpretation. Here too Stolz has offered a valuable insight. Rather than accepting the popular judgment that working mothers damage the family, "it might be more profitable to focus attention on the psychological conditions within the family, especially on the personal characteristics of the mother and the father and the kind of supervision and guidance which they provide, not only when the parents are at work but when they are at home as well."[69] While this is a more difficult challenge for the historian, it is the only way to correct the home-front record regarding America's working mothers and their children during the Second World War.

It was clear that as victory approached in 1945, child care had gained wide acceptance. Not only had enrollments in the centers continued to rise, but a large percentage of America's working women had indicated that they wanted to stay on the job after the emergency, thus presaging a postwar demand for child care. In the spring of 1945, working women, elected and unelected government officials, and child-development experts began to agitate for a national peacetime child-care policy. Memorials descended upon Congress from scores of organizations—among them, the San Francisco Board of Supervisors, the East Hollywood Club of the Communist Party, and various chapters of the Veterans of Foreign Wars, the American Legion, and the American War Mothers—in support of continuing the Lanham Act centers. Local governments, labor councils, and educational associations also sent telegrams. Although originally scheduled to expire on October 31, 1945, Lanham Act child-care funding received a reprieve from Congress—an additional $7,000,000 to keep the centers alive until March 1, 1946.[70]

The case for child care centers, suggested a school administrator in Burbank, California, would become fully persuasive only when parents and public officials put aside their political, economic, racial, and religious opinions and conceded that such centers would "assist us to win the war at home by giving our children the personal security, maximum health, social training, and the stability of character they will need." By the end of the war, many people believed that the centers had made that case—and had done so beautifully.[71]

It was clear that the Second World War experience had produced a new appreciation of the demands of child care. The *New York Times* children's expert, Catherine Mackenzie, wrote that "if day care of children in wartime has done nothing else, it has brought better understanding of the exacting nature of round-the-clock supervision of small fry, and its fraying effect on adult nervous systems."[72] The tragedy of the Second World War experience is how little carry-over value it has had. Despite the wartime successes, the United States has retreated from articulating a national child-care policy ever since.[73]

CHAPTER 6

Rearing Preschool Children

DURING THE Second World War, preschool teacher Claudia Lewis moved from New York City to eastern Tennessee. Lewis had been teaching three-, four-, and five-year-olds in a nursery school in Greenwich Village, but she wanted to broaden her perspective on childhood development by relocating to another part of the country. Arriving in Tennessee's Cumberland plateau, she soon discovered she had entered a world dramatically different from the one she had just left. The touchstone of that difference was the children themselves.

The children of the Cumberland were placid and shy. Much of the time they stayed by themselves, shunning the vigorous group play and spirited conversation so common among New York City children. Lewis had not gone to Tennessee to do research for a book, but when she "realized that there were marked differences between the mountain children and those . . . in Greenwich Village," she started "to look more closely at the structure of the community":

> I wanted to find out what kind of homes and upbringing made these children so unresisting and "easy to handle." . . . Why was there so little rebellion in the mountains? . . . What was the meaning of their outwardly peaceful, placid behavior? And why were these children so shy for months at a time? . . . Was there any relation between the . . . talent of the New York children and their energetic, self-assertive ways? Between the mediocre performances of the mountain children and their compliant, apparently untroubled behavior?[1]

While Claudia Lewis did not discover the answers to all these questions, she did describe profound differences in the childhood circumstances as well as the child-rearing traditions of these two societies during the Second World War. What proved startling was that these stark differences existed in a nation whose population—with the monumental exception of people of color—was reportedly being rapidly assimilated in the American "melting pot."

Between 1941 and 1945 Americans rushed to the nation's defense. This was a period of intense and unifying national purpose, and patriotism on the American homefront seemed to be not only forging national unity but also obliterating invidious distinctions based on social status, religion, and ethnicity. War movies featured ethnically diverse units of soldier and sailor; a Jew, an Italian, an Irishman, a Pole, or a Swede all fought together in *Bataan, Guadalcanal Diary, Sahara, The Purple Heart,* and numerous other films. By confronting hostile ethnic stereotypes in training camps and on the battlefields, these warriors got to know each other as equals. Such tolerance was crucial to the national unity required to fight a war against powerful external enemies.[2]

In addition to ethnicity, other distinctions became less important during the war, such as traditional regional and rural-urban distinctions. Undermining these were technological and demographic developments. First, revolutionary innovations were transforming communications and transportation. The spread of electricity, the automobile, radio, telephone, and motion pictures had accelerated social change and exposed dissimilar peoples to mainstream popular culture. Second, during the Second World War there was a mass migration from farms to cities. Sooner or later, most commentators predicted, the dominant culture would subsume all its variants. Indeed, within just a few years various scholars—beginning with David Riesman in 1950—wrote that a "changing American character"—a new "national character" or "modal personality"—had come to typify the American psyche.

Riesman and the other observers, however, minimized the extent to which the United States was still a land of multicultural diversity and of distinct regions. An example was the great diversity of languages spoken in the United States. Numerous Americans spoke a native tongue other than English, and there were large urban enclaves filled with people whose language was Spanish, Italian, Polish, Russian, or Chinese. Equally significant, there were distinct regionalisms in English use and dialect. In fact regional speech still thrives in America. In the mid-1980s, after forty years of television and over sixty of radio, a linguist identified twenty-seven major dialects in the country, ranging from Pennsylvania Dutch to Appalachian mountain talk, from Black English to "Brooklynese." These varieties stemmed from local eccentricities of pronunciation as well as differences in idiom, vocabulary, and grammar; but they also reflected divergent world views. For example, "Down in the mullygrubs" was a mountain expression for being "depressed"; in the South, "hung the moon and stars" meant being blindly in love; "ferhoodled" was Pennsylvania Dutch for being "mixed up"; while in African-American street vernacular, "chill out" meant "stop acting foolishly."[3]

As for a common history, one that all Americans could share, here too there were pronounced regional differences. Even during the Second World War, northerners and white southerners continued to debate the Civil War. Numerous white northerners, as well as African Americans from both sections, heaped praise on President Abraham Lincoln for preserving the Union and emancipating the

slaves, while white southerners clung to "the lost cause." Northern states celebrated Lincoln's birthday, but not a single state of the old Confederacy did so.
And in the South, there were even two versions of United States history. In 1940
Mississippi passed a law mandating different instructional materials for African-
American and white pupils; while white children's textbooks lauded the nation's
democratic traditions and institutions, all references to voting, elections, and even
democracy itself were excluded from the black children's textbooks.[4]

While the debate over American multicultural diversity versus homogeneity is
political as well as scholarly, the case for diversity on the American homefront is
overwhelming. The pitfall scholars must avoid is judging the 1940s by the social
perceptions of several decades later. No matter how much homogeneity there
might have been in the 1950s and after, in the 1940s the United States was a
country of varied regions and a multitude of subregions. Having studied hundreds of indices for all parts of the country, sociologist Howard W. Odum concluded on the eve of the war that the United States consisted of six regions, each
of which possessed distinctive physical characteristics, economic and cultural
developments, and population trends. These were the Northeast, the middle
region (that is, the states of the Upper Mississippi Valley: Ohio, Indiana, Illinois,
Michigan, Wisconsin, Minnesota, Iowa, and Missouri), the Northwest, the Far
West, the Southwest, and the Southeast. Each region, moreover, had subregions,
and some subregional differences—for example, between the Upland and Low-
land, or Yeoman and Plantation, South—were as pronounced as the interregional
differences.[5]

In a book published in 1942, the family sociologist Ruth Shonle Cavan
focused her research on "regional aspects of the family" and found impressive
differences in the factors affecting family life across the country. Some resulted
from economic disparities; for example, while annual per capita income in the Far
West was $921, in the Southeast, where many poor African-American families
lived, it was $365. If the Southeast had a larger black population than any other
region, the Northeast had a larger foreign-born population. At 75 percent, the
Northeast was also the most urbanized; in the Southeast, the figure was only 30
percent. Educational levels also varied widely; for example, the number of California's residents who had high school diplomas was 350 percent higher than in
Arkansas.[6]

The United States did undergo lasting transformation during the era of the
Second World War, but the forces of change had a greater impact on some homefront children than on others. Societal changes most directly affected girls and
boys over the age of five or six, since it was the school-age children who were old
enough to experience the war's powerful manifestations in their classrooms,
through popular culture, and by participating in homefront activities to win the
war.[7]

On the other hand, America's preschoolers generally experienced the war less
directly.[8] For America's youngest children—those under age five or six—region,

class, and culture were still the instrumental forces in their lives. These factors were most apparent in the child-rearing regimens imposed by parents. To understand the wartime experiences of all the homefront children, it is important to recognize that even in this era of rapid social change—and even at a time when child-centeredness and permissiveness were emerging in child rearing—tradition, not innovation, was the parents' main guide.

Claudia Lewis, for one, was struck by the persistence of distinctive child-rearing traditions in the Cumberlands. "Small babies are seldom out of their mother's arms, and are nursed whenever they cry. Often they are not weaned until they are well along in their second year. Children are always to be seen with their parents at buryings . . . [and] at square dances. They are never left at home or put to bed early." Lewis's description of the Cumberland children is evocative of premodern Europe during which children became part of the adult world at an earlier age. "Parents," she wrote, "do not seem to expect their children to live on a schedule that differs very much from their own. Meals are the same for all members of the family. . . . Children live a life very close to that of their parents. . . ." At the same time, these children were expected to give unquestioning obedience to elders. "They are whipped or threatened with whipping if they do not obey. Discipline is theoretically of the old 'authoritative' kind, yet the actual routine of living is far from a strictly regulated one."[9]

Obviously the Cumberland children and their Manhattan counterparts were extremely different, even though they grew up in the same nation at the same time. In this case as in many others during the war, geography had evinced its determinism, vividly demonstrating the localized and regionalized bases of a variety of cultural variations. Still, there is another dimension of region, or locale, to be considered. For scholars who study children in historical context, an important psychological perspective examines the child's "ecology," that is, the physical environment and "social space" in which children grow up. Ecological psychologists have compared childhood in varying sized communities. One significant difference is the nature of student participation. Small-town childhood, with its smaller, community-wide schools, encouraged—even mandated—greater student participation than in big-city schools. In metropolitan schools, there were not enough positions on the football team or roles in the senior play for everybody. On the other hand, an urban childhood offered superior cultural opportunities.[10]

While rural-urban differences were ecologically the most apparent, only a handful of scholars have studied these differences. But Alex Inkeles, the sociologist, has ventured some generalizations: "The greater isolation and family centeredness of rural life means a slow social awakening for the rural children, greater fear of strangers, and slower development of imagination and language skills. By contrast," he continued, "both the responsibilities early assigned to them, and their contact with animals and nature, seem to yield the rural children early and highly developed sensory motor functioning." Moreover, rural children were freer to explore their environment, and they did so "less under the immediate surveil-lance of adults. . . ." These boys and girls also had more contact with their fathers than did urban or suburban children, whose working fathers were absent from the home most days.[11] Childhood was thus a differential experience depending upon the ecology—"the real-life settings within which people behave"—whether on farms or in villages, suburbs, or cities.[12]

With ecological psychology in mind, a significant demographic trend becomes apparent: during the Second World War, while most of the country's adults were urban, most of its children were still rural. Demographers have pointed to 1920 as the year when the United States became an urban nation; for the first time in the nation's history, more Americans lived in locales defined as cities and towns than on farms and in villages. By 1940 the urban margin was more pronounced, but a majority of the nation's boys and girls still lived in areas defined as rural. As might be expected, the distribution of children was very uneven from state to state. Proportionately, in fact, some states had more than twice as many children as other states. This fact had stunning regional as well as ecological and devel-opmental implications.

Among the contemporary observers who grasped the implications of this dis-parity were the women and men who gathered in 1940 for the White House Con-ference on Children in a Democracy. "Numerically," one of the conferees explained, "[among] the Nation's children under 16 in 1940 . . . there were a good many more in the country than in the city." America's children constituted only 23 percent of the urban population, but they made up 34 percent of the farm population and 30 percent of the population living in "nonfarm rural com-munities." Variations in the fertility rate—and thus in family size—were major factors explaining the distribution of children across the United States. "The [fer-tility] rate for the rural-farm population was highest, that for the rural-nonfarm came next, and the rate for the urban population was lowest."[13]

Since farm families tended to be larger than city families, it stood to reason that agricultural states would have a larger proportion of children than industrial states. And this was indeed the case. The greatest regional differences were between the Deep South and the Southwest, on the one hand, and New England, the Middle Atlantic, and the West Coast, on the other. Even so, the range sepa-rating the states is surprising; the figures below represent the number of children under 15 years per 1000 adults.[14]

Highest Ratios		*Lowest Ratios*	
678	New Mexico	341	Connecticut
672	South Carolina	340	Washington
633	Mississippi	332	New Jersey
628	North Carolina	321	New York
619	Alabama	308	California
599	Utah	260	District of Columbia

What were the implications for the homefront children? First, families with large numbers of children tended to be not only poorer than smaller families but also more poorly educated and less healthy. Contemporary health surveys published by the Social Security Administration concluded that "the economic status of the family varies inversely with the number of children under 16 years of age in the family." Consequently, many large families waged an ongoing battle to make ends meet.[15]

The sociologist John A. Clausen has observed that large families struggle harder to raise their standard of living and to provide children with opportunities to become upwardly mobile. At its worst, large size militates against the prospect of family success in either of these areas. In contrast, children in smaller families generally enjoy greater economic resources and parental interest in school progress, and show "higher achievement motivation and academic performance" as well as greater "occupational success" later on. Family size also tends to increase the possibility that the parents will be autocratic and authoritarian. Having babies closely one after another is a financial, physical, and emotional burden; and parents in such families are more likely to use physical punishment instead of rewards to control behavior. Clausen has noted, in fact, that "differences in frequency of physical punishment associated with number of children were in general greater than differences between working-class and middle-class families of the same size."[16]

Whether based on ethnicity, social class, or physical and social ecology, it is evident that diversity, not uniformity, marked the lives of America's homefront children. Claudia Lewis had come to understand not only the significance of regional differences in childhood but also some of the developmental implications of these differences. After the Second World War Lewis returned to New York City to finish her book, *Children of the Cumberland*. Intensely curious about the differences she had observed, she pondered whether "one could picture a map of the United States in terms of its children," a developmental map that would probe "the various influences at work in molding small children." Lewis's map offered a stereotyped vision of the country, but it did so to emphasize the diversity that existed. The "East," she mused, might appear "as a great city of brick and concrete, with a small child playing in canyonlike streets. Quite a contrast would be the 'Southwest' with the little Navajo girl spending her entire time out on the arid plain, in the shadow of high blue mountains. . . ." Lewis's map also included "the Southern mountain children" along with the "children of the Northwest, who can look up daily to snow-capped peaks; children of the corn country, whose familiar 'woods' are waving seas of corn; children who live beside the real seas . . . ; to say nothing of the children of our thousands of 'towns,' with their backyard, house-close-to-house, main-street surroundings."[17]

Such a map, Lewis continued, would have to feature children and not just their physical surroundings. Indeed, "a psychologist's map would probably teem with the children themselves—eating, playing, obeying and disobeying, fighting, laughing, teasing; shy children and aggressive children; the maladjusted, and the happy and secure." Still, the "imaginary map" would be incomplete without also distributing representations of the child-rearing practices followed in the various regions and locales as "one way of getting a grasp of what the differences are, and more especially why they exist." For example, the map could identify child-rearing differences in breastfeeding and weaning, toilet training, discipline, tolerance of sexual curiosity, and other topics. Coming full circle, the psychologist's map would point out that in the southern mountains the children "are going to bed without any fuss (because the whole family goes at the same time), while up North in a well regulated modern city home the battle against bedtime may go on night after night."[18]

It is one thing to acknowledge that deep cultural differences existed in American society during the Second World War; it is quite another to understand what impact these differences had on the homefront children's lives and development. Any discussion of the relationship between culture and child development must include an understanding of culture's role in society. In 1871, the pioneering anthropologist Sir Edward B. Tylor formulated an early definition of culture, calling it "that complex whole which includes knowledge, belief, art, morals, law, custom, and any other capabilities and habits acquired by man as a member of society." More recently, however, anthropologists have offered cognitive definitions, which contend that "culture consists of whatever it is one has to know or believe in order to operate in a manner acceptable to its members." Similarly, the anthropologist Clifford Geertz writes that "culture is best seen not as complexes of concrete behavior patterns—customs, usages, traditions, habit clusters—as has, by and large, been the case up to now—but as a set of control mechanisms— plans, recipes, rules, instructions (what computer engineers call 'programs')—for the governing of behavior."[19]

In the broadest sense, then, as anthropologist Victor Barnouw has written, culture is a people's "way of life," one that is "handed down from one generation to the next through the means of language and imitation." Moreover, it is primarily the family that transmits its culture to the children.[20]

It is clear that the United States, being geographically large and ethnically diverse, was home to numerous subcultures in the 1940s.[21] Indeed, anthropologist Charles A. Valentine contends that virtually every society is a configuration of subcultures, each possessing "distinguishable lifeways of its own" and each "distinct from the total culture or the whole society. . . . The wider sociocultural system has its own coherence to which subsocieties and subcultures contribute even with their distinctiveness."[22]

But what, specifically, is the relationship between a group's culture and the socialization of its children? Anthropological research done by participant-observers in the 1940s identified significant intercultural differences in breast-feeding and weaning, toilet training, and sexual discipline, as well as in play patterns, the carrying-out of chores, initiation rites, access to formal schooling, and ideals. Focusing on the relationship between "culture and personality," these scholars were blunt in their conclusions. As one wrote: "Families influenced by different religious and political value systems differ significantly in how they see fit to train their children, and some of these value systems involve tighter and more conscious control over the details of socialization than others.... Thus what is relatively unregulated and unplanned in one group is the object of intensely deliberate socialization in another."[23]

An exemplary study of socialization differences compared several ethnic groups—for example, Zuni Indians, Mormons, and homesteaders from Texas— who occupied a single physical environment, the town of Rimrock, New Mexico. The groups in Rimrock emphasized distinctly different dominant values and types of behavior. The Zuni, for example, paid special attention to regulating aggression and promoting harmony. The homesteaders, on the other hand, focused on individualism and emphasized success, and the Mormons emphasized virtue and "rigid sex impulse control." Correspondingly, the child-rearing agendas which these parents established for their boys and girls listed these objectives first.[24]

Later on, the child's world expands to include the neighborhood and, beginning at about age five, the school. But first, as sociologists Frederick Elkin and Gerald Handel have observed, "a child is socialized into a particular subculture, not into the culture as a whole. This means that initially a child learns not the ways of his society but the ways of a particular segment of it." In addition, children find their earliest role models in the particular subculture of which they are a part. And, as Elkin and Handel have written, "since a child's sense of self is formed in large part by taking the role of others, and since his earliest significant others tend to be from his own subculture, the child's self has an anchor in a particular subculture." Moreover, if children spend their lives "associating only with those who share the same subculture," they "may have difficulty understanding the thoughts, actions, and situations of those from other subcultures." They may come implicitly to ask, "'Doesn't everyone live this way?'"[25]

But how ethnic, in fact, was the United States during the Second World War? Ethnicity ebbed and flowed in the United States between the 1920s and the 1940s. Contributing to ethnicity's decline in the 1920s and early 1930s were powerful factors, including the rise of mass culture, spread of education, migration, and the growth of large cities. But nativism and racism legislation played their parts as well. In 1924 President Calvin Coolidge signed the National Origins Act, which prohibited Asian and sub-Saharan African immigration and blatantly discriminated against continued influxes from southern and eastern Europe. Ethnic sentiments rose in the 1930s with the eruption of war in Europe, Asia, and Africa, but efforts were made during the Second World War to de-emphasize ethnic differences, as the American people pursued national war goals.

While immigration declined after 1924, immigrants still constituted a large portion of the population in 1940. The country's "foreign-stock" (first- and second-generation white Americans) dropped as a percentage of the total population, but the actual numbers remained high. In 1920, there were 36.4 million foreign-stock Americans (38.4 percent of the white population), and in 1930 the number peaked at nearly 40 million (36.2 percent). In 1940, the figure still was significant, standing at more than 34.5 million (29.1 percent).[26]

In 1940, the ten largest foreign-stock immigrant groups, by principal country, were:[27]

Germany	5,236,612	England	2,124,235
Italy	4,594,780	Sweden	1,301,390
Poland	2,905,859	Austria	1,261,246
Russia	2,610,244	Mexico	1,076,653
Ireland	2,410,951	Czech.	984,591

Prejudice and hostility heightened in-group solidarity and insularity. Studies conducted just before the war asked American college students, schoolteachers, and business people questions regarding the degree of social intimacy they were willing to grant certain ethnic, racial, and religious groups. What groups, for example, would they admit "to close kinship through marriage, to my club as personal chums, to my street as neighbors, to employment in my occupation, to citizenship in my country . . . ?" "Social distance" was not a problem for the English, Scottish, Irish, French, Germans, and other western and northern Europeans who monopolized the listings of the most acceptable groups. But as the most despised of Americans, people of color occupied the bottom rankings: Africans, Chinese and Japanese, Mexicans, Turks, East Indians. Listed just a notch above these groups were Jews and southern and eastern Europeans: Greeks, Russian and German Jews, Armenians, Russians, Poles, and Italians. Stereotypes clearly governed the responses given to these questions. Jews were considered above all to be "shrewd," "mercenary," and "industrious"; African Americans to be "superstitious," "lazy," and "happy-go-lucky"; the Irish to be "pugnacious," "quick-tempered," and "witty"; and Italians to be "artistic," "impulsive," and "passionate." The traits most listed for native-stock Americans were that they were "industrious," "intelligent," and "materialistic"; for Germans that they were "scientifically-minded," "industrious," and "stolid"; and for the English that they were "sportsmanlike," "intelligent," and "conventional."[28]

Another institution that bolstered the in-group at the expense of the "melting pot" was marriage, certainly one of the most fateful of life's choices. Marriage within the in-group served to perpetuate group culture whereas marriage "out" did not. In the 1940s, the rate of exogamy—that is, the extent to which men and women married outside of their racial, religious, or ethnic groups—was low. Racial intermarriage was rare; in fact, many states forbade interracial marriages between whites and those whose ancestry was even one-sixteenth African. According to figures cited during the war, only 2 to 5 percent of northern black marriages were to white people.[29] Intermarriage between non-Jews and Jews, especially Orthodox or Conservative Jews, was rare. In New Haven, for example, 93.7 percent of marriages involving a Jew were endogamous.[30] Less unusual was marriage between Catholics and non-Catholics, but a valid exogamous marriage within the Roman Catholic Church required a bishop's dispensation based on pledges that all children of the marriage would be baptized and educated in the Catholic religion; that the Catholic ceremony would be used for the marriage; and that the Catholic would do all in her or his power to lead the non-Catholic to the faith through prayer and example. Between 1930 and 1950 sanctioned mixed marriages totaled 30 percent of all Catholic marriages, but percentages varied widely depending upon community and region. For example, in southern dioceses with small Catholic populations—such as Raleigh and Atlanta—as many as 70 percent of sanctioned Catholic marriages were exogamous. On the other hand, in El Paso and Corpus Christi, Texas, with large Mexican-American Catholic populations, the figure was only 10 percent.[31]

Interethnic and interclass exogamy was on the rise in the 1940s. Interethnic marriages increased with the individual's length of residence and acculturation in the United States, particularly when the groups involved considered themselves to be mutually assimilable. And marriages between people of different social classes were generally acceptable if between adjacent middle classes; but family members usually expressed opposition to unions between widely separated classes. Even though fluidity was creeping into America's marriage relations in the 1940s, there was still widespread racial, religious, ethnic, and social-class opposition to intermarriage.[32]

To glean some understanding of the interaction of culture, personality, and, ultimately, society, it might be illuminating to look at a particular American group. Italian Americans, having the largest number of "enemy aliens" in the country during the war, exemplified a subculture that lived under duress on the homefront.[33] Stigmatized by other Americans as fifth-columnists and potential traitors, Italian Americans were frankly divided as to how they should respond to the prospect of a war between their ancestral land and the land they had adopted. Largely generational, the division was between the Italian-born parents and their children, who, a sociologist has observed, "were socialized not only by their parents, but also by American institutions, ... and what they were taught at home was frequently in conflict with what they had learned."[34] Prior to the Pearl Harbor attack, a leading point of contention among Italian Americans was Benito Mussolini, Italy's fascist dictator. Not all Italian-American fathers lauded Mussolini, but many did; and mothers felt caught in the middle. "We are in such ... a fix again," one recalled. "Nobody speaks to nobody. The kids, they laugh at Mussolini and fight my old man. He calls them sons of bitches and bastards. Holy Mother, they can't be in the house together. We don't even listen to the Italian music on the radio anymore. Papa shouts 'Viva Mussolini,' and then the kids they swear at him, and—oh, it's awful."[35]

The younger generation argued with their parents, many of whom read Italian-language pro-Mussolini newspapers. "A lot of them still believe Mussolini can conquer the world," scoffed one boy during the war.[36] The major concern of the immigrant generation, however, was not ideological; it was their fear that the war would magnify the biases against them.[37] Already they believed they were the victims of rampant discrimination, particularly in finding employment. They were tired of being disdained as "dagos" and "wops"; they wanted no more trouble in America. One first-generation Italian American remarked: "I came to this country for peace, and all I get is war, war, war!" It was a time to be quiet—a time to de-emphasize one's ethnicity.[38]

Convinced that the outside world viewed Italian American neighborhoods as the homes of racketeers, corrupt politicians, and subversives, Italian Americans shied away from outsiders, sometimes even including other Italian Americans who were not either family or *paesani*, immigrants who had come to the United States from the same town. During the war, there was in Italian America a status hierarchy, as northern Italians looked down upon the southern Italians, and both groups looked down upon Sicilians.[39]

Still, what bound together Italian Americans of all kinds was not only the negative stereotyping done by outsiders but also their own family and ethnicity. Italian-American families tended to share notions about sex roles and child rearing. The Italian-American family, which embraced all blood and in-law relatives, was, according to sociologist Jill S. Quadagno, "an inclusive social world in and of itself. . . . Three-generation households, including aged parents, were likely to be found at the end of the life cycle. . . . Family solidarity was the basic code of family life, and this code encompassed the parents, grandparents, aunts, uncles, cousins and godparents. And such solidarity meant that the disgrace of one member of the family affected everyone. Thus, a disobedient child was the concern not only of the parents but of the extended kin as well."[40]

Boys were reared to model their behavior after their fathers, while girls assisted their mothers in the home and were taught to be competent in women's roles as housekeeper, wife, and mother. At the same time, most Italian-American families were suspicious of and hostile to formal education for their children, viewing it as "antithetical to proper training for manhood or womanhood, involving the influence of *stranieri* (outsiders) who might interfere with *la via vecchia* (the old ways) as well as keeping young people from the more important lessons they might learn from work." Parents especially discouraged their daughters' educational interests. An old saying summed up the Italian-American attitude toward the purpose of child rearing: "Only a fool makes his children better than himself."[41]

In 1940 the number of first- and second-generation Italian Americans was 4.5 million, the second largest "foreign-stock" group. Italian-American ethnicity reached its peak in that census year, but there was little danger that assimilation would seriously, or quickly, weaken the Italian culture or undermine in-group solidarity. For one thing, Italian-American rates of intermarriage and residential mobility were the lowest of all American ethnic groups, and they continued to be so throughout the 1940s and 1950s. Of white ethnic groups, Italian Americans most often remained in the same neighborhood as their parents and siblings. For example, Nathan Glazer and Daniel P. Moynihan noted that "while the Jewish map of New York City in 1920 bears almost no relation to that in 1961, the Italian districts, though weakened in some cases and strengthened in others, are still in large measure where they were."[42]

Just as their families and neighborhoods had special value for the Italian Americans, so too did their child-rearing traditions. Herbert J. Gans, the sociologist, has studied child rearing and family patterns among Italian Americans residing in Boston's West End; his book *The Urban Villagers* is a reminder that the goal of these immigrants was not assimilation, but rather the re-establishment of their *paesani* in the cities of America's eastern seaboard. In these "Little Italies" the old ways prevailed. Unlike the child-centered families that were beginning to exemplify the American middle class (see below), the Italian-American family was "an adult-centered one." Since children were not planned, "but come naturally and regularly, they are not at the center of family life. Rather," Gans explained, "they are raised in a household that is run to satisfy adult wishes first. As soon as they are weaned, they are expected to behave themselves in ways pleasing to adults."[43]

Child rearing was the responsibility of the Italian-American mother, whose love of her children was legendary. "Her total devotion to her children," one scholar has noted, "in material and emotional support, caresses, food-giving, and in fact, supportive extension of herself makes her an indispensable figure in her children's lives."[44] While the father was the chief formal disciplinarian, the mother usually served up the punishment, "both physical and verbal: mothers slap and beat their children, tell them not to do this or do that, and threaten to tell the fathers when they come home. Indeed, to a middle-class observer, the parents' treatment often seems extremely strict and sometimes brutal." But the "continuous barrage" of threats did not vitiate parental affection, nor did the children interpret punishment as rejection. "We hit you because we love you," a mother explained. Such discipline was impulsive, but such impulsivity was permissible, Gans wrote, because the parents were "not concerned with *developing* their children, that is, with raising them in accordance with a predetermined goal or target which they are expected to receive."[45]

Reared in this ethnocentric environment, Italian-American children developed a divided self, which was also the experience of other ethnic, racial, and religious minority groups in the United States. Italian Americans harbored the same aspiration as other marginal groups—in this case, to be both a proud Italian and yet also an American with the same opportunities for success as other citizens of the country. The "two-ness" dilemma loomed large in the lives of many young Italian-American men in December 1941, when Italy declared war on the United States.[46]

"I think of myself as Italian-American," stated a young man shortly before the Italian declaration of war. "[If someone asks you what nationality you are, what do you say?] I'd say Italian descent, American born. [Do you ever think of yourself as just American?] No. [Would you like to think of yourself as just American?] No. I'd rather call myself Italian-American, because then I wouldn't be hurting nobody. If I said I was American, somebody might say, 'Why do you say that, if you're from an Italian family?' If I said 'Italian,' then they'd say, 'Why, since you were born in America?'" For second-generation Italian-American men, the most pressing issue at this time was conscription. If called, would they serve and possibly fight against their ancestral home? Almost all said they would. "Well, I know I'm a real wop," replied one. "But I consider myself an Italian-American. If there was a war, I imagine I would go for this country, since I've been raised here."[47]

Most Americans, including most historians, have an image of national unity on the home front during the Second World War. In this cheery view, the only exception was the internment of 120,000 Japanese Americans. And yet this is a simplistic view of American society and its children even during wartime. "To proceed as if group differences did not exist," wrote psychologist Marian Radke Yarrow and her colleagues, "is to ignore the cultural context in which children live, for society does not ignore differences; family customs and values and names and languages all reflect group-derived variations."[48]

In studying the ethnicity of American children and their families during the Second World War, it might be wise to remember the advice of John Dollard, the social psychologist. The individual, he wrote, must be seen "as a link in a chain of social transmission; there were links before him from which he acquired his present culture; other links will follow him to which he will pass on the current of tradition." The individual, then and now, is "one of the strands of a complicated collective life which has historical continuity."[49]

Like ethnicity, social class also had a powerful impact on both childhood and child rearing during the Second World War. Some American children obviously enjoyed greater material, intellectual, and emotional benefits than others. Income, education, and occupational status were among the crucial factors differentiating American society. Race was another, and so were region and ethnicity. Indeed, the three major cultural groupings in the 1940s were social classes, ethnic-religious groups, and racial groups. And as the family sociologist Ruth Shonle Cavan observed: "Children are born into the subculture of their parents and escape from it only under favorable conditions and usually with conscious effort."[50]

Sometimes, of course, these groupings overlapped. People of color, for example, were far more likely than whites to be poor—and being impoverished had obvious negative effects on childhood. A preponderance of African-American children in the 1940s came from lower-class families; surrounded by white racism, some black children became preoccupied with their skin color, hair texture, and other physical features, as well as with their status in a hostile society that deemed them to be inferior.[51]

Religious affiliation and social class also overlapped. A cultural geographer, James R. Shortridge, concluded after examining American religious bodies that "it is increasingly obvious that Protestant religious affiliation is more closely linked with social stratification than with official theology." Episcopalians, Presbyterians, and Congregationalists were usually solid members of the middle or upper middle class, while many evangelical churches drew their members from the lower middle class. Religion also had regional definitions. Liberal Protestantism flourished in the Northeast, but was a negligible presence in the deep South; the reverse was true for Evangelical, Pentecostal and Holiness sects.[52]

In the 1940s there was little disagreement that from the very beginning of life, class influenced a child's development in basic ways. While middle-class mothers and children enjoyed prenatal and pediatric care, such was not the case for many lower-class children who were delivered not in hospitals, but in their homes by midwives or relatives. From birth on, the middle class's ability to pay usually resulted in superior nutrition, clothing, housing, health care, and education for their children. Years later, in discussing the gross inequities that existed among American children, the historian John Demos wrote of "the reality of the 'stacked deck' versus the myth of equal opportunity."[53]

Ordinality or birth order provides another example of social class's influence. Every first-born child was for a while an only child, but then the variable of class—often in concert with ethnic factors—exerted itself. In numerous middle-class families, the first-born, especially if a son, was the one of whom the parents expected the greatest academic achievements. Indeed, it was the first-born who tended not only to perform better in the classroom but to go farther in school than his later-born siblings. But the factor of class often stood this situation on its head for those in the working class. "Among the less well-to-do . . . especially in larger families," John Clausen has written, "the first-born child is likely to have to leave school and go to work, since his parents will have several younger children to support and will frequently expect the oldest child to contribute to the family income as early as possible."[54]

Social class also influenced children's play, sexuality, and mental health, among other important areas of life. Child's play, for example, depends on role models; and as sociologist Gregory P. Stone has written, "we ought to acknowledge that one child's fantasy is another child's reality. The probability that the roles children enact in their dramas will be assumed or encountered in adult life is very much restricted by their position in the various orders of stratification—income, prestige, . . . their rural or urban residence, their 'race' or ethnicity, or their sex."[55] Likewise, in the 1940s Alfred C. Kinsey and his associates identified class differences in childhood knowledge of and attitudes toward sexuality. The distinctions were discernible in children as young as three or four, and they involved such sexual topics as "the ease or embarrassment with which such a child discusses genitalia, excretory functions, anatomical distinctions between males and females, ... the origin of babies, ... and kindred items. ..." By ages seven or eight, the differences were even greater. While the lower-class boy at that age "knows that intercourse is one of the activities in which most of his companions, at least his slightly older companions, are engaging," the ten-year-old boy "from the upper level home is likely to confine his pre-adolescent sex play to the exhibition and manual manipulation of genitalia, and he does not attempt intercourse because, in many instances, he has not yet learned that there is such a possibility."[56]

Social class was as relevant to children's lives as it was to adults' lives, influencing the quality of boys' and girls' housing, health care, nutrition, and schooling.[57] Evidence compiled during the war showed that significant social-class differences produced child-rearing differences. Two psychologists from the University of Chicago studied the varieties of child rearing then being practiced in the United States. W. Allison Davis was black, Robert J. Havighurst, white; they investigated child-rearing practices in middle- and lower-class African-American and white families. (In restricting their analysis to these two classes, they defined lower class and working class as synonymous.) Based on interviews with mothers who lived on Chicago's South Side, these psychologists reported significant differences between and among the groups.[58]

Davis and Havighurst found middle-class parents of both races to be "more rigorous than lower-class families in their training of children for feeding and cleanliness habits." They began training earlier; they placed "more emphasis on the early assumption of responsibility for the self and on individual achievement"; they continued to require "their children to take naps at a later age" and allowed less freedom in play. Thus, observed Davis and Havighurst, "middle-class children suffer more frustration of their impulses," one result being that both white and black children of the middle class were several times more likely than lower-class children to suck their thumbs. "The striking thing about this study," the authors concluded, "is that Negro and white middle-class families are so much alike, and that white and Negro lower-class families are so much alike."[59]

The central finding reported by Davis and Havighurst was that "the social class of the child's family determines not only the neighborhood in which he lives and the play groups he will have, but also the basic cultural acts and goals toward which he will be trained. The social-class system maintains cultural, economic, and social barriers which prevent intimate social intermixture between the slums, the Gold Coast, and the middle-class." Children, they explained, can learn their culture only from other people "who already know and exhibit that culture. ... *Thus the pivotal meaning of social class* to students of human development is that it defines and systematizes different learning environments for children of different classes."[60]

What makes Davis and Havighurst's research relevant is that it was contemporaneous with American children's wartime experiences and provides the best data available on the homefront children.[61] Moreover, other studies conducted during the Second World War tended to concur with Davis and Havighurst. One investigator, Evelyn Millis Duvall, studied "conceptions of parenthood" held by a sample of Chicago mothers divided among four "social status levels" and among Jewish mothers, non-Jewish mothers, and black mothers. Unlike the earlier studies focusing on just two classes, Duvall looked at four, ranging from the lower class to the upper middle class. She determined status by evaluating economic, social, and cultural factors, such as father's occupation, neighborhood, and type of housing. Duvall stressed that while certain qualities of "a good mother" (such as caring for the child's physical needs) and of "a good child" ("obeys and respects his parents") were listed consistently across class boundaries, still others were class-specific. Lower-class mothers, who listed their top priorities as "keeps house," "takes care of child physically," and "trains child to regularity," had a somewhat different conception of parenthood from the middle-class mothers whose key concerns were "trains for self-reliance and citizenship," "sees to emotional well-being," and "helps child develop socially."[62]

While ethnic culture and social class were instrumental in American child rearing during the Second World War, the war itself was unleashing forces that would transform child rearing in the United States. The 1940s was a transformational decade for both child-rearing advice and child-rearing practices. Tradition still prevailed in families where regional, cultural, and class influences were strong; but in middle-class families, permissiveness in child rearing would soon become the byword.

"Then and Now" was the title of a 1946 article that celebrated the twentieth anniversary of *Parents' Magazine*. Written by Clara Savage Littledale, the magazine's editor, the article described a radical shift in *Parents'* perspective over this period. Upon its founding, she wrote, the magazine "rode on the wave of a tremendous concern with bringing up children," and it expressed this concern by championing the rigidly behavioristic teachings of John B. Watson, who urged "habit training" in feeding, toilet training, and every other aspect of the infant's development. Watsonian training, Littledale explained, "should begin as soon as the child was born. The idea was that if you caught him early enough and trained and trained, allowing no deviations from the ideal but everlastingly hauling him up in the way he should go, why, you would never have any trouble at all. How could you, if you did the right thing every minute?" That was "then," twenty years ago, but in terms of child-rearing philosophy, it seemed light years removed. "Now" was 1946, the year of publication of a book by Watson's antithesis, *The Common Sense Book of Baby and Child Care* by Dr. Benjamin Spock.[63]

The Second World War was not only the midpoint in but also a crucial reason for the transition from Watson to Spock as the predominant influence in child-rearing advice. According to Watson's strictures, the mother should ignore the baby's demands—in her or his own best interests, of course. "No feeding five minutes before schedule because a baby cried, chewed his fists and gave every indication of hunger. No feeding five minutes late—that was enough to prove any woman a Bad Mother. Right on the dot it must be." The same rigidity applied to toilet training; "it was never too early to start that and to be firm about it." The payoff for "the good mother" was that she could "boast of never having a soiled diaper after her baby had reached the tender age of three to six months." As for affection, "If the baby cried—let him cry! All babies cried. It exercised their lungs and was good for them. But by all means don't pick him up. Don't cuddle him. . . . Never rock him, hold him as little as possible."[64] Watson did recommend two forms of familiarity and touching, one for bedtime, the other for first thing in the morning. "If you must," he wrote, "kiss them once on the forehead when they say goodnight. Shake hands with them in the morning."[65]

As cold as Watson's methods seem in retrospect, for two decades they commanded at least lip service from child-rearing experts as well as from American mothers.[66] Whether rigorously enforced or not, Watson's child-rearing advice had become the conventional wisdom of middle-class parents by the 1930s. Not everyone followed it, of course, for American culture, as child psychologist Lois Barclay Murphy observed, "is so complex and variable from one end of the country to another that observations about children in Leadville, Colorado, might not apply to children in Middletown, Indiana. Nor would the experience of children in Pittsburgh society families overlap with that of children of Iowa farmers." In addition, some parents insisted upon "tight scheduling" for their first child only to abandon this regimen later. "The first child," Murphy wrote, "was terribly important. . . . Everything had to be done just right." In addition to being reared "by the book," the first-born was likely to receive more time and care, as well as more spanking and scolding. But to do "the right thing every minute" was physically and emotionally demanding, especially for mothers. For the later-born, many parents decided that Watson's unyielding behaviorism militated against affection, spontaneity, and real enjoyment in parent-child relations. Still, Watson's guidance ranked high in influence up to the Second World War.[67]

During the 1930s, however, Watson's dogmatism began to lose its primacy, as experts urged mothers to pay closer attention to what babies "wanted," not just what they "needed." Many influences—some professional, some theoretical, some political—conspired to undermine Watson. Academically, research in child development flourished at various universities, including Iowa, Minnesota, Yale, Teachers College, and California.[68] And experts such as Dr. Adolph Meyer (director of the psychiatric clinic at Johns Hopkins University), Dr. C. Andrews Aldrich (a pediatrician at the University of Minnesota), and co-author Mary M. Aldrich offered alternatives to Watson's behavioristic theories. Meyer urged parents to develop a better understanding of how their children grew emotionally as well as physically and mentally; and in 1938 the Aldriches' *Babies Are Human Beings: An Interpretation of Growth* was published, admonishing parents to respond to their babies' needs. Indeed, warned the Aldriches, "most spoiled children are those who, as babies, have been denied essential gratifications in a mistaken attempt to fit them into a rigid regime."[69]

In the years leading up to the Second World War, Freudian theory exerted an increasingly weighty influence on theories of child development and parent-child relations. Psychoanalysts focused therapeutic attention on breastfeeding and toilet training as being integrally bound up with psychosexual development and with underlying anxieties and guilt, and they argued that forced training according to the Watsonian paradigm could produce trauma and lifelong damage to the individual. Drawing upon Freudian theory, child-rearing experts began to take cognizance of infant sexuality, the unconscious, and the oedipal conflict. Instead of recommending "tight scheduling," they urged mothers to trust their own judgment. Most of all, they told parents to provide their girls and boys with an atmosphere that was both loving and intellectually stimulating. Less rigid and more permissive child training also was the logical concomitant of the child-centered educational movement inspired by John Dewey, who stressed personal growth experience, productive citizenship, and problem solving rather than moralistic pieties and rote learning. Still another significant influence was Dr. Arnold Gesell, a pediatrician at Yale and a major voice in child-development debates, who countered Watson's extreme environmentalism with his emphasis on physical maturation as the primary determinant of child development, and who advised mothers and fathers that "the child is in league with nature and he must do his own growing." And during the war, Dr. Spock was beginning to publish portions of *The Common Sense Book of Baby and Child Care*. In March 1945, for example, his article "A Babies' Doctor Advises on Toilet Training" appeared in *Parents' Magazine*.[70]

As political attitudes changed, so too did child-rearing approaches. Politicians and parents alike worried about the effects on children buffeted first by the Great Depression and then by the Second World War. Believing that children needed to be held in troubled times, more and more parents sought to redeem "the promise of the 'secure' child in a 'world of change.'" Regimentation, on the other hand, smacked of fascism.[71]

So pronounced was the shift in American child rearing that one sociologist has declared that "the decade from 1935 to 1945 may well be called 'baby's decade' when mother becomes secondary to ... the baby's demands." This appraisal was correct, and at the center of this transition were America's homefront boys and girls. For by 1940, halfway through the transition, child-rearing advice had already changed notably. Now, only one-third of the articles recommended the behavioristic regimen, while two-thirds urged self-regulatory, "permissive" approaches. In contrast to the behaviorists' strictly enforced feeding schedules, Josephine Kenyon, writing in *Good Housekeeping*, said simply: "it is reasonable to feed a baby when he's hungry.... It is unreasonable to make him wait." Clearly the balance had shifted from "tight scheduling" to "self-regulation." By 1948 the shift was finally complete; none of the articles published that year advocated Watsonian methods.[72]

The trend toward "self-demand babies" accelerated during the Second World War. One need only compare the 1938 and 1942 editions of *Infant Care* to appreciate the change. In contrast to the 1938 edition's worried view of babies' autoerotic "impulses," in 1942 these dangerous drives were considered to be weak and almost inconsequential. The 1942 manual viewed babies as explorers who "want to handle and investigate everything that they can see and reach. When a baby discovers his genital organs he will play with them." But there was no need for mother to tie the baby's feet apart or glove his tiny hands. "See that he has a toy to play with and he will not need to use his body as a plaything." Parents should attend to their babies' "wants" as well as to their "needs." Babies sometimes cried for attention, but that was all right. "Babies want attention; they probably need plenty of it." Similarly, parents should play with their babies; amusement and fun, which earlier editions of *Infant Care* had disdained as titillating and excessively exciting, should now characterize parent-child relations. "Play and singing make both mother and baby enjoy the routine of life," stated the 1942 edition.[73]

The demands of the homefront encouraged mothers to recognize their own as well as their children's special wartime needs. "Watch out for war nerves!" warned Gladys Denny Shultz, the children's expert for *Better Homes & Gardens*, in 1943. "You'd be surprised," Shultz wrote, "how many mothers have said to me recently, 'I don't know what has gotten into my youngsters! They're fussy, impatient, and disobedient. I can't do a thing with them.' Then I say, 'Well, how about *you*? Have you been under any special strain lately?' That opens floodgates. What mother of a family isn't undergoing some special strain?"[74]

But the experts also urged mothers to recognize that their children were resilient. In June 1942, P. L. Travers, the author of the Mary Poppins books, advised the readers of the *Woman's Home Companion* that "children are tough individuals, much tougher than grown-ups when it comes to facing facts."[75] All the experts agreed that parents, particularly mothers, should be honest with their daughters and sons, letting them "know what to expect," but still seeing to it "that childhood has its full measure of happiness" and responding to emergencies "with cheerful competence." It was imperative that children, "too, in the midst of danger, shall have courage." At the same time, parents should protect their sons and daughters from hatred. "Perhaps out of this war," stated *Parents' Magazine*, ". . . will come a better world for our children. Surely it will come quicker if we can save them from hatred and fear and bitterness."[76]

In such ways, liberalized child-rearing practices became part of the homefront effort not only to win the war but to insure the peace. "Your child will be part of tomorrow's postwar democracy," wrote an expert in 1943. "Is he receiving now the training and attention necessary for the building of a future democratic citizen who will be capable of aiding stabilization of the new world order?" Parents had the responsibility of creating a better society for all of America's boys and girls. Fathers and mothers "will not only have to supply the physical and emotional needs of your own children, but you will have to bring equality of opportunity to the nation's children!"[77] The war was a democratic crusade in other ways as well, and if children learned these lessons well, they would also "learn to build the peaceful world of tomorrow...." Parents were central to the task of teaching their children tolerance; as one expert wrote: "If we parents see that democracy (the American dream if not the American reality) is based on the idea that everybody matters, whether the person agrees with us or not, we will not need to worry about the spread of intolerance and hatred."[78]

At the same time, child-rearing experts, politicians, and parents all expressed alarm that the wartime disruption of family life had fostered conditions which, according to writer Vance Packard, "could hardly be worse for baby raising." Fathers serving at distant duty stations in the armed forces, mothers working long hours in factories and shipyards, latchkey children at home alone—"all these handicaps," Packard wrote, "imposed by war and modern conditions that I've mentioned add up to a lot of perplexity for a growing child." Packard's advice: "Give the war babies a break."[79]

During the war new child-rearing approaches challenged the influences traditionally exerted by region, ethnicity, and class. There are no hard statistics to which the scholar can turn, but it is clear that there was a cultural divide in the United States, based on the child-rearing practices that resulted from diverse cultural, economic, social, and political environments. Numerous homefront girls and boys were the products of traditional child-rearing regimens, reflecting the persistence of cultural diversity; but the tide was turning. Almost without exception, the wartime advice reminded parents that "babies need mothering, the good old-fashioned kind that goes with rocking and singing lullabies...." Victory in warfare would be hollow, indeed, if the homefront children should be emotionally stunted in their development. The psychologist Frances Degen Horowitz has hypothesized that "one might look upon the Spock era ... as a kind of reaffirmation of democratic principles taken down to child rights in a reaction against fascism and authoritarianism as they were thought to have produced obedient but morally manipulable children."[80]

Child-centered approaches dominated the advice columns during the Second World War. One of the strongest statements was *The Rights of Infants* (1943) by Dr. Margaret A. Ribble, who recommended feeding and bathing as times for talk and loving play. Her prescription for the baby's emotional feeding was "a little at a time, and frequently." Published in 1946 was Dr. Spock's *Baby and Child Care*, which was the culmination of a ten- to fifteen-year process of revisionism—and which went on to sell 22,000,000 copies over the next 25 years. Surely, here was validation that a long-term change had taken place in the way parents viewed not only their children but themselves as well. Just as democracy had prevailed over authoritarianism abroad, so too democratic notions had begun to transform child rearing in America. "Belief in our children is the essence of democracy, of what we call our way of life," wrote the vice president of the Child Study Association of America in 1944. Indeed, "it is to preserve our way of life that we are fighting this war."[81]

CHAPTER 7

School-age Children Fight the War

"ABOVE ALL, as a child I felt invincible," recalled Susan Yarin, a homefront girl. "I felt that nothing would ever happen to me because I was American...." America's school-age children were certain not only that their nation's cause was just, but that the United States would prevail. "God was on our side ...," explained Beirne Keefer, a homefront boy.[1]

This was a war between good and evil; there were no shades of grey, no nuances, especially for America's school-age children. Ranging in age from five or six to puberty, these girls and boys learned well the lesson of their country's righteousness before all nations, imbibing it in their schools, churches, and theaters. Indeed, although feelings of invincibility and moral certainty were two qualities widely shared by the school-age children, another was a sense of personal self-worth, motivated by participation in the war effort. "Our pride in our country," noted a homefront girl, "made us proud of ourselves." Boys and girls collected scrap materials for recycling, and they bought War Bonds and knitted afghans. George Curtis, born in 1935, remembered that even helping his parents perform household tasks made him feel he was making "a genuine contribution" and was "a part of the overall team." Whether helping with the dishes or the farm chores, George saw the war years "as a time for children such as me to feel of greater human worth or value; to be more respected by adults, because we had the opportunity to ... contribute."[2]

Nell Thomas, whose eighth birthday fell on December 7, 1941, announced that the day was a bust for her. In fact, no one was interested in her birthday "at all, only this thing at a place called Pearl Harbor." It did not take Nell long to realize that the war would quickly become the all-consuming concern of everyone in her hometown of Greensburg, Pennsylvania. "During this whole period of my life," she remembered, "everything was focused on one thing, winning the war. Every American citizen was part of it, whether they were in the service, or an adult or a child. We were all able to contribute in some way. We were constantly reminded that ours was the best country in the world, and it was our duty to fight to preserve democracy."[3]

The experiences of America's school-age children differed markedly from those of the infants, toddlers, and other preschoolers. The world of the school-age children included not only the school but also popular culture, not to mention patriotism and participation. As the family was the focal point of the pre-schoolers' lives up to the age of five or six, so the school became the focal point for the older boys and girls. Not coincidentally, the grade-school years are the years of childhood political socialization. In the elementary schools, political scientist Fred I. Greenstein has observed, "children move from near—but not complete—ignorance of adult politics to awareness of most of the conspicuous features of the adult political arena."[4]

The grade-school years are the years of latency, pictured by Freudian psychology as a time to gain learning and social skills undistracted by sexuality; it was the calm before the storm. Likewise, educators saw it as the period when children needed to learn to become good citizens. Indeed, by legislative fiat, it was in the elementary schools that children learned about the ideals of the "founding fathers" as expressed in the Declaration of Independence and the Constitution. A New Jersey law passed in 1945, for example, mandated that these students be taught "the history of the origin and growth of the social, economic and cultural development of the United States, of American family life, and of the high standard of living and other privileges enjoyed by the citizens of the United States as will tend to instill into every boy and girl a determination to preserve these principles and ideals. . . ." By 1948, forty-four states had enacted statutes requiring "instruction on the Constitution" in the grade schools.[5]

Research done on political socialization points to the grade-school years as the formative ones. After testing 12,000 students, psychologists Robert Hess and David Easton concluded: "Every piece of evidence indicates that the child's political world begins to take shape well before he even enters elementary school and that it undergoes the most rapid change during these years," which were "the truly formative years of the maturing member of a political system. . . ." Psychologists and political scientists have found too that the grade schools are the primary agents of citizenship training and that political attitudes, most of which have been acquired by the end of the eighth grade, remain stable in adolescence.[6]

Certainly none of these social scientists has argued that children develop permanently fixed political dispositions. Childhood behavioral acquisitions, as one study reported, "may only be selectively consequential for adult behavior." Still, such acquisitions not only are durable but are "likely to remain as underlying sentiments that can be evoked in later years." One germane study, *Children and War: Political Socialization to International Conflict*, after surveying the research findings on children's conceptions of war, concluded that "war-related attitudes begin to crystallize by age seven, and by fifteen most adolescents express a fairly definite viewpoint about war and peace."[7]

Theories of child development, particularly those articulated by Jean Piaget, illuminate the learning processes of school-age children. The problem for the historian who would apply Piagetian psychology to the task of illuminating the past is that Piaget focused on internal structures and mental schemata that are unobservable.[8] Still, since the historian's goal is to identify and explain past causal relationships, the challenge is to visualize these cognitive principles in order to understand better the Second World War's impact on the attitudes and values of America's children.

According to Piaget, children's learning is a dialectical process, central to which is the quest for equilibrium. At different ages—for example, 18 to 24 months, 5 to 7 years, and 9 to 11 years—the child is "vulnerable." Increasingly, his or her understanding of the external world produces negative responses and disconfirmation. The child is in a state of disequilibrium; in Hegelian terms, his or her thesis has begun to inspire an antithesis. And as disconfirmation accumulates, the child pursues and eventually constructs a new equilibrium. With this new world view, the child then proceeds to rewrite his or her history. All of the environmental stimuli coming in take on new meaning. What has happened is that the child has experienced a vertical change—a leap upward; but it contains the seeds of its own destruction. Although equilibrium characterizes the child up to 18 months and from the ages of 2 to 4, 7 to 9, and over 11, such equilibrium is dynamic. It constantly requires correcting, just as a car requires steering to keep it on the road. But it is inevitable that the child will go too far and overcorrect, perhaps sending the imaginary car into the ditch. As the disconfirmation accumulates again, another stage begins to set in; there is a quest for a new equilibrium—and another leap upward.[9]

Piagetian psychology can help historians to appreciate "critical" or "sensitive" periods in children's development, periods in which change seems to come almost regardless of the environment. According to Piaget, up to the age of seven, children's cognitive capabilities are limited in three ways. One limitation is egocentrism, or the inability to assume another's perspective. Another is "centration," or the inability to center on more than one dimension of what she or he perceives. The third is the child's inability to perform an operation requiring reversibility. But at age six or seven, with the onset of "concrete operations," the child begins to overcome these limitations on logicality, thus inaugurating a change which, according to one psychologist, "represents a dramatic transition from illogically based thought to logically based thought." The years between nine and eleven are vulnerable ones for the child. At about age eleven, the child experiences another quantum leap as she or he develops the ability to generate hypotheses about logical relationships. This is the stage of "formal operations." "Thus," another psychologist has explained, "we now have operations on operations. Thought has become truly logical, abstract, and hypothetical."[10]

Piagetian insights can sensitize historians in other ways as well. The concept of moral development is an outgrowth of Piaget's cognitive psychology, as initially expressed in his seminal work *The Moral Judgment of the Child* (1932), and as developed by Lawrence Kohlberg and refined by Carol Gilligan.[11] Concurrent with the arrival of the stage of concrete operations at about age seven, there is a change in the child's moral sphere. There is a leap from what Piaget called "heteronomy," or blind obedience to authority, to "autonomy," or behavior based on mutual respect. An example is cooperation in game playing, in which children are able to agree on a set of rules and play a game fairly. Still, children in concrete operations tend to be inflexible in their commitment to what they perceive to be "right." At about age eleven, however, with the attainment of formal operational thought, new moral feelings arrive, and, as psychologist Lee C. Lee has explained, they "are added to and integrated with the preceding feelings," corresponding "to the newly acquired cognitive structures. Because of the ... flexibility of thought at this more advanced stage, feelings become decentered. ... The child can now grasp the concept of humanitarianism, which is the ideological stage of autonomous morality."[12]

Thus, when considering the overwhelming receptivity of America's school-age children to the wartime messages, it is essential to appreciate not only the messages but also the children's developing cognitive capacities during the war, particularly moral development. Of course, the patriotic messages were both intense and powerful. And it is only when viewing the mental structures and the wartime messages as reciprocal and complimentary that one can grasp the centrality of the Second World War to these homefront children, both then and in the values and attitudes which they have carried with them throughout their lives.[13]

In fact, there could not have been a more impressionable time for children to be exposed to the heavy doses of the patriotism and democratic ideology that prevailed on the homefront. And the place to turn to understand the transmission of these messages is not to the home—the domain of the infants, toddlers, and other preschoolers—but to the elementary school.[14]

Whatever their psychological perspective, whether cognitive, psychoanalytic, or behavioristic, the wartime experts agreed that school age was the time to learn good values and good habits.[15] Educators, having long before affirmed this, viewed the Second World War as an important challenge to, and opportunity for, the nation's schools. Soon after the Pearl Harbor attack, a group of women educators set the stage for numerous high-minded pronouncements about the responsibilities of the schools in a nation at war. Writing in *Education for Victory*, this group urged the schools "to guarantee for all children adequate *protection*, intelligent *participation*, and balanced *perspective*." Protection was obvious, including frequent air raid drills and the provision of gas masks and identification tags. But the two other categories were intensely idealistic—and propagandistic— reflecting an ideological certainty about what the United States ought to be—

and, indeed, already was, in the minds of millions of Americans. "Intelligent participation by children includes," the statement read, "understanding patriotism, citizenship, democracy...." And "balanced perspective for children" required:

1. Sensing what America is fighting for by developing an understanding of democratic ideals through daily practice in living them.
2. Seeing that America's fight for democratic principles is but one part of mankind's long struggle for freedom.
3. Knowing the real values that war cannot destroy.
4. Understanding the necessity for personal sacrifices.
5. Understanding and appreciating others by stressing likenesses as opposed to superficial differences among citizens of a democracy.[16]

All subsequent statements, whether issued by governmental bodies or educational associations or published in magazine articles, lauded these as the wartime aims for the schools.[17]

Not only was it essential for the schools to promote democracy, it also was crucial for them to eradicate hate and other antidemocratic tendencies. "Even a slight trend toward racial prejudice and national hatreds among young children should cause us concern," wrote the chief of elementary education for New York State's Education Department in 1943. "The very foundation of American society rests on the premise that individuals of different backgrounds can work together for the common good.... It is imperative that our young children shall not be encompassed by the weight of our mistakes and prejudices...." "Wars Are Made in Classrooms" was the title of a 1944 article in the *Saturday Review of Literature*, which stated that while school can be "an effective medium for the development of good attitudes and habits of life, if it remains an instrument of competitive nationalism, more wars will be made in classrooms." It was time for a change, time to drop the "autocratic classroom" that sustained "the autocratic nation," and time for school to become "a fundamental instrument for international appreciation, collaboration, and good will. Why not now," the article asked, "—now before we get involved in the passions, prejudices and power-drives which must lead to another holocaust?"[18]

And there was reason to be concerned. Sometimes in their zeal to annihilate the enemy, homefront children generalized their hatred and assaulted and vilified other children at home who were of German, Italian, or Japanese descent. If the espousal of tolerance and democracy distinguished the rhetoric of numerous school-age children, the articulation of prejudice, not to mention the total rejection of certain racial and ethnic groups, was also widespread.[19] Indeed, prejudice and rejection were not only the flip side of tolerance and democracy, they also arrived at about the same time in children's political socialization. As Gordon W. Allport, the psychologist, discovered, children's "totalized rejection" of certain groups began at seven or eight, and "by the age of 12, we may find *verbal* acceptance, but *behavioral* rejection. By this age," Allport explained, "the prejudices have finally affected conduct, even while the verbal, democratic norms are beginning to take effect."[20]

Repeatedly, homefront educators expressed apprehension that without the democratization of schooling worldwide, there would soon be another generation of fanatics plunging the world into yet another war. One such prognostication was made by an education professor at Northwestern regarding a boy named Johnny, "an active charming" seven-year-old "nibbling eagerly of every morsel of life." But "the only world he knows is a world at war," a world that constantly "emphasized the need of killing other people. He hates Japs and Germans, makes jokes about Mussolini. In his play he murders with tommy-guns. . . ." But in fourteen years Johnny would be voting. "What kind of world will he want? To grow up in hatred, suspicion, dread and terror is to become a vindictive, suspicious, anxious, terrified adult. Can children thus reared build and maintain the just, orderly and peaceful world for which we profess to fight?" It all depended on the schools: "Unless the school takes vigorous counter measures the results of this war are not worth the fighting. The world will again be run by the emotions of small, selfish, embittered, fearful men."[21]

Reinforcing the patriotism and idealism of the homefront was another formidable influence in American education at this time—the progressive education movement. In education as in other disciplines, there was usually a "conventional wisdom," and as Lawrence A. Cremin, the historian of education, has written, "by the end of World War II progressivism had come to be that conventional wisdom." Schooling that stressed the student's individual development, and worked to achieve a healthier community, was the progressives' goal. Following the philosophical guidance of John Dewey, educational policy-makers urged the wartime schools to recognize the individual needs of learners; to develop "the whole child," including personality, social skills, emotional growth, and creativity; to appeal not to coercion but to adjust "the school to the child"; and to bridge "the gap between home and school" by exposing children to "real life experiences."[22]

The ascendancy of progressive education was best represented by the pronouncements of the National Education Association's Educational Policies Commission (EPC), an influential body that included the presidents of Harvard and Cornell as well as the United States Commissioner of Education. In three notable documents published in the 1940s, two of them during the war, the EPC challenged the country to "organize a comprehensive public school system," for all children and youth between the ages of three through twenty, which not only would provide the skills necessary for earning a living but also would reform society. The grade schools should not only strive toward the "full development of all the humane and constructive talents of each individual" but also nurture "social responsibility and the cooperative skills necessary to the progressive improvement of social institutions." Such pronouncements, Cremin concluded, "signified that Dewey's forecast of a day when *progressive* education would eventually be accepted as *good* education had now finally come to pass."[23]

The schools responded favorably, and school systems and teachers across the country espoused the goals of progressive education, explaining that while it was essential to accomplish these aims in peacetime, it was absolutely imperative to do so in time of war. Indeed, what magnified the impact of progressive educational beliefs was their congruence with the popularly accepted goals of the war itself. This was a war for democracy and cooperation, and against racism and violence. And where better to teach this message, one educator wrote, than in "the school system [which] provides the only institution in our national life that can influence 'all the children of all the people.'" Teachers in a variety of situations agreed. Julia Weber, who was the sole teacher in a New Jersey country school, pondered her responsibility to teach democracy. "It seems to me," she wrote, "that in order to be fully aware of the advantages and responsibilities of democracy, to desire it, the children must practice truly democratic living in their schoolroom."[24]

Interestingly—and ironically—educational planners in the War Relocation Authority, which administered the internment camps in which lived 120,000 Japanese Americans, including 30,000 school children, saw the camp schools as an opportunity to put into practice Dewey's ideas. Not only would the educators try to mesh "school and community life as closely as possible," they also would teach the tenets of democracy and good citizenship. But as Thomas James, the historian, has pondered in his study of the wartime schooling of the Japanese Americans: How "in good faith" could these children "prepare for future life in a democratic society while suffering an undemocratic imposition of authority and deprivation of rights[?]" The democratic message in the relocation camp schools was powerful and doubtless affected the students, but it also sensitized them to violations of democracy. "If we are citizens," one student asked, "why are we in concentration camps?" "Is the United States a real democracy[?]" asked another student. And, later, an internee at the Manzanar camp observed that in his class's discussion of democracy, "we had to skip the chapter on civil liberties."[25]

Still, for the great majority of America's girls and boys attending elementary schools, there was at work a powerful congruence of the democratic message and the educational medium. But as noted above, there was yet another reason for the impact of the wartime ideology: developmentally, these boys and girls were ideal receptors—cognitively, their mental structures were set to receive. For these reasons, America's school-age children developed moral stances tightly attuned to the United States' lofty war aims. How could it have been otherwise? For example, during the war the NEA distributed 500,000 copies of "The Children's Morality Code," which admonished children to follow "The Laws of Right Living." "The Good American" was all things to all people. The code enumerated these qualities: kind, fair, reliable, cooperative, "does his duty," is loyal to family, school, town, state, country, "and above all things else ... loyal to humanity."[26]

Judging by the homefront children's recollections, the schools did stress cooperation, tolerance, and democracy. But the homefront children also remembered some stridently nationalistic teachers who, in promoting patriotism, indulged in bigotry, particularly racism, nativism, and anti-Semitism.[27] Even when well intentioned, educators could be blind to their own bigotry. In the Plainview-Beechwood school district, a war-boom town near Wichita, Kansas, the elementary-school students came from the federal projects housing the families of bomber plant workers. In teaching the "Functions of Living" to these boys and girls, the curriculum called not only for "Developing and maintaining a functioning social sensitivity," but also for "Understanding and applying the laws of eugenics to maintain and increase the nation's hereditary strength." It was a strange combination.[28]

Years later, the homefront children vividly recalled what their classrooms were like. Each day began with the "Pledge of Allegiance" and perhaps the Lord's Prayer and a prayer for those in the armed services. Jeannette Terrill, whose father was in the Navy, attended grade school in Montana. Her teacher "would inquire of the students which ones had family members in the service. . . . Then she would stand before us and pray for them and us." As for the Pledge of Allegiance, there was a war-inspired change. Prior to the war, remembered a homefront girl, "we were accustomed to placing our left hand over our heart, and raising our right hand & arm to the flag," Nazi-style. Fortunately, the extended right-arm salute was an early homefront casualty.[29]

During the war schoolgirls and boys staged patriotic plays and sang patriotic songs, the most popular of which were "Anchors Aweigh," "The Caissons Go Rolling Along," the "Marine Corps Hymn," and the Air Corps anthem that began with the rousing "Off we go into the wild blue yonder." "These were special songs to us," wrote a homefront girl. For their part, music educators urged the schools to employ music "in our present crisis to help our beloved America to bear the heavy load. . . ." The author of these words, the director of music education for the Baltimore public schools, exhorted the schools to hold "Victory Assemblies," "Victory Sings," and "outdoor flag raising and parades," replete with drum corps and military marching music. Last on his list but first in importance was the national anthem. "Music is the Handmaiden of Morale," he wrote.[30]

America's school-age children engaged in a great deal of singing. Teachers wrote the songs on the blackboards, and the boys and girls copied the words in their writing books. "For our school assemblies," recalled David Goodman of Cleveland, "we learned to sing the marching songs of all the branches of the armed services. . . ." But for Larry Paul Bauer, a Cleveland boy who attended a different grade school, there was no singing of the Air Corps song. One of his teacher's sons, a pilot, had been killed in action. "Our music teacher told us that singing that song made the teacher who lost her son very sad and sick at heart." Larry's class never sang it again.[31]

In other classrooms too, the war intruded by taking the life of the husband or son of a teacher. A first-grade teacher in Williamsport, Pennsylvania, lost her only two sons. "She walked around the room like a zombie . . . ," recalled one student; "the rest of the year she tried hard, but she looked sad." Nancy Murray attended elementary school in Dover, Delaware. One day, while in the third grade, she watched as the principal and two men in uniform entered her classroom and asked her teacher, Mrs. Wolcott, to come with them to the principal's office. There they told her that her husband of less than a year had been killed. When Mrs. Wolcott returned, Nancy remembered, "she told us what had happened, and we were all so sad for her."[32]

The war shaped both classroom discipline and the student reward system. Many teachers assigned military rank to their pupils. In a small Ohio town, remembered one girl, "our first and second grade reading groups weren't divided into redbirds, bluebirds and blackbirds, but into groups of privates, corporals and sergeants. We colored and cut out paper stripes to wear." Students wore cardboard chevrons displaying their ranks. Summer reading programs also rewarded achievement with military promotions. "We advanced in rank with every book we read," wrote a homefront girl, "and the goal, of course, was to be a four-star general by summer's end." Some teachers carried this emphasis a step further, lining students up for daily inspections of their hygiene. In Covington, Kentucky, a gymnasium teacher "insisted on drilling us just as trainees were in the military." The students had instruction in trench digging, as well as in "presenting arms, rushing the enemy, [and] falling to the ground in fighting position. . . ." She also ordered each student to make a wooden rifle and report with it to the next class.[33]

The war had a direct impact on those students whose teachers left for war. Catherine Dunlap, a student in Fort Worth, recounted that her "principal was drafted—he was also the 6th grade teacher. When I reached the 6th grade," she noted, "the former school secretary was teaching that class." Shortages were acute in rural areas and in the war-boom towns. Two thousand schools, mostly rural, failed to reopen in the fall of 1942, due to the unavailability of teachers. Oklahoma, Georgia, Kansas, and other states reported severe teacher shortages. The rural-urban disparity in per child expenditures, which had been a disturbing factor in American educational life for generations, not only persisted during the war but gave rise to the demand for federal aid to equalize educational opportunity among the states. "States with the most children," declared the *Journal of the National Education Association*, "have the least wealth; states with the most wealth have the fewest children." Led by unfilled positions in rural schools, the nation was short 75,000 teachers by the fall of 1943. Edna Boardman, who lived on a farm in North Dakota and attended a one-room school, said that her teacher was a high-school graduate who had taken an eight-week "crash course" at a state teachers' college. But if one of these graduates could not be found, then "any available adult who had finished high school was hired. . . . I even recall accepting minimally educated teachers as one of my contributions to the war effort." Shortages existed in the cities as well.[34]

Many of the best teachers left the profession during the war, for economic reasons as well as patriotic reasons.[35] School systems raised salaries, but teaching jobs were still not competitive with defense employment. A 1943 federal report showed how bad the situation was when it stated that salaries in Detroit area schools "have been increased in most cases near enough to those paid semiskilled labor so that it has been possible to prevent too large an influx from teaching positions into defense factories." Seventy percent of the teachers in New York City who had resigned, or were intending to do so, said their reason was "insufficient salary." Short of dramatically upping teachers' pay, there was only one way to fill the ranks: lowering educational qualifications for new teachers. "During the past 2 years of war," wrote Benjamin W. Frazier of the United States Office of Education, "there has been a steady lowering of standards through the issuance of war emergency permits. . . ." Such permits jumped from 4,655 in 1941–42, to 38,285 the next year, and 69,423 two years later.[36]

While many teachers who left were young and energetic, their replacements were a diverse group. "Included among emergency teachers," wrote Benjamin Frazier, were "teen-age high school students, retired teachers with a lifetime of experience, married women ex-teachers, college graduates and former students with no professional training or experience, tradesmen, and a variety of other persons." Some of the new teachers were failures, but others were successful. Margaret Dyar, who lived in a small Colorado town, fondly remembered the teachers hired from the "Emergency Certified" list. "Some were from retirement," she wrote, "with excellent abilities and some were very young lady graduates who were fun to be with on picnics and outings as well as the classroom."[37]

Whatever the talents of the teachers, the schools were invariably the focal points of children's homefront participation. On September 25, 1942, for example, the Department of the Treasury and the Office of Education launched the "Schools at War Program." On that day, 4000 school children marched along Constitution Avenue to the Department of the Treasury, where Secretary Henry Morgenthau, Jr., told them that he knew the country could "count on its 30,000,000 young Americans . . . to enlist 100 percent in our fight for freedom. . . . By participating fully in the Schools at War Program, they can tell the world: 'We Are Ready—ready for war, ready for victory, and ready for peace.'" The children's motto was "SAVE, SERVE, CONSERVE." And government officials implored teachers and students to work together to "SAVE money to buy War Savings stamps and bonds regularly"; "SERVE your school, community, and nation" by volunteering for everything from making posters for community war drives to working at the local "Child Care Center to help working mothers"; and "CONSERVE all kinds of materials, service, and money" by collecting scrap materials for recycling, repairing old clothes, toys, and furniture, "Buying wisely" and "Wasting nothing."[38]

In its first year, as 28,000 schools enrolled as "Schools at War," heroes of the war applauded the children's efforts. In his dramatic plaudit, General Douglas MacArthur ignored the girls, stating that "with the tinkling laugh in my ears of a little boy who also does his part," he sent "from far away a prayer that God's blessing may rest upon you always." Lt. Commander Edward H. O'Hare, a pilot who had won the Congressional Medal of Honor for shooting down five Japanese bombers, thanked the Schools at War Program: "It is the kind of backing the youngsters of the nation are giving ... that will move our front to Tokyo and Berlin before we finish this job."[39]

The Gary, Indiana, schools had an exemplary war program. In Gary, as historian Ronald D. Cohen has written, "the pride of the schools and the community" was a group called the All-Out Americans. Formed in September 1942 and working closely with the Junior Chamber of Commerce, the AOA initiated a massive scrap drive involving all the city's grade-school students. The students at each school elected a major and an adjutant "who together would form a city-wide organization, and they were sworn in as members of the junior civilian-defense corps...." A twelve-year-old Girl Scout served as the AOA's first colonel. "We had drives of all kinds and everyone participated," recalled a homefront girl who had been an "All-Out American." Among their tasks, the "All-Out Americans" distributed Community Chest pledge cards, "War Workers Sleeping" signs, and information about the "clean plate" campaign. They also met with the city council to discuss Gary's curfew law, helped the block mothers, and collected books to send to service people.[40]

"Children strive to exceed their elders in patriotism," wrote William H. Johnson, superintendent of the Chicago public schools. In Chicago, he explained, "they have been working hard for the salvage drive on tin, rubber, radios, and scrap metal—a typically school and home affair." More than 100,000 pounds of scrap were collected the first week. "Every child had a part in the work. Discarded scrap was brought to school by hand, in toy wagons, in wheelbarrows, in family cars, and even in trucks." The final tally was 1,500,000 pounds of scrap, including seven tons of keys.[41]

During the war schoolchildren collected extraordinary amounts of waste paper, bacon grease, and scrap metals, and they sold War Bonds and Victory Stamps worth hundreds of millions of dollars. But what is important is not the aggregate figures, but rather the individual enthusiasm and dedication that produced these totals.[42]

"Boys and girls of the United States of America," proclaimed a wartime book, "you are enlisted for the duration of the war as citizen soldiers. This is a total war, nobody is left out, and that counts you in, of course." Agleo Patri, author of this guide, explained that the Second World War was "a war for the freedom of the individual, that means your freedom." The Nazis would destroy that freedom; they would tell you where to work, what to read and think, even what to eat and wear. "We say to our enemies," Patri told the children, " 'No, never. We will fight you up and down the world . . . we will make no peace with you until you let go such a stupid idea.' "[43]

The school-age girls and boys were only too eager to participate in the war effort. Edna Boardman, who was in the first grade at the time of the Pearl Harbor attack, developed what she described as "an almost euphoric sense of being linked to something bigger than myself." Not only that, she wrote, but "I was made to feel that my . . . sacrifice was important to my nation's success."[44]

For one thing, the school-age children lived in a world that was very vivid to them. Consonant both with their psychological development and with the moral absolutes which they and their elders viewed as the goals of the war, the boys and girls objectified their wartime contributions. They were not merely volunteering a wagonload of newspapers or a wheelbarrow full of tin cans; their flattened cans, for example, were the raw materials for manufacturing airplanes, ships, and tanks. And they were purchasing War Bonds and Victory Stamps not as investments for the future, but to win the war then and there by raising money to build more weapons. Sixth-graders at the Pinckney Elementary School in Lawrence, Kansas, bought $331.15 worth of defense stamps, which, they calculated, would purchase one submachinegun, four field telephones, one tent, five steel helmets, and nine entrenching tools. "One old lawn mower makes six three-inch shells," the National Recreation Association announced. "One old flatiron makes 30 hand grenades."[45]

Children felt intimately connected to the war drives. Some were even given the honor of naming a ship. In New York State, the girls and boys who had contributed to a recent drive voted that the Liberty ship representing the state in the nation's victory fleet would bear the name of Lou Gehrig, the late baseball hero of the New York Yankees. School children in Florida named their ship the *USS Colin Kelly,* in honor of the American pilot who had gone down with his Flying Fortress in the Pacific after ordering his crew to bail out. In addition, numerous children received the plaudits of their country's leaders for their achievements in scrap collecting. In New York City, for example, four brothers earned the congratulations of Mayor Fiorello La Guardia for collecting several tons of old newspapers. The boys explained that their secret lay in their division of labor. William, fourteen, rang the door bells; Michael, ten, pushed a baby carriage for carrying the newspapers; and John and Jimmy, ages eight and three, loaded the carriage. In Los Angeles, after going door-to-door collecting aluminum pots and pans, the children took their wagons to a park where a large board stood. Painted on the board were the faces of Tojo, Mussolini, and Hitler, with holes cut out for their mouths; into the gaping dictators' mouths the children would pitch their pots and pans. In the United States during the first six months of 1943, children collected 13,000,000 tons for the schools' "Salvage for Victory" program. The Boy Scouts received special praise for their efforts; during wartime, they collected more than

3,000,000 books, 109,000,000 pounds of rubber, 23,000,000 pounds of tin, and 370,000,000 pounds of scrap metal.[46]

The school-age children collected scrap materials individually and in groups. Joan Petsch, a schoolgirl in St. Louis, won the city's award for the most scrap collected in a week. Joan had two wooden, platform wagons tied together, which she would load and then "drag, push, [and] pull back to the big scale in the schoolyard." Her grand total was 1,220 pounds of recyclables. Numerous towns and cities staged drives in which the children filled their wagons. And in New York City, a group of children had a confrontation with sanitation workers. Armed with broomsticks, the children refused to allow the city workers to cart away the three tons of metal they had collected. Backing up these ten children were fifty others. When a sanitation driver promised that the children would receive "official recognition" for their efforts, they dropped their weapons. A police officer took down the children's names; a week later, they received official certificates applauding their contributions.[47]

In their letters, the children shared memories of their homefront participation. Many told of flattening cans: first washing out the can, taking off the ends and placing them inside, then putting it on the floor and stamping on it. A homefront girl wrote that her mother taught her to say, "This is Hitler's head," as she flattened the cans. Children remembered carefully peeling the tin foil from cigarette packs and rolling it into balls for recycling. Rural children shared the experience of gathering the pods of the milkweed plant, used in stuffing Mae West life preservers. Children made important personal sacrifices. Michael McCall, a boy living in Kansas, recalled that his prized possession was a red-and-white pedal car. One day, when his mother asked where his little car had gone, he said "to war, to win the war." The next day, when he arrived at school, he saw his car sitting atop a twenty-foot pile of scrap metal. For a seven-year-old patriot, McCall wrote, "that was my proudest moment."[48]

In addition to scrap collecting, school-age children sent books, magazines, and crossword puzzles to patients in veterans' hospitals. Children also trapped rattle-snakes for venom to protect the troops from fatal bites. They raised carrier pigeons and volunteered their pet dogs for service. And they sent Christmas presents to "orphan soldiers" who had no families. Another important homefront children's activity was knitting afghans, sweaters, mittens, socks, and scarves for service people. Grade-school boys as well as girls received knitting instructions and developed their competency by knitting six-inch squares, in tan or olive green, to be made into afghans for distribution by the Red Cross.[49]

Parents, teachers, doctors, and child-guidance experts alike saw children's participation as a positive factor, which not only contributed to their happiness but enhanced their sense of personal responsibility during an anxious time. J. Louise Despert, a psychiatrist, wrote that "participation in the war effort is a powerful mechanism to allay anxiety." Sometimes, parents and teachers exploited the children's patriotism; in exhorting the children to clean up after themselves, they told them that such tidiness was their wartime duty. Munro Leaf, the author of children's books, wrote *A War-Time Handbook for Young Americans* in 1942, calling it "a guide to many ways in which we can ... do our duty as Young Citizens of the United States." These duties began at home, including being cheerful and pleasant, getting plenty of sleep, eating "WHAT you should WHEN you should," taking care not to break things, keeping clean and keeping one's room "in order just the way the soldiers in the army have to do." Eleanor Roosevelt testified before a congressional committee that "you should get your children to feel that they are making a contribution," no matter how small. One could, she said, persuade the children, who did not like milk, that "their contribution may be that they drink their milk. . . . Then they have a sense of participation."[50]

Numerous children, however, did not need to be told; seeing a homefront need to be met, they simply went ahead and did the job. After a snowfall in Maplewood, New Jersey, two boys aged thirteen and fourteen shoveled the walks of every house with a service flag in the window. They rang no door bells; they took no money; and after doing fourteen walks, they quit, thoroughly exhausted.[51]

In addition to scrap collecting, the school-age children remembered buying Victory Stamps and War Bonds. One day a week, usually Monday or Friday, the girls and boys brought in their dimes and quarters to buy the stamps, which they pasted in their stamp books. When filled, the book held $18.75 in stamps; at this point, the proud bearer would exchange the book for a $25.00 War Bond, redeemable in ten years. Some children decided to forgo school lunch in order to save their dimes for savings stamps. And given the pressure on children, it was no wonder. The government, the schools, and the media were constantly urging children to buy stamps and bonds. A girl who grew up in Davenport, Iowa, wrote that "stamp and bond booths sprang up everywhere—in schools, in stores, in public buildings." Radio announcers exhorted her to buy, as did the newspapers and popular magazines. And, she noted, "no trip to the movies was complete without a brief pitch for bonds before the main feature began." War songs also exhorted the children to buy; the most popular song was, "Any bonds today? Bonds of freedom, right when we need 'em, for the USA. Any bonds today?"[52]

To forget one's stamp money, or, worse, to be unable to afford the stamps, could be profoundly embarrassing to a child. In this regard, the homefront experience was different for children from poor families than it was for those from affluent families. Russell Grant, a ten-year-old in Kokomo, Indiana, swallowed a worm on a dare in order to win 25 cents to buy a war stamp and be, as he said, "like the rest of the kids." And when Russell's story hit the news wires, letters began to arrive with additional quarters. "My greatest fear," a homefront girl recalled, "was that I would forget my money some Friday and cause some cog in the war machine to slip." The family of means, on the other hand, could buy many War Bonds. Margaret Padgett was a seven-year-old living in Tennessee when her father bought each of his children a $1000 bond, the proceeds from which went

to build a large anti-aircraft gun. A photographer took the family's picture standing in the town square, the children gazing at the barrel and a group of soldiers
standing in the background.[53]

The War Loan Drives in the schools were top-down as well as bottom-up
affairs. An example was the Third War Loan Drive, which began in the fall of
1943 and concluded on the second anniversary of Pearl Harbor. School districts
announced that they were planning to "Back the Attack" by setting targets for
their drives. Assuming a quota of $10,000,000, the Detroit Public Schools
enrolled the teachers as salespersons and had them certified as issuing agents
under the supervision of the Federal Reserve Bank. Obviously, the pressure to
buy was great. The grade schools asked each student to write a letter home
explaining "what his class is doing to speed victory, how he is saving and earning
money for War Stamps, and why we should buy all the War Bonds we can." The
children had to bring their letters back to school after securing their parents' signatures.[54]

That fall the national drive stressed the new "Triple Threat" jeep campaign.
The schools' bonds would purchase either the familiar land-going jeep, the
amphibian jeep, or "Quack," or the flying jeep known as the "Grasshopper." The
government set a goal of 20,000 jeeps by December 7. Smaller schools had an
alternative to entering the buy-a-jeep campaign. For these schools, there was the
Junior Triple Threat Campaign, with the scaled-down goals being motor scooters,
parachutes, and lifeboats. School children tripled their quota in Milwaukee, and
in Philadelphia they bailed out a lagging city-wide drive by organizing a "commando attack" during the campaign's last week. Smaller schools were proud of
their accomplishments as well. There were only five students in Rauville, South
Dakota, near Watertown, but they sold $3000 worth of bonds in a single week
and were awarded a "Diploma of Achievement" by the Treasury Department.
During 1944, nationwide bond sales in the schools financed 2900 airplanes,
33,000 land jeeps, and 11,600 amphibious jeeps. Moreover, during the summer,
when schools were not in session, other children's organizations stepped in,
among them the Junior Brownies, Junior Scouts, Girl and Boy Scouts, Camp Fire
Girls, YWCA and YMCA, the Boys Club, and church and social groups. Probably the children's proudest moment occurred on June 9, 1944, when a representative of President Franklin D. Roosevelt accepted a bullet-scarred Japanese propeller shot down by American forces. Engraved on the blade was the children's
record of $510 million worth of war equipment paid for with their stamps and
bonds.[55]

America's school-age children also worked in victory gardens, either with their
families or classmates. Joan Petsch wrote that, on Tuesday afternoons, she and
her schoolmates "filed out of the classrooms" with child-size shovels and buckets
and marched three blocks to plant vegetables in their victory garden.[56]

Magazines provided gardening instructions for parents and children. *House &
Garden* featured sketches of a garden grown "by two youngsters five and seven

years old." Divided into two identical plots, the garden "promotes competition," with "great rivalry to see who will grow the better tomatoes" and pumpkins. "Here comes the enemy!" the article warned, the enemy in this case being insects and weeds. As for the "pint size gardeners," those between three and eight, the article suggested they "will not have such a precise idea of what a garden is," so they should haul water or "tend a few tomato plants." At the Thomas A. Edison School in Tampa, Florida, 219 pupils cultivated a large garden plot, growing cabbages, carrots, turnips, lettuce, beets, and other vegetables, "all destined for meals in the lunchroom serving 300 children a day." The children brought tools and "hauled fertilizer from home in little wagons," turned the soil, "and planted everything themselves." In Chicago, park officials set aside 30,000 plots, each 5 by 12 feet, for youngsters in grades five through eight "to plant vegetables and a few flowers." Here, and throughout the nation, children were tending their victory gardens.[57]

In recognition of the many patriotic activities of America's homefront girls and boys, the Congress in 1944 adopted House Concurrent Resolution 104 expressing thanks for their "contribution to the victory effort." Patriotism was the wartime passion that motivated these children; indeed, this was a time, recalled a homefront girl from North Dakota, when "*not* being patriotic was unthinkable." "I became fiercely patriotic," recalled a Nebraska farm girl, "and participated in every war effort that I could." And a homefront boy remembered that in his hometown of Florence, Alabama, "there was such a unity of spirit that hardly anyone questioned any war-winning effort...."[58]

Another form of children's homefront involvement was at times oblique, but always present: government rationing. Among the memories recounted by the homefront children, the topic of rationing appears with great frequency. Unlike father absence, for example, or mother's employment outside the home, government rationing affected everyone in the United States, including the children. In January 1942 President Roosevelt had signed an act establishing the Office of Price Administration (OPA). Later that year, the OPA issued the first war ration books, providing coupons for sugar, then coffee and gasoline. The OPA instituted point rationing in early 1943, with meat, fats and oils, butter, cheese, and processed foods. Later it added shoes. Local rationing boards set quotas, allocating coupons for each family. The homefront children remember the wartime deprivations well. Their letters complain about shortages of bicycles, gasoline, butter, nylon stockings, cigarettes, bubblegum, pencils with erasers, and, most of all, sugar, meat, and shoes.[59]

In many of these letters, however, there is evident pride in "making do." For some children, these shortages were merely continuations of the Depression shortages caused by falling incomes; "getting along with less," wrote one girl, "was something I already knew." Naturally, many shortages hit low-income Americans the hardest. A homefront girl who grew up in a coal-mining town in the Ozarks recalled the shortages in her life: "shoes, coffee, soap, sugar, cooking oils, gas, tires, elastic, tin, and chewing gum." War for her meant further "hardships," especially after her father was drafted. Many children remember the substitute foods resulting from rationing—"Marshmallow Ice Cream," for example, which was "full of ice crystals and crunched as you ate it," concocted in the absence of sugar. Ann Johnson, who lived on the Washington coast, gathered oysters as a meat substitute for her family's meals. Her mother then traded meat stamps for extra sugar stamps—for making jams and jellies. Other homefront children reported that instead of eating customary cuts of meat, they ate tongue or tripe, and that instead of enjoying a new pair of summer sneakers, they could find only huaraches in the stores.[60]

The homefront children, however, also told of the sharing, and the kindness, that was inspired by the rationing system. Neighborhood grocers helped out in a number of ways. At Lutz's Grocery Store in St. Louis, Mrs. Lutz, with the permission of her customers, kept the neighborhood's coupons at the store. "She would give us stamps that we would use in exchange for some we did not use." Mrs. Lutz also used any extra sugar allotments to make grape jelly, a couple of jars of which she gave to every customer. Other grocers held back items, such as meat, sugar, coffee, and even cigarettes, for families that had been doing without. These grocery stores were important neighborhood institutions whose functions transcended the strictly commercial. One clerk taught honesty: "Mama would send me with the black billfold and ration books" along with the shopping list, recalled a homefront girl; the clerk then "took out the money and put the change back inside, tore out the right ration stamps, and warned me not to use any of it on the way home and not to stop and play with anyone."[61]

Many of the homefront children have fond memories of rationing. They remembered mixing the yellow pellet into the oleomargarine to give it color. They remembered receiving wonderful presents—bubblegum, for example, which children seemed to covet more than any other rationed item; or, as one girl recalled, her "best Christmas present ever," when she was age six, "my *very own* 1 pound of real butter!" And they have humorous memories. Because of shortages of elastic, children wore tie underpants. "We tied our pants with a bow on the left side," recalled a homefront girl. But the tie came loose easily, and several of the homefront girls remembered the mortification they felt when it finally gave way and their underpants fell down.[62]

The homefront children also learned valuable lessons from rationing. Whenever they saw people queuing up, they knew they were to join the line, whether or not they knew what was at the end of it. "Mother instructed my sister and I," recalled one girl, "to always get in a line if we saw one" because at the front of it might be butter, coffee, or sugar. The first thing some mothers did when they arrived at the supermarket was to put a child in each line. "I stood in line with my older sister many times to get what we needed," wrote a girl; sometimes the girls waited two or three hours, but "we had to be there when the trucks came in . . .

because the items would be gone immediately." Finally, a girl living in Manhattan went to daily Mass with her sister. On the way home they walked past the Grand Union supermarket. "If there was a line, one of us would get in it, and the other would go home to tell my mother."[63]

Moreover, the homefront children remembered the generosity of relatives and friends. One grandmother, who was a domestic in a wealthy San Francisco household, visited her grandchildren in the San Joaquin Valley, traveling by bus. She would deboard with her suitcase filled with treats, including chocolates and coffee. In a Milwaukee family, Aunt Rose was a hoarder, but she shared her larder with her family. In 1938, sensing that war was imminent, she "started laying in supplies": shoes, shortening, sugar. (In 1964, when Aunt Rose gave up housekeeping and came to live with her niece, "we found three cases full of Crisco still stockpiled in her vegetable cellar.")[64]

But as many positive recollections as there were of sharing and generosity, there were even more that were negative. Shortages caused violence, and the children witnessed behavior that upset them. A homefront girl in a small town in Missouri, for example, had gone to the store for bread, but when she came to the bakery section, she saw two women fighting over a loaf. In their struggle, they "tore it in half. Bread went everywhere." Another girl remembered that at Christmastime, her mother had taken her butter and sugar coupons, which she had been saving for holiday cookies, and gone shopping. But she left her bag of groceries unattended for a few moments, and someone stole it. More disillusioning and upsetting, however, was the personal dishonesty children witnessed, including questionable behavior by close relatives. Under the pressures and temptations of rationing, parents not only hoarded things, they shopped in the "black market," participating in the illicit trade in rationed goods that took place under the counter, out the back door, or from the trunk of an automobile.[65]

Obviously, not all parents were black marketeers or hoarders; many actually condemned these practices. "My parents were very offended by this," remembered a homefront girl, "and had many discussions with us kids about the moral choices involved." What is remarkable, nevertheless, is the number of homefront children who remember with chagrin the violations committed by relatives. Sometimes the offense, while minor, caused consternation to a child. In one household, a great aunt routinely gave her coupons for shoes to a grandniece. When the school-age girl then took one of these coupons, she was "always afraid that someone . . . would ask me my name," but her mother told her "to lie & give my aunt's name."[66]

Again, some illicit activities appeared justified, such as the father who shopped in the black market for liver, which the doctor had recommended for the man's anemic daughter, but which he was unable to obtain legally. Rationing was a touchy business. A daughter noticed that even though her father had already used up his allotted gasoline coupons, he continued to produce more coupons. "He became very angry when I asked him where they came from"; in fact, he had traded his liquor coupons for gas stamps. Another father outsmarted his family, boasting of the steaks and rib roasts which he had bought "on the Black Market." After the war, he confessed to his wife and children that it was actually horse meat—and legal—but he knew they would not have eaten it had they known.[67]

Less laudable, of course, were parents' unpatriotic and sometimes illegal activities. One father, a truck driver on the East Coast, returned from a trip with a 100-pound bag of sugar. He swore his wife and children to secrecy and hid the bag under a bed at the grandparents' farm, several miles away. In one family, it was the grandmother who, remembering the shortages during the First World War, hoarded sugar. She "bought it and bought it," but never used it; "it just piled up in an upstairs room. After the war we took hammers to it. . . ."[68]

What proved most upsetting to children was unpatriotic behavior by mothers. One mother, after illegally obtaining a 25-pound bag of sugar, made an otherwise innocent remark—to the wife of the chairman of the local rationing board—about her plan to make strawberry preserves. Realizing her mistake, "she was so petrified that she hid the sugar," first, in the baby grand piano, next in a closet, and, finally, fearing arrest and public humiliation, she flushed it all down the toilet. Another mother, who always seemed able to find cocoa to serve to family and guests, "credited her good luck . . . to the fact that she had arrived at the warehouse in advance of the hoarders." When suspicions rose that she was the hoarder, she reacted angrily, and "it very soon became clear that further discussion of the matter would be to our disadvantage. . . ." Obviously, the effects of rationing varied according to the idiosyncrasies of the adults in children's lives. Thus, although rationing touched the lives of all the homefront children, it was usually the parents or the grandparents who defined the experience by their example. But however they experienced the rationing system, the girls and boys remembered it well. Wrote a homefront girl: "I think rationing will remain with us for the rest of our lives."[69]

In terms of wartime political socialization, one figure towered over the children's homefront. That figure was Franklin Delano Roosevelt, who encouraged the children's patriotism and participation. Whether glorified or despised by their parents, the President's name was a household word. "Dominating this good old time," observed Willie Morris, the writer, "was the image of FDR—his voice on the radio, his face with the dark rings under his eyes on the newsreels." To Morris, a homefront boy in Yazoo City, Mississippi, "FDR was the war itself. . . ."[70]

Roosevelt's was the voice which the homefront children heard over the radio, though some children admitted that they confused his voice with that of the news commentator H. V. Kaltenborn. The children knew, though, that whether it was Roosevelt or Kaltenborn on the air, they had to "be quiet." The homefront children recalled that "Roosevelt's talks on the radio" were "totally reassuring" and made him seem "like a 'god' to us kids." Diana Bernal, a Mexican-American girl living in San Antonio, recalled one "fireside chat" in which Roosevelt's "resounding voice" called for national unity in fighting the enemy. But the eight-year-old girl had a special reason for remembering this particular radio talk: "I had just learned English fluently and was proud to have understood every word he said."[71]

If "you weren't for Roosevelt during the war," remembered James Covert, a homefront boy in Portland, Oregon, "you were not patriotic. He was the man who was going to take us through the war, and anybody who seemed to be against him was sort of un-American." So pro-Roosevelt were some of the school-age children that they defaced the yard signs of the 1944 Republican presidential candidate, Thomas E. Dewey. A boy who lived on Chicago's South Side wrote that a "Vote for Dewey" sign, with a picture of the candidate on it, adorned a yard half a block from his school. "Every kid that walked by on the way to school would spit on it."[72]

Of course, not all the children agreed. After all, Roosevelt was re-elected in 1944 with his smallest voter margin yet, drawing just under 55 percent of the popular vote. "My parents detested FDR," recalled Paul Alexander, the son of a Methodist minister. "They blamed him for all the problems." In other families, there were noisy and bitter disputes. One homefront girl described "the arguments my father had with my grandfather over how the war was progressing. My father thought Roosevelt walked on water and my grandfather hated him!" Eleanor Roosevelt also inspired wide-ranging reactions, from adulation to hatred.[73]

Whatever they thought of Roosevelt, most of these homefront children thought that he would, as one explained, "always be there, always be our President." Indeed, for the school-age children, he had always been the President. "He was president when I was born & I was almost 12 when he died," wrote a homefront boy. "I was shocked the day we learned that President Roosevelt had died," remembered a girl from Dover, Delaware. "I had assumed that he would be president forever. . . . I was really afraid. I thought nobody else would know how to be president."[74]

Like the distinct memories they had of Pearl Harbor earlier and of John F. Kennedy's death years later, for numerous homefront children time stopped on April 12, 1945, when they learned of Roosevelt's death. They remembered the tears and sobbing of their parents and grandparents and the crying of people in the streets. A girl in San Francisco was crying when "a woman walked over, crying as well, and hugged us." An eleven-year-old girl thought her father had gone "into shock—I never saw him with such an expression on his face. And then he started to cry," something she had never seen him do before. A homefront boy, Beirne Keefer, saw the tears in his parents' eyes as they told him; then "I went to my room," he remembered, "laid across the bed, and cried." For homefront children who had never experienced a death in the family, Roosevelt was, wrote one homefront girl, "the first person I mourned for in my life!" "He was the first person I knew who died," remembered Howard Kleinberg, who was a twelve-year-old in the Bronx, New York. "And it turned out that everyone else knew him. And that brought me into a direct relationship with the rest of the world." Few children suffered greater grief than Albert Sheppard, a boy in Warm Springs, Georgia, who used to stand near the railroad tracks and was thrilled, his mother recalled, when the train carrying the President passed by "and the President would wave at him and my young son would wave back." Albert was very sad and stayed home from school the day Roosevelt's body began the train trip north. That afternoon, Albert carefully wrapped "a favorite little black train engine that was made of cast iron . . . in a velvet cloth and buried it and mounded the dirt up on top of it and stuck a little American flag and put flowers on the top of the mound of dirt." Every day for weeks, he visited the grave, and despite his mother's entreaties to dig the engine up and begin playing with it again, "he refused and for all I know it is still buried [there]. . . ."[75]

The homefront girls and boys listened to the radio accounts of Roosevelt's funeral train as it proceeded to his home at Hyde Park, New York. Musical dirges accompanied the broadcasters' descriptions. Lining the tracks were grade-school students holding red roses and singing "God Bless America." Howard Kleinberg wrote that at five in the morning, he felt his father shaking him. "Get up," he said, "we're going to say goodbye to the president." They walked to a bridge spanning the Harlem River; under the Bronx side of the bridge ran the railroad tracks toward upstate New York. "There we stood, just the two of us, me and my dad. Through the gloomy dawn, a train light appeared. . . . Slowly it came to a point just under the bridge. . . . I could make out the last car, with its circular windows through which I saw a flag-covered casket. I looked up at my dad. He was saluting—and he was crying."[76]

Collecting scrap materials, buying War Bonds, observing the rationing require-ments, sharing the grief over their President's death, and, most of all, imbibing in the nation's intense pride and unity—thus did the homefront children participate in the victory. Such participation was especially important to the school-age chil-dren—those who, beginning at the age of six or seven, had entered a new stage of life. Jean Piaget called it concrete operations; Sigmund Freud called it latency. Guaranteeing the school-age children's political socialization was the great out-burst of patriotism that consumed Americans of all ages.

"We all became so patriotic," recalled a school-age girl. Aside from families of conscientious objectors, contrary opinions were very rare; in most places, dissent was unheard of and even unimaginable.[77] As a result, the school-age children's recollections echoed each other. "We were all so Proud then," wrote a homefront girl from Gary, Indiana; like other children, she knew without a doubt that "we would win." We were "so proud to be Americans," wrote another girl, and that pride "prompted us to do everything we could to hold onto our wonderful way of life." "Through it all," agreed still another girl, "there was this deep sense of pride and togetherness and incredible patriotism." "Everyone was patriotic to their very being," observed a homefront girl from Newark, Ohio. ". . . Everyone sacrificed and worked for one cause—America and to keep our freedom. . . . Everyone rallied around the flag."[78]

So intense was the wartime pride and unity that some of the homefront children have insisted that a "bonding" took place among the boys and girls who had lived "in an historic time within a certain age group" and had "shared similar problems in stressful times." The school-age children shared a sense not only of unity and invincibility but also of moral and military superiority. "I was an American," wrote a homefront girl from Springfield, Massachusetts, "and on the right side."[79]

CHAPTER 8

Children Play War Games

Whistle while you work,
Hitler is a jerk,
Mussolini is a weeny,
And Tojo is a jerk.[1]

Three blind rats, three blind rats,
Hitler, Benito and the Jap,
Started off with a yip and a yap,
And ended up with their tails in a trap.
Three blind rats.[2]

Slap
The Jap
With
Iron
Scrap
Burma-Shave[3]

WHETHER CHANTED while jumping rope or skipping on the way to school, or read from the ubiquitous Burma-Shave signs by the side of the road, for children rhymes became an important connection to the war. In their rhymes and jingles, children put words to age-old tunes as well as to those of more recent vintage, such as "Whistle While You Work" from Walt Disney's 1937 film *Snow White and the Seven Dwarfs*.

Related to rhyming and chanting was another of the children's homefront activities—telling war jokes. One of the homefront girls recalled that a popular joke among the boys was: What did the Germans say whenever a baby was born? "'Hotsy-totsy! Another little Nazi!'" The war inspired even little children's wit. A four-year-old boy was telling his little brother the Christmas story, but when he described Mary and Joseph's futile search for a place to stay, he changed the inn-keeper's words to: "No, there's no room here. Don't you know there's a war on!" Mocking Hitler was popular among boys, who would comb "their forelocks slantwise across their foreheads, put two fingers between nose and mouth to create the Hitlerian mustache, raise right arms in salute and goose step around the playground screaming, 'Heil, heil.' ...." Sometimes the jokes backfired. Jon Presnell, a boy in Enid, Oklahoma, and a pal were walking to the movie theater when they decided that they would pretend to be newsboys and began shouting, "EXTRA, EXTRA, WAR ENDS." Front doors started opening, and people began streaming out of their houses. With that, "and realizing we had pulled off a joke in very bad taste," the boys took off running and did not stop until they were inside the theater.[4]

Humor was developmental, like political socialization. "From an early age," the psychologist Martha Wolfenstein has written, "children avail themselves of joking to alleviate their difficulties. They transform the painful into the enjoyable, turn ... the envied bigness and power of adults into something ridiculous, expose adult pretensions, make light of failures, and parody their own frustrated feelings." The humor of little children was often nonsensical, such as endless rhyming for its own sake. But Wolfenstein found "a sharp change in the style of joking from the age of five to six, a shift from improvisation and joking fantasies to the learning and telling of ready-made jokes." Wolfenstein, whose orientation was psychoanalytic, related the change to the onset of latency; but cognitive psychologists, in the tradition of Jean Piaget, have agreed that there is a fundamental developmental shift at about this time, with the arrival of the stage of concrete operations. "For Piaget," psychologist Jerome L. Singer has written, "symbolic play goes on from about 18 months to the age of 7 when it gradually disappears ... as the child ... is transformed to the more overt play of games with rules."[5]

Like rhymes and jokes, war games and war play changed with the age of the child. "Whereas older children can talk about their fears with varying degrees of ease," wrote Dorothy W. Baruch, an education professor, in 1942, "younger children can many times play out their fears more readily than they can talk about them." Early in the war, Baruch observed young children at play. She saw two three-year-old boys clamber into upturned wooden boxes. "We're airplane men and the Japs are fighting us," one of them shouted. They're going to kill us." His eyes were wide and frightened. "They're going to shoot us," he continued. "They're going to shoot us. Hurt us and shoot us. Dead." The other boy picked up the chant; soon both were shrieking. Then the first boy picked up a wooden block and lifted it high in the air. "It's a gun," he shouted. "Hey, look. Two Jap bombers are going by." He aimed his gun: "Boom, boom! I shot them down." The other little boy joined in, and, Baruch wrote, there ensued "an orgy of shooting them down. Two small boys have conquered, not Jap bombers, but their own fear."[6]

Like the preschoolers, the school-age children also played war games that involved fantasies about killing, but theirs demanded a level of verisimilitude unknown to the younger children. For grade-school children, attention to detail was essential in everything from uniforms to weapons. Conrad Burton, a school-age boy whose family lived on the outskirts of Aurora, Indiana, recalled that the nearby woods "included all the necessities of a battle ground." He and his friends dug fox holes and built an air-raid shelter "by putting boards over a large hole leaving space for the entrance and covering the roof with dirt from the hole." The children shouldered wooden rifles, and Conrad was proud to carry "a wooden 50 Cal machine gun complete with tripod," which he had received as a Christmas present. He also wore a khaki Army belt with pockets for a first-aid kit, pretend bullets, and maps. A quarter-mile from his house were railroad tracks, and behind the tracks lay the Ohio River. "Many a day," Conrad mused, "we 'shot-up' the German train going by and fired at Japs across the river trying to invade our position."[7]

Like Conrad and his friends, school-age children across the country played war games with improvised weapons, for commercially manufactured toy guns were scarce. (Indeed, most toys made of metal, rubber, and other rationed materials were in short supply during the war, including electric trains, doll carriages, and rubber balls.) But adults made guns out of wood and cardboard, and the children fashioned their own weapons from tree branches, clothes pins, and rubber bands. War gadgets also came from cereal boxes, on the backs of which were patterns for cutting out walkie-talkies. Children craved guns of all kinds; in 1942, when New York City's postmaster allowed journalists to examine 1000 letters to Santa, one reporter commented that "the children seemed to consider their favorite patron saint as an important part of war production." Children asked for machine guns; one asked for a "tommy gun and a baby brother." Of the children's items in short supply because of the reallocation of supplies to war production, none seemed more precious than a bicycle. Used bikes were occasionally available, and in some cases parents would combine parts from several aged, non-functioning units to come up with one operational vehicle. Sadly, the shiny new bicycles of which the homefront children dreamed were casualties of the war effort.[8]

It was natural that children's play would turn from "cops and robbers" and "cowboys and Indians" to war games. "One side would be the Americans," wrote Larry Paul Bauer, who lived in Cleveland, "the other side the 'dirty Japs' and Nazis. In our scenario," as in those of all other homefront children, "the Americans always won in the end." Birthday parties featured war games too. A homefront girl who lived in Berkeley remembered parties where the children threw beanbags at a cardboard cutout of Hitler, which they then "stomped to bits." Moreover, children in all parts of the country were shouting "Bombs away" or "Bombs over Tokyo," as they pretended to be airplanes. "Arms outstretched and vocal cords straining, we raced around pretending we were bombers or pursuit planes. . . ."[9]

N. Scott Momaday, the Native-American writer, lived in Hobbs, New Mexico, as a homefront boy. War fantasies filled his head: ". . . I'm in a Bell P-39 okay no a Flying Tiger okay sons of the rising sun this is for my kid brother ha gotcha oh oh there's a Zero on my tail eeeeeeooooooooooow lost him in the clouds just dropped down and let him go over me and climbed up oh he can't believe it he's in my sights crosshairs there Tojo that's for the Sullivans well Chuck you can paint four more Zeros on old Sally here no I'm okay honorable colonel we must stop Momaday he comes from nowhere from the sun I tell you he's not human they say he's an Indian that he wears an eagle feather has the eyes the heart of an eagle he must be stopped...."[10]

"Shoot those Japs." "Bomb those Nazis." Parents disagreed over the value of their children's war play. So too did the child development experts. "Some are distressed by it and feel it should be discouraged," wrote a professor of education, but "others declare it a perfectly natural way for youngsters to release pent-up emotions in wartime." The debate over what to do "when play goes warlike" peppered the pages of parents' magazines as well as the conversation of concerned adults. Arthur L. Rautman, a clinical psychologist, warned that "habitual war play" was "a symptom of a basic neuroticism, and not ... desirable or whole-some...." More than that, it was a sign that "our care and guidance have failed; we may know that we have not provided him with the basic security he requires." The last thing in the world these children needed, Rautman wrote, was "encour-agement in their preoccupation; nor do they need more war games and realistic war-play objects to make their neurotic dramatizations more vivid." What they required, rather, was "protection against overstimulation." As alternatives to war games, Rautman urged parents to engage their children in War Bond and scrap drives, and instead of realistic toys, he recommended toys that stimulated the imagination, such as a stick, which could serve as a walking stick, an airplane, or "a gun to shoot buffalo." But "a fancy, painted wood-and-tin tommy gun" was "good for only one thing—as a tommy-gun to kill the Japs."[11]

Other experts expressed far less concern about the detrimental effects of the war games. "Don't be too shocked," wrote Ethel Gorham, the author of *So Your Husband's Gone to War!*, "if you find your children whooping about after Japs, killing countless Germans, [and] sinking submarines...." Child psychologists tended to agree that this "harmless outlet" was preferable to the children "dam-ming up." John J. P. Morgan, a psychology professor, concurred, adding that it was "only natural for children to build their play around military themes in war-time." But, Morgan cautioned, "the danger lies not so much in the games children play as [in] what they see and hear concerning the viciousness of our enemies." These stereotypes were worse than the war games in developing hatred; in fact, these were the same tactics "used by the Nazis to develop hatred among children for everything non-German."[12]

Among those recommending war games for children, none did so more enthu-siastically than experts in the field of recreation. One suggested that children "Cel-ebrate with a 'this-is-the-army' theme" party. The schedule of events began with flag raising and proceeded to recruiting and induction and featured basic training, which consisted of calisthenics, the obstacle course, and target practice with beanbags. Next was reconnaissance duty, which in this case meant searching "not for an enemy," but "for a treasure"—in this scenario, a watermelon hiding in the woods. Finally, there was mess followed by the lowering of the flag. "Tell your fun seekers," concluded one of the experts, "that they are in the army now and it's the G.I. way of doing things!"[13]

As for parents, few forbade their children from playing war games. Most seemed to agree that children's war games were educational and "a natural outlet for the emotions"; that they were "merely adaptations of old games"; and that they satisfied "the child's natural urge to imitate adults and to hero-worship older brothers off to war." With this in mind, a writer in *Recreation* magazine recommended the following activities: children could run an obstacle course "patterned after the Commando courses"; take a "half-hour trek" guided by a compass; build a snow fortress and wage snowball fights; construct a cardboard periscope and set up a lookout post; or play "The Prisoner and the Soldiers" modeled after "the old game of Hare and Hound." The children could also play "Blitzkrieg" in which "each player takes his turn swinging on a rope over a ground target and dropping beanbag bombs on the bull's-eye"; "I See a Spy" in which one of the children runs ahead of the others and "camouflages himself with whatever materials are available"; and "Miniature Tanks" in which the children convert cardboard boxes into "a pliable tread-like oval. . . . The drivers crawl into their box-tanks, the commander signals his comrades, and away they go on hands and knees."[14]

A key issue in children's war games was, Who would play the enemy? For one boy, the answer was simple: "The same wimps (of which I was one) who got maneuvered into being the Indians." A rarity was the child who wanted to be a Nazi; a boy later confessed that he did so because he enjoyed speaking in a German accent. Unheard of, however, was the child who wanted to play a Japanese soldier. Market research done in 1943 indicated that Japan was "a stronger emotional symbol of the enemy" than Germany, and many children, taking their initial cues from adults, viewed the Japanese as not only exceedingly cruel but also "ungodly, subhuman, beastly, sneaky, and treacherous." But there were ways to mobilize an enemy force; to fill these ranks, the older children would select the younger ones. It was smaller children too who were designated the prisoners of war.[15]

There was, however, a second major source of enemy soldiers and prisoners: girls. This was not surprising. Psychologist Rosalind Gould has noted that when "an all-boy group engaged in some killing, shooting fantasy" suddenly encounters a girl or a small group of girls, the results are characteristically the same. "At such times, the boys attempt to make contact with the girls only as their 'victim.'"[16]

Sometimes girls, either through their own perseverance or as result of having an older brother, did gain entrée into the all-boy fighting forces. Older brothers were helpful in other ways as well, teaching their sisters to play with toy soldiers as well as how to identify airplanes. It seemed that an only girl in a neighborhood stood a better chance of "playing war" with the boys; and in areas where there were few children, girls often got their chance. Still, while large numbers of girls clamored to play war games, many never got the opportunity to do so. To a certain extent, segregated war games were an extension of the gender segregation that already prevailed on school playgrounds. Moreover, those girls who did play war games with the boys seldom did so on equal terms. For one thing, explained a homefront girl, "Girls tended to be killed early on in the war." Another added that while she enjoyed being an officer, it did not do her much good because she "usually got to be the enemy officer who was captured."[17]

"Girls were nurses," remembered a homefront girl who lived in Long Beach, California, "or if pushy, maybe a spy, but not soldiers, certainly not officers." A homefront boy's recollection was the same. "My pal across the street had a younger sister . . . about 7 or 8," wrote Larry Paul Bauer. "She insisted on playing with us boys. Otherwise, she would throw a tantrum. So we made her the 'nurse,'" a profession which she pursued with determination, carrying a doctor's bag with a toy stethoscope along with pieces of candy, which she dispensed to the "soldiers" as pills.[18]

Why was it that—in vacant lots and fields across the country—the boys fought the war as soldiers, risking imaginary wounds, while it was generally the girls who were the nurses treating them? The answer begins in infancy; by the time children begin social play at two or two and a half years, it is the boys who engage in aggressive behavior that is sometimes intended to do physical harm. Not only do boys play differently at an early age, but also, as psychologists have observed, they prefer playing with other boys—while girls prefer playing with other girls. The developmental psychologist Eleanor E. Maccoby has reported that once these preferences become evident, they "increase in strength between preschool and school and are maintained at a high level between the ages of 6 and at least 11." Maccoby observed that girls are averse to the boys' insistence on competition and dominance, and that, finding it frustratingly difficult "to influence boys" to change their behavior, they become wary of social play involving boys.[19]

Although Maccoby made her observations in 1990, psychologists during the Second World War also wrote about gender differentiation in play. In 1943 Lois Barclay Murphy, a psychologist who studied preschool children in nursery schools, reported that children's play began to diverge at about age four; "girls play more domestic games and boys give up playing family in favor of fireman or engineer." For boys, "a strong identification with father is usually part of this transition, accompanied by exaggerated experimenting with aggressive expressions of their new masculine role."[20]

Social scientists have diverged in their explanations of gender differences in aggressive behavior. "The greater aggressiveness of the male," Maccoby and her colleague Carol Nagy Jacklin have written, "is one of the best established, and most pervasive, of all psychological sex differences." A 1986 study reported that in no behavioral category were women more aggressive than men.[21] While psychological researchers explain that these differences are "learned as aspects of gender roles and other social roles," other social scientists vehemently disagree. "That belief," the anthropologist and physician Melvin Konner has written, is "a tenacious modern myth," which "becomes less justified with every passing year. . . ." Gender differences in aggressive behavior, Konner explained, resulted not from social learning, but were "intrinsic, fundamental, natural—in a word, biological." Konner did not claim to have unlocked the entire explanation of male aggression, "but it seems increasingly clear," he noted, that testosterone, the male sex hormone, "is a key."[22]

In addition to this variant of the nature-nurture debate, there is the related gender issue of power. Clearly, in the children's wartime games, it was the boys who not only initiated the warfare but also selected the opposing sides.[23] Although one cannot entirely reconstruct the children's homefront battlefields during the Second World War, the historian can benefit from the field of psychology. Possibly the girls' play as nurses ministering to the wounded derived from what Carol Gilligan has called "the ethic of care." Gilligan has written that boys and girls see moral dilemmas differently and construct different kinds of solutions. For example, while males might see clear-cut answers to all problems, with solutions producing winners and losers (as on the battlefield), females might respond in more ambivalent ways in an effort to see that all are cared for and that none are left out. Women's "moral imperative," Gilligan has written, "is an injunction to care, a responsibility to discern and alleviate the 'real and recognizable trouble' of this world." The homefront girls' letters repeatedly give examples of the ethic of care. A homefront girl wrote, for example, that she and her friends "played with our dolls as if we were in a war zone"; but their dolls were refugees, not soldiers. "We were constantly moving our 'families' to safe areas," she explained.[24]

From the homefront girls' memories, it is clear that they acted less in accordance with the ethic of care than in conformity with what was culturally acceptable in the eyes of their parents, teachers, and peers. The "feminine role," the sociologist Mirra Komarovsky has written, is something which girls learn. Komarovsky related a postwar study of seventy-three homefront girls in which thirty recalled experiences when "their dispositions ran counter to the stereotype of femininity." They "disliked dolls sufficiently" to have had conflicts with their parents and grandparents, who made them feel "queer" or like "little 'freaks'" because they preferred climbing trees or playing with toy soldiers, marbles, or electric trains. "Pressures were exerted upon the girls to select girls' toys and to be more restrained, sedentary, quiet, and neat in their play than their brothers or boys in the neighborhood," Komarovsky wrote. And "woe to the boy who likes dolls!" One girl recalled playing dolls with a little boy, but "one day his mother walked in and saw the boy cradling a doll. She laughed and called him a 'sissy,'" and his father was "quite annoyed" at the boy.[25]

In the end, then, girls could consistently play warriors only if they played with other girls or played alone, usually beyond the reproachful eye of parents. One girl remembered that she and "a couple of girl friends . . . used to crawl on our bellies, playing commando," and another that she and a girl friend used their walkie-talkies in playing war. Her friend's "large backyard became the South Pacific, or Germany. . . . Sometimes, we were wounded and turned into nurses, sometimes we died and the other turned into the captain performing a full burial at sea." Nancy Morris, a girl in Amelia, Virginia, recalled that playing alone was the surest way to attain command. An avid movie-goer, she "transferred the war action" from the screen to her backyard where she flew missions over Japan in her airplane. Using two sawhorses "with boards across and a bombay in the middle," she "could pretend to be the captain. Rank was very important," Nancy explained, "—I was never just a private."[26]

Girls played war games with their dolls, some of which were in uniform and wearing the patches of the various service branches and commands. Valerie Drachman, in Chicago, named her stuffed bunny "Private Philip" and dressed him in khaki and attached the Army Air Corps insignia. Other girls had paper doll Wacs and Waves; one remembered playing "hours on end pretending I was either in the Waves or Wacs," and another wrote that her "cut-outs changed from Shirley Temple to men & women in uniform." There were paper doll "handsome military men" as well as "pretty Wacs and Waves." The cutout paper uniforms were sex-specific, with the male dolls dressed "in proper attire for combat," while the women were nurses "in medical uniforms to heal the wounded."[27]

In their everyday activities, however, the homefront girls shared many of the boys' enthusiasms. For example, both showed a fascination with military aircraft, and girls also collected and traded the airplane cards from packages of Wings cigarettes. Both also made balsa-wood models of the airplanes, which had rubber-band engines to propel them, and they cut out and assembled the model planes that came on the backs of Pep cereal boxes. Moreover, girls and boys alike pieced together jigsaw puzzles that pictured airplanes; there were Japanese Zeros and German Messerschmitt 109 fighter planes and Stuka dive-bombers as well as American and British aircraft. One of the children's most popular was the American P-38 Lightning, which had a high rate of climb and a top speed of over 400 miles an hour.[28]

Another wartime passion shared by homefront girls and boys was drawing pictures of airplanes, ships, tanks, and battle scenes. Jan Heinbuch, a girl who lived in Arlington, Virginia, "spent many hours drawing American, German, and Japanese aircraft dropping bombs. . . . I still have a few of these drawings," she wrote, "and am amazed that they were drawn by a little girl." In their obsession with airplanes, the children spent hours poring over pictures of them. "I got all the airplane magazines & books I could afford," a homefront boy in Chicago remembered, "& could identify anything that flew." Numerous homefront boys and girls remembered drawing airplanes. Like the boys, the homefront girls also made model airplanes. Emily Cropper, who lived in Coronado, California, reported that her favorite model plane was "the super fast Lockheed P-38 Lightnings" not only because she thought they were "exciting and beautiful," but also because they were "probably the fastest propeller planes during the war. . . ."[29]

Probably the major gender difference in drawing airplanes and war scenes was that the boys were far noisier about it. At her grade school in Chicago, recalled a homefront girl, the boys "made noises to depict what was happening," using "a lot of gutturals and sibilants" as they drew pictures of ships with their big guns firing, of "planes with bombs and smoke coming out . . . as they dove into the sea," and of torpedoes shooting from the tubes of submarines and blowing up enemy warships."[30]

During the war, children even dressed for their part, wearing child-size military and naval uniforms. Boys wore cast-off uniforms of all kinds; "khaki became the normal garb for young kids my age," recalled one boy. Their favorite item of headwear was an aviator's cap. "I dressed as much like a soldier as I could," wrote one of the boys, combining uniforms given him by family friends in the Army Air Corps with relics from the First World War, such as the gas mask that hung from his shoulder. Army outfits seemed to be the most popular, but boys also wore naval uniforms. George Curtis, who lived near Portland, Oregon, used to wear an Air Corps uniform to school. During recess the boys played war games, and he remembered that one day all the boys agreed to wear their uniforms so that "we could play war in grand style."[31]

But what about the homefront girls? Many also wanted uniforms, including combat gear, but relatively few ever got them. Girls used their imaginations and improvised. Some wore pea jackets, like the sailors. A second-grade girl in Lyons, Kansas, wore a brown skirt and blouse, her Wac uniform, "and I marched around proudly in it. I probably didn't look very military—I was so little the skirt kept sliding down—but I felt truly patriotic." Some girls were resentful because it seemed that only boys received uniforms. One described her little brother's blue worsted naval uniform, with its gold braid and brass buttons. He was "*always* dressed in nautical style," she noted, but "I was *never* dressed in anything military." But sometimes, when there were siblings, both the son and the daughter received uniforms. A four-year-old girl found a Wac uniform under her Christmas tree while her brother got a soldier's outfit; and another brother and sister were gleeful one wartime Christmas when they both opened boxes containing Army officer uniforms. In the eyes of most girls, however, a Wac uniform was somewhat second-rate; their strong preference was for a Wave outfit. "I wanted to be a Wave when I grew up," recalled a homefront girl, "because their uniforms, especially the caps, were *much* prettier than the Wacs." Susan Anzovino, who lived in Cleveland, had a sailor dress, and when her mother made her "a navy cape with a gold satin lining and a gold star on each side of the stand-up collar," she was immensely proud to be taken for a Navy nurse.[32]

It was not just the uniforms, however, that inspired the homefront girls; many also idolized the women who had volunteered to serve their country. "We were fascinated by the Wacs and Waves," stated one of the girls, "for they were obviously making huge sacrifices not staying home to be Mothers and housewives. . . ." For their courage and adventurousness, the women in uniform served as role models—indeed, as inspirations—for the girls. During the war there were organizations of Junior Wacs and Waves. Girls also attended "Junior Wac Camps" where they were given military ranks: "The youngest girls and the first to come to camp were called buck privates, but as they proved their helpfulness, they were moved up the scale to private first class, corporal, and sergeant."[33]

On the other hand, boys derided the Wacs and Waves. Peter Filene, the historian, has observed that having women in the armed forces was "a startling phenomenon for Americans, evoking ridicule or suspicion" of the women themselves. Thus, while the homefront girls generally greatly admired the Wacs and Waves, the boys had a perverted view. Boys on the verge of puberty, who were developing a pervasive curiosity about sexuality, thought they were immoral. One boy, who was twelve when the war ended, reported that he "heard all kinds of stories" about how the Wacs and Waves "fucked their way through the war." One Friday evening, he saw a woman Marine at Jewish services and wondered whether she "got her share." Similarly, at Mass one Sunday in Duluth, Minnesota, a priest railed against the lack of morality of women in the Army and Navy, and, recalled a homefront girl, he was "really down on girls having anything to do with the military. As he finished, the mother of . . . a Wac got up and walked out." Her defiance of the priest "really shook many people up."[34]

Generally, the homefront girls and boys adopted the sex-typed behavior which parents and teachers, not to mention peers, believed appropriate to their gender and age. Some homefront girls complained about their exclusion from the battlefield, but sex roles were a fact of life, and arguing against them was usually unproductive. "It was naturally assumed," wrote a homefront girl, that "the boys would serve as soldiers." As for the girls, wrote another, "we were the support group. We were the nurses, the nurturers, the reasons the boys wanted to come home." The "boys were always the soldiers, sailors, and marines, and the girls were the nurses," recalled a homefront girl. But the important thing, she wrote, was that all the children were patriotic.[35]

The girls knew they ran a risk in playing war games since, as one recalled, "the neighbors frowned on a girl doing so." Numerous homefront girls did so anyway. As for the boys, some reported that during wartime they not only learned to cook and clean house, but also to knit squares for afghans to be sent to the soldiers. In Cleveland, the contributions of Larry Paul Bauer, a grade-school pupil, were in the domestic sphere; he ran the household in the absence of his parents, both of whom did defense work. His father was on the shift from 7 a.m. to 5:30 p.m., his mother from 3 p.m. to 11:30 p.m. Larry Paul learned to cook, and he made dinner for his father and himself; he also cleaned the house and did the ironing.[36]

Over time, what the homefront boys and girls learned was which dreams and expectations they could afford—and not afford—to have. And while there were exceptions, it was not androgyny, but rather sex-typed behavior that has had the greater historical significance in the lives of the homefront children. Earlier, there were discussions of two important ingredients in gender roles: first, biological and psychological differences in aggressive behavior; second, male power in determining boy-girl relationships. In the view of historian Carroll Smith-Rosenberg, gender is the "most fundamental ordering of . . . power arrangements within the society." And in children's play, these arrangements were apparent at age three. Yet there is a third ingredient: identification, or the means by which children internalize values and, in the process, develop an identity.[37]

To the question: "How does one develop an identity?" there is no simple answer. For one thing, the term identification is Freudian in origin, and few developmental psychologists have given it serious study. Moreover, the mechanisms of identification are complex, involving power, status envy, and fear of both punishment and withheld love. Identification, the historian Cushing Strout has observed, is "a vexed concept in the literature of psychological theory, particularly with respect to the process, or mechanism of psychological forces that generate emulative tendencies in any individual." But the historian need not unravel these complexities; for the historian, Strout contends, it "is sufficient . . . to observe specific emulating or internalizing of a given model's actions, standards, or expectations and to see if it is appropriate . . . at a given stage in the life-cycle."[38]

Identification is a powerful element in children's play. In playing house, school, or war, children perform the roles they believe they will fulfill in adult life; these roles generally are gender-specific. But in child's play, a major difference is that girls who assume the roles of mother, teacher, or nurse have a far better understanding of what these jobs entail than do the boys who pretend they are warriors. Thus, as sociologist Gregory P. Stone has noted, "the dramatic play of children in our society may function more to prepare little girls for adulthood than little boys."[39]

America's homefront boys dreamed that one day their heroics on the battlefield or in the air would epitomize their manhood and bestow the crowning achievements of their lives. Interestingly, the homefront girls agree that it was the boys who—as they contemplated future battlefield challenges—bore the heavier emotional burdens during the war. "It seemed to me even then," recalled a homefront girl, "[that] their attitude was 'more serious' about the war." Other homefront girls concurred. "I felt the war was more a boy-thing," wrote another. "Boys were taught to be brave & called cry babies if they dared cry from fear." Gayle Kramer, a girl whose father was serving in the Pacific, remembered feeling that adults placed "an awful lot of pressure" on the boys whose fathers were absent. "'Well, John,'" she overheard a grown-up telling a boy in her neighborhood, "'now that your father's gone, you're the man of the family.'"[40]

All this added up to a heavy burden for many homefront boys. They worried: Would they succeed at heroism, or would they bolt under fire? Could they withstand torture? Would they betray their country if their Japanese captors jammed bamboo slivers under their fingernails, or if their German torturers broke their bones? "I used to think I would have to be a soldier when I grew up," observed Fred Humphrey, a boy born in 1939 whose father was overseas in the Army, and "I worried whether I would be 'up to it.'"[41]

While some boys were dubious about how courageous they might be, others expressed bravado. "All the boys devoutly wished we were old enough to enlist . . . ," explained one. "I was afraid the war would be over with before I was of age," recalled another. In fact, some of the boys could not wait. In 1941 Allen Van Bergen, a four-year-old in Burlingame, California, entered a barbershop and got a military crew haircut on credit. He then went next door to a confectionery where he charged a dish of ice cream. After walking twelve blocks to the railroad station, he boarded the Del Monte Limited. He never made it, though. Allen had confided to the conductor that he was traveling to New York to join the Navy, and the conductor had contacted the boy's mother. But for a four-year-old, Allen had had quite an adventure.[42]

In Arizona a twelve-year-old boy strode into a recruiting office and announced, "I want to join the Army and shoot some Japs. Sure, I'm 17 years old. You enlist men 17 years old, don't you? I don't need my mother's consent. . . . I'm a midget." Some of the boys did succeed in enlisting. In 1942 the Marines issued an honorable discharge to William Holle, of Eau Claire, Wisconsin, who had enlisted the year before at the age of twelve. Another boy, Jackie MacInnes, of Medford, Massachusetts, said that he "liked the Navy fine." A thirteen-year-old, Jackie had taken his older brother's birth certificate to the enlistment office in Boston. When the consent papers arrived for his parents to sign, he forged their signatures and reported for duty at Newport, Rhode Island. "Everything was going fine," stated a newspaper report, "until he wrote a letter home," prompting his parents to come pick him up. But at least one young enlistee escaped detection during the war; and more than four decades later, in 1988, President Ronald Reagan signed a special bill granting Calvin L. Graham, a Marine veteran injured at Guadalcanal, the disability benefits to which he was entitled, but which the Navy had denied him because he had lied upon enlisting at the age of twelve.[43]

In reflecting upon boys' and girls' wartime experiences, a homefront girl observed that "the obvious difference was the expectation that boys would go to war as soon as possible" while girls would stay behind to tend the victory gardens. The girls "naturally assumed," explained another, that if the war lasted long enough, "the boys would serve as soldiers." But a third girl wrote that the major difference would not become apparent for twenty years, that is, in the mid-1960s when a "boy would . . . be reminded, 'Your dad served honorably in WW II—shouldn't you go uncomplainingly to Vietnam?'"[44]

While adults frowned on girls playing war games with the boys, there was nothing but approbation for girls and boys working together in scrap collection drives and savings bond campaigns. Indeed, just as the war widened the gender gap, so on occasion it also narrowed it. "We both pulled the wagon to collect scrap metal and papers, we both tended victory gardens and collected pop bottles to get money for candy and Liberty Stamps," recalled Marlene Larson, who lived in Whittier, California, and whose best friend was a boy. Significantly, in many communities it was girls who captained the paper and scrap collection drives, and who, recalled one homefront boy, were "more active than the boys in war stamp and bond drives and in the Red Cross program."[45]

Because of their shared goals, some of the activities of the Girl Scouts and the Boys Scouts converged. Historian Mary Logan Rothschild has observed that from its beginning, in 1912, the Girl Scouts had pursued "two main programmatic themes." One was domesticity, including "the teaching of traditional domestic tasks for women"; the other was "a kind of practical feminism which embodies physical fitness, survival skills, camping, citizenship training, and career preparation." To these themes, the war added a third: participation. Both Boy Scouts and Girl Scouts took classes in first aid, and together they collected scrap materials, worked as Civil Defense spotters, raised victory gardens, and placed flags on the graves of departed service people.[46]

In its pages during the war, *Boys' Life*, the Boy Scout journal, exalted patriotism and glorified the warriors—"Real Supermen," it called them—but it did not praise war. Although the Boy Scouts boasted that many war heroes had been Scouts, most notably Captain Colin P. Kelly, Jr., the group itself proclaimed that it was "not a military organization. The ideals for which it stands are the antithesis of militarism. But Scout training produces men—men of character and decency, men who cooperate for the common good, men who have known Freedom. . . ."[47]

While the Boy and Girl scouts worked jointly toward a variety of goals, the major difference was that the Girl Scouts sometimes engaged in domestic tasks disdained by the Boy Scouts, such as rolling bandages and making hospital beds for wounded veterans. In the homefront Girl Scouts, according to Rothschild, "traditional domestic tasks . . . became cast as war work," so that canning, child care, and household cleaning to reduce the pressures on working women became girls' contributions to the war effort. In this way, the girls became "Deputy Women."[48]

It was not the Girl Scouts, but another organization, the Camp Fire Girls, that promulgated a policy unequivocally based on gender roles. Indeed, no other program so illumined the sex-typing imposed on the homefront children. In 1943 the Camp Fire Girls established a two-year training course to prepare its members for life in postwar America. Its basic premise was that these girls would have to help shoulder the enormous responsibility for nursing the wounded war veterans back to health, including even financially supporting these men during their recovery for as long as it might take. The Camp Fire Girls' national council reasoned that since the girls would have to care for the veterans, they "must be prepared for the fact that many men will return physically handicapped. . . . The girls to whom they return must be prepared to help them psychologically and financially. . . ." For the benefit of these girls as well as of the "girls who may not marry if the casualty list grows," the Camp Fire Girls urged "a greater emphasis on preparation for jobs in which the girls may become self-supporting." While good advice generally, the organization's reasons for articulating this program must have aggravated the homefront girls' uncertainty about their future.[49]

Whether resulting from socialization, testosterone, or power, gender differences in the homefront children's activities were striking. And because so few adults questioned the validity of sex-typing, they did not hear the girls' wartime lament that their gender had cheated them. "Oh, how much I wished I'd been born a boy!" exclaimed one girl, who "was convinced that just being a male made one brave and heroic." Another girl lay awake at night thinking about what it would be like to sit in a foxhole, eating out of a mess kit, drinking from an Army thermos, and smoking a cigarette as she watched the bombs explode: "How I wished I was a boy, so I could grow up and do all those things." Girls also felt short-changed by government rationing, which permitted children with paper routes to purchase bicycles, without recognizing that mostly these were boys. "During the war," wrote one girl, "I hated being a girl. The younger boys could have bikes; the older boys got to join the service. . . ." The girls could do neither.[50]

One undaunted homefront girl was Patsy Fisher, a twelve-year-old who lived in Racine, Wisconsin. Patsy decided that she needed to write the President. "Dear Mr. Roosevelt," she began. "I am just a common ordinary school girl and so are my friends. But we feel that all the women in the woman[']s army need assistants to help them." The girl assistants would wear uniforms and, after receiving the same training as the Wacs, would be shipped to battle stations overseas. "If you say O.K.," Patsy concluded, "and we get shipped across we may be in the same predicament as some of the men but we don't care. . . . We looked at all the angles. We want to give up everything we have for our country[,] even our lives if we have to."[51]

CHAPTER 9

Children's Entertainment:
Radio, Movies, Comics

"**M**OM, WHAT WAS on the radio before the war started?" one of the homefront girls asked—and hers was not an isolated question. Children's wartime radio adventure programs were integral to their homefront experiences. So too were their comic books and music, as well as movies, cartoons, and serials, all of which had a memorable impact on these boys and girls. Radio was an important source of their war news, and they came face-to-face with war's horrors in the newsreels and on the pages of *Life* magazine. The homefront children have reported that popular culture amplified the impact of the war, causing them to be both thrilled and terrified during the years from 1941 to 1945.[1]

"I tried to never miss 'Captain Midnight,'" wrote a homefront boy. "He was constantly chasing down Nazi spies." In reciting the Captain Midnight oath, the children pledged "to save my country from the dire peril it faces or perish in the attempt." Other radio adventure shows were *The Shadow, The Green Hornet, Gangbusters, Jack Armstrong—The All American Boy, Hop Harrigan, Dick Tracy, Don Winslow of the Navy, The Lone Ranger, Tom Mix, Sky King, Terry and the Pirates,* and *Superman.* Many radio heroes were not much older than their audience—heroes such as Jack Armstrong, a high school student, and Hop Harrigan, "America's ace of the airways," an eighteen-year-old aviator who fought the enemy and constantly escaped capture while dodging bullets. For their part, the radio villains were truly despicable. Jack Armstrong battled the Vulture and the Silencer, while Captain Midnight's arch rival was Ivan Shark, who was first "a Russian mastermind" in 1940, but during the war became part of "the oriental peril." While some heroes operated abroad, others such as Superman and the Green Hornet tracked down spies and saboteurs at home.[2]

America's radio adventure shows had a moral tone that was similar in its righteousness to the homefront children's war games. Good confronted evil and justice prevailed. Focusing on the need to defeat the heinous Germans and Japanese, the radio shows exhorted children to collect scrap materials, buy War Bonds, and plant victory gardens. Listeners to *Dick Tracy*, for example, took the five-point pledge to combat waste, vowing "to save water, gas and electricity, to save fuel oil and coal, to save my clothes, to save Mom's furniture, to save my playthings." These girls and boys had the satisfaction of having their names placed on the Victory Honor Roll, which the show's announcer guaranteed would be read by General Dwight D. Eisenhower at his headquarters. In 1943 more than a million children joined the Jack Armstrong Write-a-Fighter Corps, pledging to write once a month to a service person as well as to collect scrap and tend their victory gardens. Most of all, the show's messages reaffirmed the nation's patriotism. Every week during the war, western hero Tom Mix fought spies and saboteurs on his show, and on V-E Day, May 8, 1945, he told his listeners: "We've shown Hitler and his gang that we know how to lick bullies and racketeers, but we've still got a big job to do . . . fighting the Japs."[3]

The impact of radio varied depending upon the child's age. According to a study done in the 1940s, it was between the ages of four and seven that a child began "to take an active, continuous interest in radio entertainment," including requesting particular programs and listening intently. By age six the child had become "an habitual listener," who could not only identify the characters but imitate them, and who "could enter freely into fantasy and whimsy" and use radio as "an exciting realm for imaginative wanderings." At age nine or ten, however, interest in fairy tales flagged before the escapades of radio's adventure heroes; and as children's desire for "more realism or pseudorealism took precedence," they demanded "greater verisimilitude in situation and character."[4]

Homefront children, who listened to radio an average of fourteen hours a week, had their special times to tune in. On weekday afternoons between about 4:30 and 6:00, they listened to *Jack Armstrong, Captain Midnight,* and other shows. And they expressed their gratitude by eating "Wheaties, the Breakfast of Champions," or drinking "Ovaltine," the drink for "young, red-blooded Americans." They also bought the premiums offered by the shows, such as Captain Midnight's "Code-o-graph" for decoding messages from the Secret Squadron. "You are the keen-eyed fliers of tomorrow," stated one message, "[you are] the skippers of . . . atomic powered ships that will girdle the world. . . . America needs you—healthy, alert, and well trained to guard her future. . . ." Of all the radio heroes, Superman was not only the most popular, but probably the most imitated by children, because the show had "considerable physical action and fighting in exciting, fast-moving episodes." An overnight sensation when introduced in 1938, Superman vigorously fought the enemy, whether mobsters or spies. He also battled social problems, such as racial intolerance and juvenile delinquency. Superman could fly at supersonic speed, and his other extraordinary gifts included long-distance hearing and X-ray vision. Larger than life, he provided assurance of righteousness. Parents, however, were not as taken with *Superman*. They "find it noisy," an article in *Child Study* reported, but this is a "factor which children seem to enjoy."[5]

Children had other favorite listening times as well. A popular time for boys was Friday evening, with the boxing matches followed by *The Bill Stern Sports Show*. Both genders enjoyed the Saturday morning shows *Grand Central Station* and *Let's Pretend*, which told fairy tales about princes rescuing "beauteous maidens" from "witches, dragons, gnomes, dwarfs, and other mythical fauna." There were plenty of popular family shows too, and everyone gathered around to listen to singers such as Bing Crosby and Kate Smith and comedians such as Fred Allen, Jack Benny, Bob Hope, and Edgar Bergen. Also popular with families were musical programs and mystery shows as well as a new genre, exemplified by *The Aldrich Family* and *Meet Corliss Archer*, that probed teenagers' problems with love, school, and their parents.[6]

During the war, radio entered the children's lives in another way: it conveyed the war news. So central was the radio news that one homefront girl asked her mother: "What did the news have to talk about when a war was not going on?" Or, queried another girl: "Has there always been war? Has there ever been other news . . . ?" A survey conducted in November 1942 estimated that three-fourths of America used the radio as its major source of information about hostilities. Rural and urban children alike remember listening to the war news; and everywhere the decorum expected of them was the same: Be quiet. "When the news was on," one girl remembered, "we did not make a sound." And they recalled the seriousness of their parents as they bent forward to listen; sometimes the room was dark except for the glow of the radio. Among the popular newscasters were Gabriel Heater, H. V. Kaltenborn, and Walter Winchell. Some of the broadcasts, such as those by Edward R. Murrow from London, were live transmissions, and after listening to the news, families would update their war maps by moving pins to indicate the locations of battles. Children in Cincinnati had their own news program, with the station supplying world maps and tiny flags on pins; each afternoon, the "announcer would read the war news and tell us where to insert the proper flag pin."[7]

Wartime radio clearly stimulated the children's imaginations, so much so that parents, teachers, police officials, and child-guidance experts expressed concern that radio overstimulated children's fantasies and even encouraged them to commit antisocial acts. This was not a new fear; apprehension about radio violence had risen in the 1930s. And during the war, radio critics pointed out that the crime shows featured murder, sabotage, robbery, arson, assault, and drug peddling. Sound effects amplified the commission of these crimes, "with nothing left to the imagination," complained one writer. "One hears the thud of the lead pipe against the head of the victim as well as the resulting crashing of the skull and his blood-curdling shrieks for help and mercy." Another critic argued that some of the homefront children suffered from an "addiction" so acute that at night "they dream about killings and have fears of kidnappings." But there was a gender gap. The chief concern of radio critics was the medium's impact on boys; the debate only marginally addressed girls' fantasies and the possibility of radio's negative impact on their behavior. An article entitled "His Ear to the Radio" voiced fore-boding about the future of the boy who, after listening to his radio shows, "may sit brooding over the terrors, refuse to eat his meal, fight against going to bed; and when he does get to sleep, toss restlessly until morning."[8]

Still, the girls as well as the boys listened to the radio adventure and crime shows during the war; in fact, gender seemed to have little impact on children's programming preferences. While investigators paid little attention to gender dif-ferences in children's wartime listening habits, one study—of children in the fifth through seventh grades in New Rochelle, New York—did conclude "that, on the whole, girls were almost as interested in crime and mystery shows as boys." But this study also observed that while boys and girls both "liked stories of the lives of successful men and women," it was the boys who "had a tendency to look upon the detective or the 'superman' as a hero while the girls preferred stories presenting the 'good mother' or the 'nurse' or the 'faithful wife' as a heroine." Putting these conclusions side by side, it seems likely that the girls and boys lis-tened to the same programs, but heard different things.[9]

Whatever their mission, most of the radio "superheroes and supersleuths" shared one characteristic: they were male. There were few women adventure heroes; women starred not in these shows, but on the daytime serials such as *Stella Dallas, Portia Faces Life, Our Gal Sunday,* and *Ma Perkins.* One of the few crime drama heroines was Little Orphan Annie, and while some male heroes, such as Jack Armstrong and Captain Midnight, had girls for sidekicks, they did in the same way that the Lone Ranger had a faithful Native American by his side and the Green Hornet had a loyal Filipino valet.[10]

Gender differences were evident in all aspects of American popular culture. During the war, for example, advertising and popular music were two of the loud-est boosters of American patriotism, not to mention two of the fundamental shapers of sex-role images. Advertising both reinforced patriotism and exploited it. Ads urged Americans to do their part, to sacrifice for the noble cause. Symbols played a vital role in mobilizing support for the war, and in the hands of the adver-tising industry, no symbols were as powerful as those evoking sacrifice. "'I Died Today,'" stated an ad for the 5th War Loan Drive in 1944. Depicted is the face of a dead American soldier, who asks: "'What Did *You* Do?'" "Who gets hurt if you buy gas in the black market?" asked another ad. "Will you give a pint of life insurance?" was an appeal for blood donors. But the primary goal of the wartime ads was to sell the products—everything from breakfast cereals and liquor to cig-arettes and vitamin pills; so the advertising industry manipulated the symbols of patriotism and sacrifice to fuel the consumer culture. "What was offensive in peace time," wrote Raymond Rubicam, a leading advertising executive, "was ten times more so in war."[11]

Regardless of product, the drawings and photos in the wartime ads appealed to people's patriotism. Ads were filled with pictures of tanks, airplanes, and other war equipment, as well as with scenes of air, land, and sea battles. There were also scenes of the homefront—pictures of busy factories and farms and of families praying together for the safe return of loved ones. Likewise, patriotic symbols adorned the ads, such as Uncle Sam or his hat, the eagle, a minuteman, the "V" sign, the colors red, white, and blue, and, of course, the flag. Numerous patriotic jingles and slogans were targeted at women. "In times like these serve hearty breakfasts," stated an ad for a pancake mix. "Uncle Sam's workers must be well fed. Do your part for national defense right in your own kitchen." An ad for hats urged women to wear "Victory colors—let your hat be brave, bright, a banner of your courage." A store advertising its "lower prices" announced: "Thrift is no longer a private virtue. It is a patriotic duty." And should they forget, a cosmetics ad reminded America's women: "Your first duty is your beauty—America's inspiration—morale on the home front is the woman's job."[12]

As for the music of wartime, it too was intensely patriotic. The homefront children sang all the service songs, and they thrilled to hear Kate Smith sing "God Bless America." They either listened to the radio music or played 78 rpm records. A boy in Brooklyn delighted in cranking up his grandmother's RCA Victrola, placing the needle on the record to play "Remember Pearl Harbor" or "We're Gonna Kill the Dirty Little Jap," and then marching around the Victrola. If not rousingly patriotic or jingoistic, the songs were sentimental ballads reflecting the sadness and loneliness of couples separated by the war; most focused on the women who waited while their men fought abroad. "Don't sit under the apple tree with anyone else but me," stated an exceedingly popular number, "'til I come marching home." Many children knew the words to all the songs: "I'll Never Smile Again," "I'll Walk Alone," "Sentimental Journey," "When Johnny Comes Marching Home," "The White Cliffs of Dover." After patriotic songs and sentimental ballads, the third type of music listened to by the homefront children were novelty songs, most of them silly, with lyrics like "Send up in smokio the city of Tokyo, Show the Nipponese that Uncle Sam does not jokio!"[13]

Whether broadcasting music or adventure shows, the radio was a link between the children and the homefront. And because numerous homefront girls and boys have expressed pride in being the last pre-television generation, it is difficult not to speculate about the childhood effects of the homefront radio as opposed to the postwar television. The homefront child generally listened in silence to the war news, while the parents discussed the news among themselves. The parents also answered the children's questions, describing the images which the boys and girls then internalized. With television, however, the parents ceased to be the major interpreters of events, the main filterers of the news. As the anthropologist Margaret Mead observed, "for the first time the young are seeing history made *before* it is censored by their elders."[14]

Even without television, however, American childhood during the war was anything but innocent, for the children did have access—direct and immediate—to powerful visual images, in which no adult stood between image and child. One example is the pages of *Life* magazine, another the newsreels. *Life*'s impact could be shattering. Its photographs compelled numerous girls and boys not only to reorder their sense of what was right and wrong, but to recalculate the outside limits of human horror. "*Life* did more than anything else," wrote Ed Morris, a boy in Aaron's Fork, West Virginia, "to give me an adult's perspective of the war." At the outbreak of war, Ed's family did not have a radio or newspaper subscription, but their copy of *Life* arrived weekly in the mail. Decades after the war, the images of *Life*'s photographs—"fiercely gruesome photos," one homefront girl called them—still evoke deep emotion. For one girl, it was a photograph "that showed a very young boy standing on his stumps . . . that brought home to me the horrors of war." For another, it was a picture "of a Japanese officer beheading a blindfolded American flyer. I see it in my mind very clearly," she wrote forty-seven years later.[15]

Not surprisingly, *Life*'s photographs of the Holocaust deeply moved the home-front children. "Most of my life," wrote a homefront girl, "I have been haunted by a photograph from *Life* magazine." It showed a young woman "standing in a Betty Grable-like pose, looking over her shoulder. Her skirt was raised, rather like a 'pin-up girl,' but, instead, she was displaying horribly disfigured legs, the result of medical experimentation." For another girl, the Holocaust photographs "exceeded anything we could have imagined—even in our wildest nightmares."[16]

The other source of uncensored visual images was the newsreels, which one reporter called "a sort of *Life* magazine made animate." The newsreels were "our only view of the war," recalled a homefront girl. Not only did they bring "The Eyes and Ears of the World" to the audience, but, one girl wrote, it was through them that "the war became very real. . . ." In black and white, the children saw scenes from air, land, and sea battles. During the war, about three-fourths of the newsreels showed military or naval hostilities or war-related activities; much of the combat footage was shot by professionals trained by *The March of Time*, *Fox Movietone*, and Hearst's *News of the Day*. Early in the war, there was strict government censorship of both the newsreels and the combat photographs in *Life*; for example, it was a year before the government released footage of the attack on Pearl Harbor. Fearing that civilian morale was flagging, however, the government later allowed the release of films and photographs of atrocities that would shock the people in an effort to redouble their commitment to the war effort.[17]

"The most terrifying thing I can remember," wrote a homefront girl, "was our weekly walk to the movie theater." She was afraid of what she would soon see; shown in between the double feature, the newsreels gave her "terrible nightmares. It was apparent, even to a child, that this was not part of a movie but reality." When a newsreel came on, she remembered, some people began to cry, others shouted at the images of Hitler and the Japanese, and "some were terrified—as I was." It was through the newsreels that children learned about the Holocaust. These films were "very upsetting," remembered one girl. "There were scenes of rooms full of eye glasses. . . . Pictures of gas chambers . . . [of] skeletal people. . . ." Some theater managers edited out the Holocaust footage; the manager of Radio City Music Hall said that he did not want to risk "sickening any squeamish persons in the audience which is usually family trade, mostly women and children." But Frances Degen, a girl in New York City, saw the unabridged black-and-white newsreels at her local theater. Years later, in 1967, she was in Israel when war erupted. She thought to herself that this was not at all like war because it was in color, not in black and white. It was "really a lovely day," she reflected, with blue skies and bright sunshine.[18]

Of course, the homefront children paid their dime or twelve cents to see not only the newsreels but two full-length features, a serial, the previews, and several cartoons. And the children saw westerns and musicals as well as war movies. In 1943 Hollywood, perceiving that the public was tiring of war movies, began concentrating on "escapist" films, particularly musicals and comedies.[19] At some theaters the children viewed classic films funded by the government, notably the documentaries in Frank Capra's *Why We Fight* series as well as John Ford's *The Battle of Midway* (1942), William Wyler's *Memphis Belle* (1944), and John Huston's *Report from the Aleutians* (1943) and *The Battle of San Pietro* (1944). They saw government-sponsored public service films and cartoons; the best-known of these cartoons was Walt Disney's *Der Fuhrer's Face*, starring Donald Duck, which gave the nation a hit song based on the unlikely topic of flatulence: "We heil [Bronx cheer], heil [Bronx cheer], right in der Fuhrer's face." Not to be left out, Minnie Mouse did her part for the war in *Out of the Frying Pan and into the Firing Line*, which exemplified Disney's sex-typing and was consonant with the sex-typing evident in all forms of popular culture. Thus, Minnie's task was not to confront Hitler, but to show "why it was important for housewives to save fat. . . ."[20]

Movie attendance skyrocketed during the Second World War. Some people wanted escape, and with fat pay checks, they could go to the movies several times a week. In response, Hollywood studios released 1500 films during the war; one-fourth were combat pictures. Weekly attendance in 1942 was an estimated 100,000,000—at a time when the national population stood at 135,000,000; the previous high had been 90,000,000 tickets sold per week, set in 1930. In numerous theaters, the films ran from Thursday through Saturday, changed on Sunday, with the new films continuing through Wednesday night. Double features were widespread. As people of all ages responded to these offerings, box office receipts soared, doubling from $735,000,000 in 1940 to $1,450,000,000 in 1945.[21]

The homefront children said they were going to "the show," and the most popular was the Saturday afternoon matinee. Going to the movies was an experience in independence; the children either went to the theater with their siblings or playmates or were dropped off by a parent or a grandparent, to be picked up later. In either case, they enjoyed being there without adult supervision. Most ushers were only teenagers, and reports were frequent of unruly boys and girls. Children screamed in their seats, and they ran up and down the aisles making machine-gun sounds, whistling, and pretending they were bombs falling. The theater could be a rough place. Writer Willie Morris, a homefront boy in Yazoo City, Mississippi, described the Saturday matinee in his hometown as always "crowded, noisy, filthy, full of flying objects; one of the Coleman boys from Eden had his eye put out when somebody threw a BB." A theater manager in El Paso, Texas, complained that "from the beginning, the boys and some of the girls . . . decided they were going to run things as they pleased." Sometimes, he said, "when told not to roam the aisles, to quit talking, . . . they would sneer or spit in your face."[22]

"Comics—Radio—Movies," asked an article in *Better Homes & Gardens*, "What are they doing to our children? And what should parents do about them?" Although this article tried to be inclusive, it failed to discuss another of the homefront children's favorite entertainments: the movie serials. Children thrilled to the serials, featuring such freedom fighters as Captain Marvel, The Spider, Batman, Spy Smasher, and Special Agent X-9. Most of these heroes were variations on the Superman theme. In *Captain Marvel*, the boy hero Billy Batson became "the World's Mightiest Mortal" simply by shouting the magic word "Shazam"; the first fifteen adventures of this serial were released in 1941. *Spy Smasher*, who "in reality" was Jack Armstrong's twin brother Alan, started in 1942. *The Black Commando* also debuted that year; he trapped the leader of the alien spies in America by offering to provide the secret formula for synthetic rubber. The next year, with the release of another new fifteen-chapter serial, *Batman* began to pursue "Dr. Daka," the "Japanese superspy" who was conspiring to seize control of the United States. "You're as yellow as your skin!" Batman railed at Dr. Daka. Also in 1943, *The Masked Marvel* took to the screen in pursuit of "Sakima," a Japanese spy trying to sabotage America's defense industries. And in 1944, *Captain America* donned his red, white, and blue tights to do battle with the Nazis.[23]

As they had with radio, parents and educators worried about the psychological effects of the movies. But there seemed to be far less concern about the movies, because parents believed they could monitor the movies more closely than radio. The film Production Code, adopted by the Motion Picture Producers and Distributors in 1930, reassured American parents; in addition, some states and communities had their own censorship boards that licensed films and removed scenes and dialogue considered objectionable. Radio, on the other hand, could be listened to, almost surreptitiously, by children lying in their darkened bedrooms at night. All the same, the Production Code, which was in effect until 1968, had a curious view of what was permissible. On the one hand, it allowed arson, dynamiting of trains and buildings, and murder, so long as "the technique of murder" was not "presented in a way that will inspire imitation." On the other hand, nudity was "never permitted," including "nudity in fact or in silhouette, or any licentious notice thereof by other characters in the picture." As for sexuality, "excessive and lustful kissing, lustful embraces, suggestive postures and gestures, are not to be shown." Finally, "miscegenation (sex relationship between the white and black races) is forbidden."[24]

Perhaps there should have been greater concern about the movies' effects on children, particularly the psychological results of the violence portrayed on the screen. Certainly many of these films were gruesome. As Bosley Crowther of the *New York Times* wrote in 1943: "It is one of the ironies of warfare that such dubious films . . . manifestations which, in peace time, would be regarded as outside the pale—can be palmed off as righteous inducements to a true war consciousness." These violent and horrifying films agitated and frightened children. More than radio or the comics, concluded a survey of psychiatrists, it was the "visual experience" that had "the greatest and most lasting impact."[25]

The writer Scott Momaday remembered that the children in his New Mexico town "would not let go of the movies, but we lived in them for days afterwards." After seeing a swashbuckling adventure film, "there were sword fights in the streets, and we battered each other mercilessly." Girls and boys alike took fantasies of violence with them as they left the theaters and headed for the playgrounds and vacant lots. But other children were terrified, and took their emotions home. According to one wartime study, girls especially found the films to be too vivid. "I felt I was actually *there*," remembered a homefront girl, "and I would be terrified for days afterwards." Another girl recalled a "totally horrifying film" in which a Japanese pilot fired bullets through the windshield of an American airplane, and "the starburst in the windshield is repeated in the starburst of the [American] pilot's goggles and the trickle of blood that ran down his face is indelibly imprinted in my memory."[26]

Radio and movies were relatively new media in children's lives; books comprised a more traditional medium. Considering the nation's appetite for war-related material, publishers released surprisingly few war titles for children. And librarians' lists of recommended children's books were relatively devoid of war-inspired stories. There was a canon in children's literature, and its guardians were active in exposing a new generation of readers to it, war or no war. Among the magazines that published librarians' lists were *The Nation, Horn Book, Newsweek, Journal of the National Educational Association*, and *The Atlantic*; they also appeared in publications of the Children's Bureau and various child welfare agencies and associations. Frequently throughout the war, the *Library Journal* and *Publishers Weekly* also selected children's books to recommend for all reading levels. These magazines recommended that children read the adventures of Mary Poppins, Homer Price, Lassie, and other fictional characters of the day as well as the classics, such as *The Adventures of Huckleberry Finn, Little Women*, and *Treasure Island*. They recommended animal stories, fairy tales, and biographies of George Washington, Davy Crockett, Abraham Lincoln, Julia Ward Howe, George Washington Carver, and other great Americans.[27]

Tales of military and naval glory from the Second World War never displaced the older traditions in children's literature. New titles did appear, like *PT Boat, Bombardier*, and *Navy Gun Crew*; and there were anti-fascist adventures such as *Shadow in the Pines*, about a boy in New Jersey who foils a Nazi plot. But unlike the radio and the movies, most new children's books did not spotlight warfare's blood and gore. They tended to be more on the order of Theresa Kalab's *Watching for Winkie*, a book for grades three to five, which tells of a carrier pigeon that saved the lives of four Royal Air Force aviators.[28]

Much of the credit for the ongoing innocence of children's books was due to Edward Stratemeyer. Stratemeyer, in 1910, founded a syndicate specializing in children's books, which during the next sixty-five years originated over 100 series, featuring Honey Bunch, the Bobbsey Twins, the Hardy Boys, Nancy Drew, and Tom Swift, and published more than 1200 titles. According to the novelist Bobbie Ann Mason, a fan of Stratemeyer, he was "keenly attuned to what turned children on, and he gave them the action-jammed adventures of ideal, wholesome, all-American characters who lived in ideal places. His own favorite," she noted, "was the Rover Boys, a series he wrote personally, with love." Many girl readers, however, preferred Nancy Drew, "the first official girl sleuth."[29]

Homefront children have also paid special tribute to John R. Tunis, who was the author of a series of children's sports novels published during the war. Tunis's baseball players—with names such as Roy Tucker, the Duke, Highpockets, Sourpuss, and Rats—were a diverse group racially, religiously, and ethnically; this was precisely the point. Tunis told exciting sports stories. His stories were quaint too, about a hot hitter who "whangs the old apple" and "thumps the old tomato" and about grown men who affectionately call each other "lads" or "chaps." But his novels were bold.

"What made Tunis's novels particularly fine," Toby Smith, a journalist, has written, "was that they were less about sports than about social concerns," notably, the need to confront racism and anti-Semitism. One of the heroes of Tunis's football story *All-American* (1942) is Ned LeRoy, a black wide receiver subjected to bigotry at his school. In this novel, as in the others, tolerance triumphs. In *Keystone Kids* (1943), a Jewish catcher, Jocko Klein, joins the Dodgers only to be taunted as "Buglenose" by the fans and opposing players. Jocko is an excellent player, but he becomes discouraged. "He won't last . . . ," says a team veteran. "The bench jockeys will get him. You'll see. They'll ride that baby to death. Besides, these Jewish boys can't take it. Haven't any guts." In time, the fans get on Jocko: "Yer yeller kike, look out there or he'll pare your beak offa ya." But the manager Spike takes Jocko aside, telling him that the club "can win if you play like you can. . . . Once they see you're a scrapper, they'll be for you, all the way." And naturally Spike is correct. At last, the bigots confront their intolerance, unite to overcome it, and witness "the re-birth of their team." The new spirit is evident in the postgame nods in Klein's direction. "Nice catching, Jocko. . . ." "You sure handled those pitchers, Jocko. . . ." In 1945, a leading children's book editor praised Tunis's "craftsmanship and his ability to make the reader *feel* and like characters before he hits—and hits hard—at intolerance." Children eagerly read Tunis's other wartime books, *World Series* (1941) and *Rookie of the Year* (1944), and they went to the library shelves for a prewar publication, *The Kid from*

Tomkinsville (1940). Toby Smith spoke for several decades of readers when he paid this tribute to Tunis: "He made a world—where winning and losing and bad bounces all have a place—come alive . . . for an 11-year-old."[30]

Some children's wartime heroes owed their existence not to writers of fiction, but to cartoonists and animators. Whether animated cartoon heroes or comic book heroes—and there was a great deal of overlap—these drawn images were integral to children's experiences. Donald Duck and Minnie Mouse starred in both the cartoons and the comics, as did Bugs Bunny, Mighty Mouse, and others. The plots varied little from one medium to the other, and the messages were invariably the same. Death to the enemy, or at least a severe maiming, was usually the goal. In *Bugs Bunny Nips the Nips*, the rabbit hero is on a Pacific island where he sells Japanese soldiers "Good Rumor" ice cream bars that have grenades hidden inside. "Here y'are Slant Eyes," says Bugs as he hands out the treats.[31]

Many of the comic books, however, demanded a level of sophistication uncalled for in animated cartoons. As with cartoon and radio heroes, several leading comic book stars also were "born" just prior to the outbreak of the Second World War. In June 1938 *Superman* made his debut on the cover of the first issue of Action Comics. *Batman* arrived the next year. Many of these heroes could fly on their own, but with the Pearl Harbor attack, others began to pilot airplanes. As exemplified by *Hawkman, Blackhawk*, and other comic-book aviator heroes, the Second World War was "a time when the evil and the airborne were both simplified in the comics' pages."[32]

The homefront children began reading comic books at age five. From six to eleven, as the children's comprehension of the comics progressed from simple description to the interpretation of events, many boys and girls became fixated on the comics. At age twelve, their fascination peaked. Incredibly large numbers of children read comic books; of the homefront children between the ages of six and eleven, 95 percent of boys and 91 percent of girls "read comic books regularly." And they had a large selection during the war when there were 150 different comic books selling 20,000,000 copies each month.[33]

Just as there was no discernible gender difference in the comics' mass readership, so, concluded a researcher for the Child Study Association, "the comics appear to have an almost universal appeal to children . . . regardless of I.Q. or cultural background." The "Captain Marvel Club" had 575,000 members, and in 2500 classrooms during the war school children learned to read from "Superman workbooks." To offset the perceived evil influence of comic books, in 1942 the publishers of *Parents' Magazine* brought out the first issue of "True Comics," "designed deliberately to look like other comic magazines," *Parents' Magazine* explained, but "free of vulgarity and slang" and "the trashy horror fiction." With monthly sales of comic books so high, however, these sanitized comics did not cut substantially into the sales of the others.[34]

Although most of the superheroes of the comics were men and boys, there was a notable exception—and for numerous homefront girls, she was a magnificent hero: Wonder Woman. A girl in Toledo, Ohio, Gloria Steinem was seven when Wonder Women made her debut in 1941. Prior to that time, her heroes had performed "superhuman feats," but all had been men. In watching these heroes, she wrote, "the female child is left to believe that, even when her body is as grown-up as her spirit, she will still be in the childlike role of helping with minor tasks, appreciating men's accomplishments, and being so incompetent and passive that she can only hope some man can come to her rescue.... 'Oh, Superman! I'll always be grateful to you....'" But then Wonder Woman came to the rescue. She was "as wise as Athena and as lovely as Aphrodite, she had the speed of Mercury and the strength of Hercules." She also was an Amazon and had honed her superhuman skills in training with her sisters on Paradise Island, their home. Like the male heroes, Wonder Woman had marvelous gadgets. She had a golden magic lasso which she threw with unerring accuracy, and with her bulletproof bracelets, she reacted with incredible quickness to stop speeding projectiles. And she flew on her missions in an invisible airplane, which was also a time machine.[35]

Since this was wartime, Wonder Woman was a patriot who fought Axis spies, and she did so in her red, white, and blue costume. Like her male compatriots, she could be jingoistic and racist. But clearly there was a difference: Wonder Woman's message was feminist. She says in one episode that she could never love "a dominant man," while in another episode it is a woman villain who proclaims that "girls want superior men to boss them around." In one story, Wonder Woman rescues Prudence, who in the process discovers her self-worth and agency: "I've learned my lesson. From now on, I'll rely on myself, not a man." Although written during the war by a man, William Marston, Wonder Woman's plots revolved around "evil men who treat women as inferior beings." This, Steinem wrote, was exhilarating because "in the end, all are brought to their knees and made to recognize women's strength...." Recalling Wonder Woman years later during the rebirth of feminism, Steinem observed that hers was a precursory "version of the truisms that women are rediscovering today...."[36]

In an era when women were breaking stereotypes on the labor front and in the armed services, Wonder Woman did not originate in a vacuum; but she was a rare exception in wartime popular culture. On most of the children's radio adventure and crime shows, as well as in their comic books and animated cartoons and their movies and serials, men invariably were the heroes, the action figures, the ones who took risks. Women's role, on the other hand, was usually that of supportive nurturant, as evidenced by the title of one radio serial, *Mary Noble—Backstage Wife*. As it was with the sex-typing in children's war games, so the prevalent sex-typing in popular culture assigned girls to the role of nurse and boys to the soldiering.

Admittedly, there were exceptions. By their example, working women, dressed in slacks and making their own decisions, clearly undermined gender roles in the minds of their children. So too did accomplishments of the 350,000 women who joined the armed forces. Nevertheless, sex-typing was a powerful force in America throughout the Second World War.

The wartime films exemplified both the depth and persistence of gender roles. One paradigm was the romance film in which, a homefront girl wrote, the "wives and sweethearts were at home waiting for the men to return, fearing that they would not." One of the most memorable of these films was *Since You Went Away* (1944), in which the husband and father of two daughters was missing in action and presumed dead. It was Christmas, and Claudette Colbert, the wife, unwrapped the present her husband had left her. It was a music box that played "Together." Sitting there alone, she listened to the song. Suddenly, the telephone rang. It was her husband; he was alive. "There was not a dry eye in the theater," remembered a homefront girl.[37]

The movies also contained lessons in "subtle sexuality," or how to "trap" a man. Personified by Betty Grable and Rita Hayworth, these films, according to a study of women in popular culture, "portrayed images ... which were at great variance with the working girl image popular with audiences of real working women." Produced with the serviceman in mind, these features often had a chorus girl as the female star; into "the wartime musical melee danced the sweet little sex kitten who, underneath the chorus-girl figure, was all mother love." Herein lay major future conceptual problems for both the girls and the boys. Interestingly too, several significant 1940s films about independent women were released not during the war, but before or after it, such as *Woman of the Year* (1941) and *Adam's Rib* (1949), both starring Katherine Hepburn. The import of the wartime films was that they strengthened rather than weakened Americans' attachment to traditional roles for women. America's women participated in the national defense, but in popular culture the institutions which they defended were their traditional domains: romance and marriage, the family and the home. As for working women, the film version was that they joined the labor force entirely out of patriotism and only for the war's duration.[38]

The messages of these films, as well as of much of the wartime popular culture, were simplistic. In reply to questions about what it was that the United States was fighting for, the answer in several movies was "Pumpkin pie!" Or as an aviator in *Thirty Seconds over Tokyo* (1944) put it, "When it's all over ... just think ... being able to settle down ... and never be in doubt about anything." Just as the films indulged in sex-typing, so too they spun myths about American society. "One big family—that is America," stated one film character. "We all see eye to eye," asserted another. America was "a happy home." Multiple cultures, some of them oppositional, existed in wartime America. Even while patriotism and shared national goals helped to shape popular culture, homefront unity often papered over deep racial, ethnic, religious, and class divisions. Nevertheless, on key ideological issues, the war's imprint was deep and lasting. With few dissenters, Americans trumpeted the glories of their country, boasting that it was the greatest nation in the world—and getting better all the time.[39]

Because it engaged in myth-making as well as in the presentation of "ideal types" whether based on racial or gender stereotypes, much of the wartime popular culture distorted the past, present, and evolving future of the United States. For one thing, as was evident during the 1943 race riots, the country was not a big happy family. Also with regard to gender roles, conceptions derived from popular culture were misleading, telling America's girls and boys that they lived in a sex-typed world and that they should enjoy it.

It is clear that these attitudes and values were perpetuated by the movies about the Second World War made after 1945. "A lot of the perspective on the war was taught to the children . . . after the fact through the movies made *after* the war," a homefront boy has written. These films also romanticized the war, just as they deepened society's obsession with gender roles.[40] Indeed, one of the war's legacies for the homefront children was a system of representations in mass culture about what properly constituted the male sphere, and what the female sphere. And this ideology, which exercised its hegemony in subtle but overpowering ways, helped to govern the homefront girls' and boys' behavior for years to come. Moreover, these images of men's bravery and women's sacrifice were the same ones the postwar baby boomers imbibed until well into the 1960s, when the feminist movement issued its challenge to sex-typing.[41]

Although scholars can debate whether or not films and other forms of popular culture are "social artifacts" that reflect trends in society, there is no denying the instrumentality of these images in the minds and lives of Americans. Forty years after the war, a homefront girl, Jean Bethke Elshtain, wrote a book entitled *Women and War*. Hers, she explained, is "not-a-soldier's story; it is, instead, a tale of how war, rumors of war, and images of violence, individual and collective, permeated the thoughts of a girl growing up . . . in the 1950s. . . ." "Born in 1941," she wrote, "I knew the war only at second hand. . . ." Her encounter "as a child and citizen-to-be with the larger, adult world of war and collective violence" was, to a significant extent, she observed, "filtered down to me through movies. . . ." Through them, she and countless others, boys as well as girls, learned lessons— and myths—and developed idealized, sex-typed notions, some of which in fact had been around for ages in epic tales of good versus evil. And in the absence of new story lines, Elshtain recalled, the film images of the Second World War told of "the Just Warrior and the Beautiful Soul."[42]

CHAPTER 10

The Fractured Homefront:
Racial and Cultural Hostility

T HERE WAS A central contradiction on the American homefront during the Second World War: Amidst enthusiastic national unity there existed deep racial, ethnic, and cultural animosities that occasionally exploded into violence. Although historians have recognized this contradiction in studies of the 1943 race riots and of the government's internment of 112,704 Japanese-Americans, they have barely scratched the surface in exposing the fractures that rent the homefront. Buried have been countless stories of the hostility suffered by children because of their race, ethnic heritage, and religious beliefs. Suffering most were America's children of color, but also affected were its Jewish-American children, children of German and Italian descent, Mexican-American children, and children of religious pacifists and nonconformists.

This chapter exposes the persistence of prejudice during the war. The victims were children, but so too were many of the bigots. For these reasons, it is important to understand the influences that fostered hatred and meanness in wartime America. This chapter begins with six vignettes that are emblematic of the gap between lofty ideals and ugly realities on the homefront.

Wanda Davis and the Jehovah's Witnesses: Nine-year-old Wanda Davis experienced hurt and dismay in 1943. Her beliefs were those of a Jehovah's Witness. The Witnesses were unpopular, Wanda remembered, because they refused to "worship an emblem of the state" and thus would not salute the flag or pledge their allegiance to it. The Witnesses' enemies were legion: governments throughout the world, including Nazi Germany; Roman Catholic and Protestant clerics; the American Legion and Veterans of Foreign Wars; vigilante mobs; and the police officers who openly sided with mobs assaulting the Witnesses. Attacks occurred as the Witnesses engaged in streetcorner distribution of the *Watchtower* magazine and even as they met in their Kingdom Halls to pray.[1]

In Winnsboro, Texas, in December 1942, O. L. Pillars and a number of Jehovah's Witnesses were handing out magazines when a mob approached. As the

Witnesses packed up and began to leave, they noticed that parked in the street was a "sound car," with the local Baptist minister sitting behind the wheel. "He started ranting and raving about how Jehovah's witnesses would not salute the flag," Pillars recalled. "He told how he would be happy to die for Old Glory, and that anyone not saluting the flag should be run out of town." As soon as Pillars and the others had passed the automobile, "we looked ahead to see another mob coming right toward us." The city marshall intervened by arresting the Witnesses. And that night he permitted lynchers to enter the jail and drag the prisoners outside. "I was taking a terrible beating," Pillars remembered. "Blood was gushing from my nose, face, and mouth. . . ." Beaten senseless, then revived by douses of cold water, Pillars would not relent, and every time he refused to salute the 2-by-4 inch flag which the vigilantes had brought with them, they struck him again. Finally, they put a hangman's noose around his neck, threw the rope over an extended pipe, and began to pull it taut. "As I was lifted off the ground," Pillars said, "the rope tightened and I lost consciousness." What he did not learn until he woke up later in a hospital bed was that the rope had broken.[2]

Although the Witnesses were known for their refusal to salute the flag, it was the school-age children who suffered the cruelty inflicted by classmates and teachers. In 1943 the Supreme Court reversed an earlier ruling and upheld the Witnesses' position. Writing the majority opinion, Justice Robert H. Jackson compared the flag-salute requirement to the totalitarian practices of America's wartime enemies and asserted "that compulsory unification of opinion achieves only the unanimity of the graveyard."[3]

Teachers did not need the mandate of the Supreme Court, however, to impose their will on nonconforming students. In 1943 Wanda Davis was a third-grader in a two-room schoolhouse in Tulare, California, in the San Joaquin Valley. Each room had four grades with about thirty students, and Wanda was the only one in her room not to salute the flag. Her teacher became angry and threatened her. Salute the flag, she ordered the girl, or she would swat her with the wooden paddle she kept by the side of her desk. Wanda obeyed, but she felt horrible, having betrayed her religion. That evening, she asked her mother what could she do for redemption. "Pray for forgiveness," was the answer. Wanda prayed, but she never forgot; and this incident reinforced her lifelong religious commitment.[4]

Rick Ceaser: Born in 1936, Rick Ceaser lived in Detroit during the Second World War. His father had come from Sicily in 1912; his mother was a second-generation Italian American. Even though his mother was a native-born United States citizen, the federal government ordered her to appear before an official to attest to her loyalty; Rick accompanied her on this mission. They spent "long hours waiting in line to get this done." He remembered that his mother became "very frustrated since she had been born in Detroit, and annoyed by the questions she was being asked. I remember feeling like they thought we were on the side of the Italians, and it scared me."[5]

The Ceasers lived in a Jewish neighborhood, and at school, Rick recalled, "I was always *relieved* to be taken for a Jew. I remember once when a big Jewish holiday was coming up and my teacher had asked all the children who would be absent that day to raise their hands. Only about three or four of us were not Jewish. But when I kept my hand down, the teacher asked me, 'Richard, won't you be celebrating the holiday?' When I told her I wasn't Jewish, she asked me what I was. I told her I was Italian and all the kids *booed!* This was in 1944, I was seven. . . ." Rick's tribulations and anxieties persisted. "Another time," he remembered, "some Jewish boys were waiting in a field that I had to cross to get home from school. . . . I was walking with a Jewish girl, and the boys threw my books in a mud puddle and scolded the girl for being with me. They called me 'Mussolini' and chased me home." To avoid further incidents, Ceaser purposely failed the second grade "because," he recalled, "I wanted to get in a class that didn't know me. I remember telling my mother and teachers that I was having trouble seeing—and it worked!—the next semester I had a new pair of glasses and new classmates!"[6]

Rick Ceaser's homefront experiences were crucial to his childhood development. Living as a member of a disdained group reinforced his introversion: "I had few friends during that time except for a family of Italians that lived around the block." He remembered that he was proud of his two older brothers who were serving in the armed forces; worried about his parents who fretted constantly about his brothers; "ashamed of being Italian, and afraid the bombs would drop soon. And yet I didn't really know what the war was all about." Rick was not alone in his insecurities. Ethnic children, particularly those of German and Italian descent, felt at risk in their schools and neighborhoods.[7]

Helen Baldwin: While this is a short story, it is sadly revealing. As a girl Helen Baldwin moved with her family to a small fishing village on Cape Cod. Newcomers stood out there, particularly Helen's bearded father, an artist, who often set up his easel on the beach. "Briefly put," Helen recalled, "my experience was that of being the child of a man who was widely . . . considered to be a German spy." When he painted on the shore, the townspeople whispered that he was "drawing maps for German submarines." Indeed, "everything my father did was interpreted as bearing out the idea that he was a spy." Helen and her family became outcasts in the village: "Gradually a vicious, even dangerous, attitude developed toward us all. . . ." Bitter memories of this period have stayed with Helen, alerting her to "how easily hostility can arise from prejudice, especially when whipped up by war fever." During wartime especially, the outsider was a suspicious person.[8]

Racism and Riots: Newcomers of all kinds often encounter suspicion and hostility. A notable wartime example were the migrants who came from the southern Appalachians to Detroit and other midwestern cities. Even more despised than the "hillbillies," however, were the African-American migrants. During the war social scientists conducted surveys of racial opinion in the United States. Donovan Senter, who worked for the Bureau of Agricultural Economics of the Department of Agriculture, took a train trip in 1942 from Dallas, Texas, to the Florida panhandle. "Just two ears listening," he reported, but what he heard was "a rising din of racial hatred." Whites expressed anger bordering on rage that African Americans had obtained defense jobs, were receiving decent salaries, and refused any longer to bow to white supremacy: "No nigger should get the money a white man does. . . . We won't tolerate them niggers workin' around a white woman. . . . They get impossible when they get a little money. . . . I'd kill them. . . ." One southerner's solution was to draft only black men "and let them get shot before the white men. Why should they take all of the white men in the army and put the niggers . . . in good jobs in the defense plants to make good money[?]" Another white man explained the southern point of view: "I like the niggers when they do what they are told and stay in their place, but they better stay in their place. Would you like for the nigger to rule in the North? Well, we aren't going to have them rule here in the South."[9]

It was not surprising, then, that during the war upwards of a million African Americans took a harsh look at their southern surroundings and decided to migrate to the North and West. But to the migrants, the mentality of many white northerners seemed disturbingly similar—and, in its own way, just as ugly and uncompromising. This was evident in 1942 in Detroit. A few miles from Rick Ceaser's apartment house, a federal project to house African-American war workers and their families was under construction. Named for the black abolitionist and feminist Sojourner Truth, the project was scheduled for occupancy on Saturday morning, February 28.

Polish Americans in the neighborhood next to the housing project prepared for battle, and violence erupted. At 6 a.m. on moving day, according to an eyewitness, "automobiles with horns blowing drove throughout the Polish section, arousing people to come and defend their rights. The Polish came, by the hundreds. The first Negro truck to appear was destroyed. . . ." Hysteria prevailed: knives flashed, volleys of rocks and stones flew through the air, a shotgun roared, an overturned car was set ablaze. This was a war for territory; motivated by both hatred and desperation, blacks and whites battled each other. But blacks bore the brunt of the injuries. Because of the biased police response, most of the people hospitalized were black, and of the 109 rioters held for trial for carrying concealed weapons or disturbing the peace, only three were white. In the days following the rioting, white people predicted victory. They would be waiting, one man warned, if blacks tried again to move in; "the minute it starts, there will be hundreds of people blocking all the streets like there were last time. The police are on our side, too—they don't want to see them get in any more than we do." Outside the project, whites set up a picket line led by three boys aged seven and eight. For his part, a discouraged African-American boy said, "I'm a Jap from now on."[10]

Sixteen months later, in June 1943, massive racial violence erupted in Detroit.

In three days, thirty-four people died—twenty-five blacks and nine whites—and over 700 were injured. In the downtown area white rioters, most of them young men, pulled black people off streetcars; they overturned cars driven by blacks, savagely beat the occupants, and set the cars on fire. Armed with knives and clubs, mobs of between 100 and 400 whites chased individual blacks down streets and up alleys. As one streetcar with a black conductor entered an area of rioting, a drunken white man approached and saw the conductor: "Here's some fresh meat. Fresh meat boys. The conductor's a nigger, c'mon some fresh meat." Another white mob chased a black youth down the street. "They probably caught him" said a white woman, "—they were gaining ground when I saw them." As before, the conduct of the police was shameful. "Those police are *murderers*," wailed a young black man. "They were just waiting for a chance to get us. We didn't stand a chance. I hate 'em, Oh God how I hate 'em."[11]

Other riots erupted that summer—242 racial battles in forty-seven different cities, including Harlem, Philadelphia, Mobile, Alabama, and Beaumont, Texas. Six African Americans died in Harlem in August; at least 400 people were injured. But in Los Angeles in 1943, there was a different kind of riot; in that city, mobs of up to 1000 white soldiers and sailors hunted for zoot-suiters, most of whom were Mexican-American youth. The zoot suit, which was also popular among African Americans in various cities, consisted of a long draped coat, key chain, and "Porkpie hat"; the pants were high waisted, with baggy legs and pegged cuffs; a duck-tailed haircut completed the ensemble. Whites seemed to view this manner of dress not only as a threat but as an insult that needed to be avenged. Thus, the Los Angeles city council responded to the violence not by dispersing the mobs, but by declaring the wearing of a zoot suit to be an act of "vagrancy," which was a misdemeanor justifying arrest. When finally rescued from frenzied mobs, therefore, the bloodied zoot suiters found themselves not in ambulances headed to a hospital, but in paddy wagons on the way to jail.[12]

There was little justice in Los Angeles that summer. A grand jury looking into crime and violence explained that "'Mexican' youths are motivated to crime by certain biological or 'racial' characteristics...." This "would be laughable," lamented educator George I. Sanchez in 1943, "if it were not so tragic, so dangerous, and, worse still, so typical of biased attitudes and misguided thinking which are reflected" throughout the United States.[13]

Diana Bernal: In numerous towns, cities, and farm communities, Mexican-American children worried about their personal safety on the homefront. Since many were very poor, their lives already were precarious: health care was substandard, schools were segregated, and prejudice against Mexican Americans was virulent in the Southwest and other parts of the country.[14] Diana Bernal, a homefront girl in San Antonio, remembered her father's difficulty in locating work; her mind flashed on the sign she had seen frequently during the war: "No Mexicans Hired." She could not "eat in the same restaurants as other people," nor swim in the same pools. Feeling insecure because of their second-class status, some Mexican-American children feared that the hatred unleashed against the nation's enemies might spill over and harm them and their families. "At times," Diana wrote, "I could picture ourselves in a concentration camp. I would question why the Japanese Americans were placed in camps. Are they going to do the same to us because we are Mexican?" Theresa Negrete, a Mexican-American girl living hundreds of miles away in Scottsbluff, Nebraska, also had a fear of internment. Twice during the war, servicemen on leave had "torn up" the Eagle Cafe, which was owned by a Japanese-American family. "There was so much propaganda against the Japanese that," Theresa confessed, "I did not like them. . . . When I'd see them . . . I would actually stick out my tongue at them. We were told they were killing our boys, and they wouldn't let us forget with songs like 'Remember Pearl Harbor,'" which urged Americans to "remember Pearl Harbor as we did the Alamo!" And Theresa, fearing that "maybe other people were remembering the Alamo when they saw me," worried that after the authorities had rounded up all the Japanese Americans, they would come next for the Mexican Americans.[15]

Executive Order 9066: On February 19, 1942, President Franklin D. Roosevelt took up his pen and signed Executive Order 9066, thus stripping more than 112,000 Japanese Americans on the West Coast of their freedom. Within little more than a month, the Army had taken these men, women, and children from their homes in California, Oregon, and Washington, and locked them behind barbed wire in isolated locations in the Mountain States and the Southwest. Most of the internees were young; more than three-fourths were under twenty-five, and 30,000 were schoolchildren. Morever, almost 6000 children were born in the camps. Their numbers—when added to the infusion of 1,037 Japanese Hawaiians and several hundred people of Japanese ancestry from Panama, Peru, and elsewhere—swelled the total to about 120,000, including 77,000 who were United States citizens. The large number of children and youth skewed the population in the camps. And as young adults left the camps for military service or college, the remaining population was increasingly split between children and older people.[16]

Some Japanese-American fathers did not accompany their children to the camps, but these father absences were of a different kind than was usual during this war. "On the very day of my eldest daughter's 11th birthday, February 21, 1942," recalled Masao Takahashi, "I was roused from my sleep very early in the morning. The FBI, along with four Seattle policemen, searched my house, ransacking closets" before taking him away to a detention center run by the Justice Department. He thought he would be released in time to share his daughter's birthday cake later that day. "However, when we were stripped naked and thoroughly inspected, I was shaken. . . . After about a month and a half, my family came to the train station when a group of us were transferred to [the detention center at] Missoula, Montana." Takahashi was "allowed a few minutes to walk to the fence and to say goodbye to them. I was at a loss to find comforting words," he remembered with sadness tinged with bitterness forty years later. "Boarding the train, I heard my daughters crying out, 'Papa, Papa.' I can still hear the ring of their crying in my ears today."[17]

Charges of criminal behavior were never brought against Masao Takahashi; indeed, no Japanese American was convicted of espionage, treason, or sedition. Even their alleged "crime," disloyalty to the United States, was not against the law. The real issue was the racist presumption that the Japanese Americans were part of an enemy race; on that basis, they were rounded up by the Army and put in prison camps. "It was really cruel and harsh," recalled Joseph Y. Kurihara, a veteran of the First World War. "To pack and evacuate in forty-eight hours was an impossibility. Seeing mothers completely bewildered with children crying from want and peddlers taking advantage and offering prices next to robbery made me feel like murdering those responsible. . . . I could not believe my very eyes what I had seen that day."[18]

One interned mother was Yuri Tateishi. On the day that she and her family were evacuated from Los Angeles, her three-year-old son developed the measles. She covered him up, but a nurse noticed the spots and took him away. All the way to Manzanar, the internment camp in northern California to which the Tateishis were being taken, "when I thought about how he might wake up and be in a strange place, with strange people, I just really broke down and cried." He stayed at the hospital for three weeks before rejoining his family. Other internees were Mary Tsukamoto, her husband, and five-year-old daughter. After staying in temporary barracks in Fresno, California, until October 1942, the Tsukamotos were taken by train to the relocation camp at Jerome, Arkansas. Upon arrival, Mary Tsukamoto recalled, her daughter "cried for a whole week—she cried and cried and cried. She was so upset, because she wanted to go home; she wanted to get away from camp. Adults felt the same way," she admitted, "but we weren't children and so could not dare to cry. I remember I always felt like I was dangling and crying deep inside, and I was hurt." Violet de Cristoforo's daughter, who was born in the temporary camp at Fresno, developed double pneumonia, and "by the time we got to Jerome, Arkansas," Violet remembered, "the ambulance was waiting for her and she was taken off to the hospital," where she stayed on and off for the next year. Several times the baby almost died; many nights, Violet and her two other children, aged seven and five, stayed at the hospital with her. "I didn't try to blame it on the government," she said; "I didn't try to blame it on any-body—but the self-recrimination: What have I done to deserve this? Why? Why? I love my children dearly. I haven't done anything wrong, and why?"[19]

Anomalies abounded, as did cruelties. Not only were Japanese-American fam-ilies interned; so too were Japanese orphans, including babies for whom "loyalty" was scarcely an issue. And so were crippled Japanese-American children, who were taken away from their doctors and nurses and relocated along with every-body else. There also was the issue of children of mixed heritage. Some Alaskan Indians who were part Japanese were sent to assembly centers, including children ill with tuberculosis. A religious missionary reported to the Children's Bureau that at an assembly center in northern California, two children aged six and eight were taken from their non–Japanese-American mother and placed in an internment camp with their father, who did not want them. She also wrote about a twelve-year-old boy whose Japanese father had died in Tokyo when the boy was not yet one. The boy's mother, who was Caucasian and a United States citizen, "brought him back [as] a little baby and raised him 100% American. . . ." But now, "he has been taken from her and put in [a] Japanese Camp—twelve years old, a delicate boy who knew no Japanese people."[20]

Memories of the internment are still vivid to these homefront children. Nine years old at the time, Tatsuko Anne Tachibana remembered the evening of December 7, 1941, when FBI agents seized her father Chikamori Tachibana as a "dangerous enemy alien." "He was taken away . . . in a dog-catcher, after our house was searched." The government sent him to the detention center at Missoula, where, Anne heard, he and the other men were being held "to exchange [for] hostages and prisoners." Anne's thirty-one-year-old mother, with four children between ages one and nine, had "to close up the farm . . . and entrust our valuables and possessions to friends." The Army then took the family by train to the Poston internment camp located on an Indian reservation on the western Arizona desert. Anne remembered the train ride: "The guards all around on the crowded train—we were allowed only one suitcase apiece—the shades had to be drawn at all times—the temperature rising with each mile—and the worst was not knowing our destination." What greeted them were "the wooden tarpaper shacks all lined up in the desert with barbed wire fences and lookout towers with guards all around." Their address was Block 43, 12-D, which was a single room furnished only with cots and mattresses. "I can remember the oppressive heat, dust storms blowing right through our shacks. . . ." Toilets and dining facilities were communal; privacy was almost nonexistent. Diarrhea from contaminated water was epidemic, and Anne remembered that several old people died "from sickness and the . . . heat."[21]

The Japanese-American internees were sent to arid and desolate spots. Although the names were evocative—Rivers, Arizona; Heart Mountain, Wyoming; Topaz, Utah; Manzanar, California—the camps themselves were bleak and demoralizing. Still, every morning in the camps' grade schools, these homefront children, imprisoned because of their race, pledged allegiance to the flag—"one nation, indivisible, with liberty and justice for all."[22]

The message of these six vignettes is that racism, nativism, and religious bigotry were all examples of homefront prejudice. But at what age did children's racial and ethnic hostility arise? In research done in the 1940s, psychologist Marian Radke Yarrow and her colleagues investigated children's group stereotypes and prejudices. "Local neighborhood patterns and family group memberships," they wrote, "are among the important sub-cultural differences which influence the responses." In a neighborhood marked by tensions between Catholics and Jews, for example, "the children show a heightened awareness of these groups." And the most important influences on children's attitudes are adults' "values and interpretations of the social world."[23]

As the psychologist Gordon W. Allport explained in his classic study *The Nature of Prejudice*, the development of prejudice parallels the development of other kinds of attitudes, values, and, ultimately, behavior. He concluded that "identification" is the psychological process by which children learned prejudice, for it conveys "the sense of emotional merging of oneself with others. . . . A child who loves his parents will readily become depersonalized from himself and 'repersonalized' in them." But, Allport continued, "it is not only affection for the parent that may lead to identification. Even in a family where power is dominant over love, the child has no other model for strength, for success in life, than his parents. By imitating their conduct and mirroring their attitudes, he often can gain approval and reward from the parents." In addition, the child-rearing regimen adopted by the parents, along with the home environment, also shapes prejudicial attitudes. Allport suggested that even if the parent did not barrage his or her child with negative stereotypes—even if prejudice was "not *taught* by the parent"—it could still be "*caught* by the child from an infected atmosphere. . . . Children who are too harshly treated, severely punished, or continually criticized, are more likely to develop personalities wherein group prejudice plays a prominent part."[24]

At age six, Allport observed, the child is still in "the first stage of ethnocentric learning," which he called "the period of *pregeneralized* learning." The child knows that certain groups are deemed hateful, but he or she cannot yet generalize beyond these "linguistic tags." "He does not quite understand what a Jew is, or what a Negro is, or what his own attitude toward them should be. He does not even know what *he* is—in any consistent sense." Allport quoted an observer who had watched three young children sitting at a table reading magazines. Suddenly one of the boys said: "Here's a soldier and an airplane. He's a Jap." The girl said: "No, he's an American." Then the boy shouted: "Get him, soldier. Get the Jap." "And Hitler, too," chimed in the other boy. "And Mussolini," said the girl. "And the Jews." Then the children started to chant: "The Japs, Hitler, Mussolini, and the Jews! The Japs, Hitler, Mussolini, and the Jews!"[25]

Allport's views accord with those of cognitive psychology that at age seven there is a perceptual transformation: the emergence of the stage which Jean Piaget has called "concrete operations." According to Allport, at seven the child enters "a second period of prejudice—one that we may call the period of *total rejection*." Able to generalize, the child "now has the adult category in mind—she will undoubtedly reject all Negroes [or Jews], in all circumstances, and with considerable feeling." Total rejection "seems to reach its ethnocentric peak in early puberty. . . . The paradox, then, is that younger children may talk undemocratically, but behave democratically, whereas children in puberty may talk (at least in school) democratically, but behave with true prejudice."[26]

Also at work during the childhood years are powerful environmental factors. As the child's world expands to include school, other attitudinal influences begin to rival those exerted by the parents, siblings, extended family, and neighbors. And children learn prejudice in these new settings. During the Second World War, popular culture was a major influence on children. Another was the school, for it was there that the homefront children found not only teachers and books but also peers. By exploring these influences in the context of children's lives, this chapter looks at America's racial, ethnic, and religious minorities in wartime.

The people most despised in the wartime America were the Japanese. The hatred which the homefront children possessed was frightening at times, but perhaps understandable, considering that their anti-Japanese attitudes reflected the parental, cultural, and peer influences that surrounded them. With the Pearl Harbor bombing, many of the homefront children witnessed their parents' hatred. They heard the epithets shouted by their fathers and mothers: "The little slant-eyed bastards," growled one father. "Dirty Japs," "dirty yellow Japs," and "Little Yellow Bastards" were other denunciations. Some children stood helplessly as their enraged parents smashed to pieces any toys "made in Japan." One homefront girl, a first-grader, had won a race; her prize was a doll made in Japan. But her mother destroyed it. Another homefront girl had a toy piano she "dearly loved"; but her father took it to the basement and "chopped it all to pieces with an ax." Fathers and mothers were on the prowl, and it did not matter that the objects they destroyed were their children's prized possessions. In one case, a neighbor boy noticed a little girl's set of china dishes and started smashing them. Usually, however, it was parents who broke the toys. Sometimes their rage was understandable. When a five-year-old girl's big brother was reported killed in the Pacific, her parents gathered up everything made in Japan and smashed it, including the girl's set of three white ceramic dogs, which she treasured and kept in her playhouse. When she tried to glue them back together, her mother caught her and was very upset.[27]

Another homefront mother implanted deep fears in her daughter with warnings about "the yellow peril." The family lived on the Pacific coast, in Eureka, California, and the mother warned that the Japanese might invade. Moreover, she implied that, if they did, they would rape her and the other women: "She asked if I wanted a sister with yellow skin and slanted eyes." Certainly the behavior of adults directly influenced the children. Once, a family relative who was about to go overseas dropped by to say goodbye. A three-year-old boy dressed in a sailor suit asked where he was going. "To whip some big Japs," he replied. "Well, I can whip the little ones," said the boy.[28]

The homefront children saw disturbing examples of anti-Japanese hatred. Signs appeared in restaurant windows: "No Dogs or Japs Allowed." One father referred to dried peaches as "Jap ears." In trying to understand this anti-Japanese hatred, a Japanese-American homefront girl, Tatsuko Anne Tachibana, has discerned a truth that applies to the wartime treatment not only of the Japanese

Americans but of other racial and ethic groups as well. The causes of Japanese-American mistreatment were twofold, she wrote: "racism and fear." "I was very afraid of anyone with slanted eyes . . . ," wrote a white homefront girl, and other homefront children admitted they shared this fear.[29]

Hatred and fear had numerous progenitors during the war, and one of them, popular culture, functioned to amplify both. A homefront boy remembered that the comic books portrayed the Japanese "as loathsome buck-toothed little yellow savages, but cunning devils." Movie images were just as inflammatory and perhaps more vivid and long-lasting. Some children found the war films to be terrifying. "I don't know why," wrote a homefront girl, but "the Japanese frightened me far more than the Germans," and movies about the Japanese "frightened me terribly." Several homefront children recalled seeing films about "chattering, monkeylike Kamikaze pilots," in which the pilots made suicide crashes on the decks of American naval vessels. Such images caused a homefront girl to question what motivated this fanatical self-sacrifice. Such "alien" behavior reinforced her belief that the Japanese enemy posed a horrifying threat. "And so," wrote a homefront girl, "I became afraid of the Japanese. They . . . had to be stopped." "There wasn't any question that the Japanese were the 'enemy,'" wrote another girl: the "stereotypical slant-eyed 'Jap,' the march to Bataan, babies impaled on bayonets, horrible tortures, kamikaze madmen, treacherous spies. . . ."[30]

Some children did learn tolerance from their parents. One set of parents, hearing their daughter sing "some limerick about dirty Japs," sat her down to talk. They told her that the Japanese were human beings with feelings, and "that Japanese children were loved by their parents just as I was by them. . . ." Some children had Japanese-American classmates who suddenly disappeared. A kindergartener's little friend was gone one day: "he was Japanese. I worried about him." One homefront girl developed "much empathy" for her pen pal, Yooko Mayekawa, who was interned at the Heart Mountain camp and wrote about her family's life in the barracks. Some children wondered why the Army had taken their friends away. "It was for his own protection, dear," explained one mother. But some homefront children had unanswered questions.[31]

Although Hitler and Mussolini were powerful emotional symbols of the enemy, the common people of Germany and Italy usually were not. The war against the Japanese was different. It "became kind of a racist fight," a homefront boy remembered, "whites against the yellow race. . . . In Europe it was a little different. You felt that Europeans were good people. They just followed the wrong leaders." Another perceived distinction, explained a homefront girl, was that while the German soldiers were "fathers/sons/brothers," the Japanese had "sprung into being family-less" and, as a result, were "unspeakably evil, vicious, & sub-human." Thus, all the Japanese were enemies, no matter whether they were leaders like Hirohito or Tojo, kamikaze pilots, jungle soldiers, or even United States citizens of Japanese ancestry. Typifying this perspective was an advertisement for lapel buttons that appeared in *Newsweek* in 1942. The button for the German enemy displayed a picture of Hitler with the words "Wanted for Murder" printed above his head. The button for the Japanese read: "Jap Hunting License—Open Season—No Limit."[32]

"I remember being called a 'Jap' in grade school, which hurt a lot," remembered Tatsuko Anne Tachibana of an incident that occurred before her family was placed in an internment camp. The "Japs" were customarily the wartime enemy on the school playgrounds and in the fields and vacant lots, and children's war fantasies involved dog fights with Japanese Zeroes and dodging Japanese machine-gun fire during imaginary landings on Pacific island beaches. Also in school, recalled a homefront girl, children drew "cartoons of villainous looking Orientals with knives hidden behind their backs." Indeed, sometimes the victims of anti-Asian prejudice were not even of Japanese ancestry, but were Chinese or Filipino. One homefront boy remembered a Filipino classmate in his grade school in Cleveland: "We would chase him down the street, yelling 'Jap! Dirty Jap!'" Although the image of the Chinese "as our wonderful allies" was highly positive, even if condescending, *Time* magazine deemed it necessary to publish an article on "How to Tell Your Friends from the Japs." After listing differing physical characteristics, most of them specious, *Time* asserted: "Those who know them best often rely on facial expressions to tell them apart: the Chinese expression is likely to be more placid, kindly, open, the Japanese more positive, dogmatic, arrogant."[33]

In retrospect, the extent of anti-Japanese sentiment among the homefront children was not unexpected. But the depth of these hatreds, as well as their longevity, has been remarkable; for even in the 1990s, numerous homefront children still resent and even hate Japan. If "I had to list in order of preference ten cultures," wrote a homefront girl, ". . . Japan would be last." In the 1990s there has been a new Pearl Harbor: Japanese imports. "I cringe," confessed a homefront girl, "every time I read or hear about Japan buying or financing something in America."[34]

Surprisingly, only a handful of non–Japanese-American homefront children wrote to express deep personal regret about the internment of so many of their fellow citizens. Janet Scholtes, who lived in Elkhart, Indiana, during the war, said that she kept a diary in which she recorded her hatred of the "Japs" and her jubilation that an atomic bomb had destroyed Hiroshima. She had confided to her diary that the bomb was a victory for "the whole *civilized* world." Writing in 1990, however, she reported that she was "ashamed that the Japanese-Americans were incarcerated" and chagrined by her diary entry: "Now I wonder how civilized it was of my country to drop a bomb that indiscriminately wiped out hundreds of thousands of civilians."[35]

Not surprisingly, the anti-ethnic sentiments of the war years found their way into homefront children's play. Here is one taunting jump-rope rhyme:

> Red, white and blue,
> Your father is a Jew,
> Your mother is Japanese,
> And so are you![36]

Children mimicked adults in using rhymes to malign other people, especially people of color and Jews. Sometimes Jews or African Americans appeared together in these rhymes, as in this piece of doggerel which circulated during the presidential election campaign of 1944. In it, President Roosevelt addressed his wife Eleanor Roosevelt:

> You kiss the niggers,
> I'll kiss the Jews;
> And we'll stay in the White House
> As long as we choose.[37]

Although perhaps lumped together politically, Jews and African Americans represented the two extremes of racial and ethnic stereotyping in the 1940s. "The Jewish stereotype is to be sharply distinguished from the Negro stereotype in two respects," wrote Carey McWilliams in *A Mask for Privilege: Anti-Semitism in America* (1948). "In the first place, the Jew is universally damned, not because he is lazy, but because he is *too* industrious; not because he is incapable of learning, but because he is *too* intelligent—that is, too knowing and cunning." On the other hand, McWilliams explained, African Americans exemplified "the basic stereotype of subordinated groups." As such, this stereotype was similar to those held by the Germans of the Poles, by the English of the Irish, or by Californians and Michiganders of the "Okies," "Arkies," and "hillbillies." Thus, the African-American stereotype held that "the group is lazy, shiftless, irresponsible, dirty, can't learn and won't work, competes unfairly, lowers living standards and property values, has excessively large families, and is 'incapable of assimilation.'" And in children as well as adults, such negative stereotyping led to "total rejection."[38]

Although the Japanese Americans were probably the most intensely hated people of color in the country during the Second World War, hostility to people of Japanese ancestry was very much an abstraction, since there were so few in the population and fewer still who had not been imprisoned in relocation camps. Interaction had little to do with this stereotype, and except on the West Coast, this hostility was generally of recent vintage. In 1935, for example, a group of African-American students at Howard University were given a list of eighty-four adjectives for a study of stereotypes. For each racial and ethnic group, the students selected the ten adjectives which they believed were the most descriptive. For the Japanese, the top three descriptors were "intelligent," "industrious," and "tradition loving." Faced with the same task in February 1942, a group of Howard students put these adjectives at the top of their list: "sly," "treacherous," and "extremely nationalistic."[39]

Clearly, however, the most widely hated people of color were African Americans. More so than for any other homefront group, the persistently negative stereotyping of African Americans led to generalized, and total, rejection. Negative stereotypes abounded, whether in advertising, cartoons, or the movies. And on such radio shows as *Pick and Pat* and *Amos 'n' Andy*, the African-American characters (played by whites) spoke in a dialect spiced with malapropisms: "I'se regusted," exclaimed the Kingfish. "Ain't dat sumpthin'," responded Andy.[40]

Also during the war years, racial rumors spread throughout white neighborhoods and frightened and angered the residents. According to one rumor, black people were in a state of rebellion and were gearing up to "take over." Numerous whites reasoned that African-American troops would return not only with experience in warfare but also with dangerous aspirations of "social equality." According to the rumors, blacks were stockpiling weapons, notably icepicks and razors, for the seizure of power. No white women would be safe. A rumor circulated in Boston that African-American "troops are organizing Eleanor Roosevelt Clubs ... to plot the overthrow of the whites." In the South the rumors were similar. "Nearly everybody below the Potomac," journalist Virginius Dabney observed in 1943, "has heard of the 'Eleanor Clubs,' or the 'Daughters of Eleanor,'" composed of "colored cooks and maids who have vowed to abandon domestic service as degrading. Their slogan: 'Not a maid in the kitchen by Christmas.'" In Tyler, Texas, a homefront girl was "warned not to be in the 'city square' on Thursdays because of 'bump 'em' day," when black people would "see how many 'whites' they could bump off the sidewalks. Needless to say," she reported, "... I was terrified...." In the white South, there always seemed to be fears of impending interracial warfare. "After we win the war," grumbled a white southern man in 1942, "there is going to be a racial war and we'll clean out these niggers."[41]

African America was on the move during the war. Indeed, the black population was more migratory than the white population; during the 1940s, almost 1.5 million southern blacks relocated to the North and West. In 1940, only 48.6 percent of African Americans were urban dwellers, and 77.1 percent still lived in the states of the old Confederacy; in 1950, the figures were 62.4 percent and 68.1 percent. On the West Coast, for example, between 1940 and 1945 Los Angeles's black population swelled from 75,000 to 135,000, and San Francisco's from 4800 to 20,000. Smaller communities registered even bigger proportional changes; in the state of Washington, Vancouver's black population jumped from 4 to 4000, and Bremerton's from 17 to 3000.[42]

Upon arrival in many defense areas, African Americans entered communities where black newspapers and civil rights organizations were unflinching in their advocacy of racial equality.[43] Still, enormous problems awaited the migrants in the war-boom communities. There was hatred. Whites, a city official in Cleveland explained, perceived "the in-migrants as a menace from the outside." Certain ethnic groups, such as Polish Americans in Detroit and Italian Americans in Chicago, were particularly hostile to the newcomers, whom they feared and hated.[44]

Moreover, the influx created severe health, housing, and education problems. The health of African-American children was bad anyway; of 10,000 white males and the same number of African-American males born in 1942, the estimate based on current statistics was that there would be 500 more white boys alive at age ten. And wherever housing shortages existed, conditions for black families were the worst of all. During the war, the Tolan Committee of the House of Representatives, which investigated "national defense migration," held hearings in war-boom locations across the country. Virtually everywhere, the story was the same. The "Negroes," Dr. Abel Wolman of the Maryland State Planning Commission told the Tolan Committee, "as always, suffer most acutely from these shortages because of their economic status or because of restrictions..., or because they are the first to have difficulty finding reasonable facilities at a reasonable price."[45]

A related problem—blaming the victims—was striking during the war. For example, a member of the Tolan Committee asked the chief health officer of the District of Columbia whether he did not agree that black people "prefer to live in these overcrowded conditions, and that they would take the finest dwellings in the world and convert them into a shambles." "I hardly think," the doctor replied, "it could be said that they prefer to do that." Detroit's housing director concurred: "The housing shortage has put pressure on the families at the bottom of the economic heap. Welfare families are being evicted...." Reports from other cities were the same. In Baltimore, an estimated 200,000 African Americans, or about 20 percent of the city's total population, lived in three districts whose total area comprised fewer than four of the city's total 78.6 square miles. On the other hand, these districts had the city's highest death rates for tuberculosis, pneumonia, and meningitis. A black man in Baltimore whose family had been evicted complained to a reporter on the *Afro-American*: "I was forced to move here in a basement unfit to live in, but I can't do no better, so please come and see where me and [my] wife and five children got to live."[46]

Not the least of the newcomers' problems was that southern whites had migrated to the same areas. The black people, moaned a white southerner in Detroit, "want to get the white under and they are going to try to do that. I would give the last drop of my blood I have before I would be a slave to a Nigger." For their part, northern blacks viewed the white southern newcomers as hopelessly bigoted. "They are too prejudiced," explained a twenty-year-old automobile mechanic in Detroit. "They have got their own ideas about what a colored man should do and say and speak and sit and stand and walk.... So every time they try to tell you what you should do, there is a fight."[47]

African-American children encountered hostility in their neighborhoods and schools. The victims of racial epithets and physical assaults, some black children also were assigned to segregated schools or were seated in the back rows of their classrooms. In Hastings, Nebraska, for example, the black pupils sat alongside their Native-American classmates at the back of the room, and these students were "made to hang their coats & all outerwear in separate areas in the coat room." At noon, while the white students "ate lunch in the large library room, the Indians & Blacks ate on the floor in the auditorium." Finally, the school nurse inspected the children for head lice, which she and other school officials feared "would become an epidemic, originating from the Indian & Black students"; not a single case was found.[48]

Evidently, white children were learning racism first-hand, and like their parents, they too expressed the fear that the black children were "taking over." A homefront girl born and raised in Richmond, California, deplored the arrival of both southern whites and blacks. "It was ... very overwhelming," she wrote, "to have the black race thrust upon us in school...." Her school was on the brink of a race war because the "extremely aggressive" black children sought to "take over [our] playgrounds, areas that were never broached before."[49]

Despite the racism and deprivation which the African-American children endured on the homefront, there were offsetting benefits. For one thing, there was hope that blacks' participation in the war effort would enhance democracy at home. For another, there were friendly white classmates as well as hostile ones. When Calvin Campbell's father, a Tuskegee graduate, took a job at the Puget Sound Naval Yard, he moved his family from Tullahoma, Tennessee, to Bremerton, Washington; Calvin was seven at the time. He remembered that a white boy, a recent arrival from Alabama, called him "a nigger and a white friend of mine beat the hell out of him. Later on," he added, "someone else called me a name and I took care of him myself." Calvin felt proud when Jesse Owens came to Bremerton to show his track prowess and beat all comers. He remembered happy times at home too, because whenever his father saw a new black arrival to Bremerton, he invited him to join the family for dinner. Later, Calvin reflected that "moving to Bremerton" had been an "eye opener for me as everybody had something in common ... and there was not a lot of room or time for racial tensions...."[50]

African-American girls and boys were immensely proud of the black men and women who were serving in the armed forces. "Yellow and Black is no match," wrote a twelve-year-old African-American boy, "... the Negroes shall win the war." In the *Pittsburgh Courier*, *Chicago Defender*, and other black newspapers, children read about black war heroes, beginning with Dorie Miller, who was awarded the Navy Cross for heroism at Pearl Harbor; his citation read that "in the face of serious fire during the Japanese attack," he shot down four enemy planes while a machine gunner on the battleship *Arizona*. Lawrence Streeter, an African-American boy in Birmingham, Alabama, deeply admired the black men he saw in uniform, but none more so than the Tuskegee Airmen. These black aviators, who had received flight training at Tuskegee Institute, served with distinction as fighter and bomber pilots; the 332nd Air Group, for example, composed of four all-black squadrons, fought in Italy, France, Germany, and the Balkans, participating in dive-bombing and strafing missions. So proud was Lawrence Streeter that in

1990, as scoutmaster of the Post 606 Tuskegee Junior Airmen in Benton Harbor, Michigan, he took nine African-American boys 183 miles to Detroit to meet his, and their, heroes, the Tuskegee Airmen, at their annual awards banquet.[51]

Native-American children also had their heroes, the best known of whom was Ira Hayes, a Pima Indian and member of the United States Marine Corps who fought in the Vella LaVella, Bougainville, and Iwo Jima campaigns and gained immortality as one of the six marines in the photograph of the raising of the United States flag on Mount Suribachi. The children were proud of the decorations for valor which Native Americans had won, including two Congressional Medals of Honor. The Marines developed a special relationship with Native Americans, whom they saw as valuable warriors; and to teach the Navajo recruits who did not speak English, the Marines established special language classes. Well known to Native Americans were the Navajo "code talkers" who served in the Marines. After receiving training in radio communications, these Navajo joined the assault forces that landed on beaches in the Pacific. Scurrying ashore, often under hostile fire, they quickly set up their radio equipment and, using a special code based on Navajo, began transmitting information between advance units. The "code talkers" also made vital contributions by relaying communications between support units and headquarters, reporting sightings of enemy forces, and directing shelling by American detachments.[52]

Native-American children participated in the war effort in various ways. Some joined their families in the massive wartime migration. The superintendents at various reservations told an investigator that "they had seen not only the steady flow of young men leaving for duty in the armed service, but they saw whole families pick up their most essential possessions and leave the reservations for war work." Like other homefront children, Native-American girls and boys played war games; they collected scrap materials and bought Victory Stamps; they attended the Saturday afternoon movies. And at the theater, as Scott Momaday, the Native-American writer, observed, "we cheered to those wonderful newsreels in which, out of nowhere, a Zero . . . suddenly swerved into our sights. . . . The whole field of vision shuttered with our fire," as Momaday and the other children loudly shot the airplane down with their imaginary machine guns. At school too the war dominated events. "Along with reading and writing," Momaday recalled, "we were taught to hate our enemies. Every day, after we had pledged allegiance to the flag, we sang . . . 'Let's remember Pearl Harbor, as we do the Alamo.' . . ." Ironically, despite Momaday's fierce patriotism, classmates called him "Jap." Momaday was very proud of his ethnic origin, but one consequence was that "nearly every day on the playground someone would greet me with, 'Hi 'ya, Jap,' and the fight was on."[53]

At the same time that Native-American children shared in homefront experiences, there was great diversity in their lives. For one thing, child-rearing practices varied from tribe to tribe, as Erik Erikson, the psychiatrist, demonstrated in his

1945 study of the children of the Sioux and the Yurok. Second, while some of the children attended boarding schools, others did not. Scattered throughout New Mexico and Arizona, for example, the Navajo children attended boarding schools as far away as Oregon. These children were forcibly taken away from their parents by the Bureau of Indian Affairs; their heads shaved and their names changed, boarding-school children were subjected to a strict regimen, including military drills and uniforms. In contrast, other Native-American boys and girls attended day schools near home. Third, among many Native Americans, the forces unleashed by war accelerated modernization and "Americanization"; but among others, they reinforced tradition. On the one hand, Native Americans migrated to the shipyards and aircraft factories and learned valuable skills as machinists, assemblers, and electricians. Others became soldiers and, as a Navajo leader put it, "got a glimpse of what the rest of the world was doing." On the other hand, the war caused a revival of American Indian traditionalism, including prayers and dances. With the outbreak of war, for example, the people of Santa Ana Pueblo "left their homes and went secretly to their ancient shrine. There, in their former home, long since abandoned, the entire pueblo remained for one unbroken month in secret prayer." Among the Navajo, too, there was a revival of war chants and dances, as men of the tribe left for the armed services.[54]

Another exemplar of intertribal diversity was the range of differing attitudes toward participation in the war. For example, as observed by two scholars who studied the Navajo and the Zuni, "The attitude of the Navaho toward the war situation contrasted sharply with that of the Zuni. Although one could not describe the attitude as one of complete enthusiasm and whole-hearted co-operation, there was not the disinterest in and reluctance to go to war that was so characteristic of the Zuni." The receptions accorded the returning veterans also varied. While the Navajos regarded their veterans as "returning warriors," the Zuni "did not look upon this war as any concern of theirs. They were concerned only with the safe return of their veterans which they believed depended upon the prayers of their priests." When the Navajo veterans came home, their return was ritualized in three ways. First, there was the ceremony of Blessing Way, which was also performed prior to the soldier's departure, "to invoke positive blessings." The second Navajo ritual, for those who had come in contact with dead German or Japanese soldiers, was Enemy Way, "the traditional ceremony for dispelling the harmful effects of alien ghosts." In the third ritual, the Navajo veteran took a sweat bath with older male relatives. For the Zuni, however, there was only one ceremony, a purification rite known in English as "bad luck get rid of it," in which a male waved a cedar bark over the head of the returning veteran. Finally, in the postwar tribes, the Navajo veterans "have tended to be regarded as potential forces for constructive change," while the Zuni veterans "have tended to be regarded as forces for destructive change and have been forced to accept the traditional Zuni values or to leave the pueblo."[55]

For Native-American as well as for African-American children, the war was a

mixed and sometimes confusing experience. Pride abounded in their races' contributions to victory, and wartime participation promised opportunities for an
equitable share of the nation's wealth as well as for democratic access to its institutions. But for most of these homefront girls and boys, these dreams did not
materialize, subverted by racial stereotypes as well as by their classmates' animos-
ity. Sometimes it was an epithet—"Hey chief," "Hey nigger"—followed by: "Get
your ass out of here!" "Go back to where you belong!" Children of color quickly
learned that the war had failed to eliminate racial hatred and discrimination. For
Freda Jones, an African-American girl who was nine years old when the war
ended, the realization came when she and the family of a friend visited an uncle
at an Army base. "There was flags flying, soldiers marching," and the black children were thrilled to be offered hot dogs and malted milk shakes. But there was
a problem: "We had to go to a back room." And, she wondered, "Is this what
this black man was fighting for?"[56]

There was some fear that Italian- and German-speaking Americans might commit
sabotage and espionage. On both coasts, the government arrested "dangerous"
enemy aliens and relocated Italian Americans and German Americans away from
vital defense areas. But whereas over 112,000 Japanese Americans were relocated
from the three states of California, Oregon, and Washington and placed in internment camps, the number of alien enemies interned in the entire country was about
5700. Mostly of German and Italian descent, these internees, unlike the Japanese
Americans, were arrested first and charged with crimes before being imprisoned.
One reason for the difference in treatment was there were far more of these ethnic
Americans, and the reality was that they possessed political and economic power
that the Japanese Americans did not. Many were assmilated and, above all, they
were white and thus not members of an "enemy race."[57]

Nevertheless, many homefront children of German or Italian descent shared
the apprehension that in time they too would be seized and jailed. A German-
American girl who lived near an ammunition depot in Hastings, Nebraska, wrote
that she "secretly harbored a fear" that "if the government could haul away
the Japanese, as a threat to our national security, . . . the government could also
round up my family, friends, and neighbors. . . ." Some German- and Italian-
American families did receive visits from FBI agents, deepening children's fears
that the government would imprison them. One set of parents planned to get a
divorce in the event that happened, with the three children to go with their non–
German-American mother. One of the daughters wrote that her parents tried to
keep their plan secret, but "I don't think my parents realized that we children
knew this." And a father in Tahlequah, Oklahoma, fearful of having his family
interned, told his children, "Don't dare tell anyone you are German. *Never.*"[58]

German-American children from all over the country reported their families'
determined efforts to de-emphasize their ethnicity. "You are an American, speak
English," admonished one mother, who told her children "to be quiet about our

heritage...." "Hush, speak English," another mother warned, "or we'll get in trouble." And when a three-year-old girl boasted that her family was German, her mother upbraided her: "We are NOT German! We are AMERICAN." Some German-American families masqueraded as other nationalities—Swiss, for example. And in numerous German-American communities, the Second World War marked the end of speaking German at home and at church.[59]

German-American boys and girls were the subjects of taunts. "We were called 'Jew haters,' or Nazis," remembered a homefront girl who lived in Dumont, New Jersey. And a homefront girl who lived in New York City reported that she and her German-born mother became "Enemy #1 to many people in the neighborhood...." Neighborhood children beat up the girl, and "Mom was called a Nazi." Other children were chased home from school; "It got to the point where my Mother had to walk me to and from school...." Sometimes the children fought back, including an eight-year-old girl who was taunted by a boy "to the point where I couldn't stand it any longer. I pounded on him, there was blood in the snow...." For other homefront children as well, the burden of being German-American became intolerable, making them feel ashamed and isolated. Barbara Fritz was five and in catechism class when a nun turned to her and said: "With a name like yours, you should pray harder."[60]

Italian-American children also suffered the taunts of schoolmates. The boys were routinely called "Mussolini." And a thirteen-year-old girl in New York City, Viginia De Carlo, wrote President Roosevelt that she was "just a plain girl who lives just like ordinary children" and has "no very important troubles but one: ... I live on a block with mostly all Irish people. And they call me a guinea and an Italian wop." The roots of this hostility preceded Pearl Harbor. Ronald H. Bayor, the historian, has examined "neighbors in conflict" in New York City from 1929 to 1941. Focusing on the Irish, Germans, Jews, and Italians, he has found that "group conflict was significant during this period" and that all of the groups participated. Ethnic stereotypes were widespread; those of Italian Americans depicted them as artistic and passionate, impulsive and quick-tempered, and very talkative, not to mention physically dirty, lazy, and unreliable.[61]

In general, however, Italian-American girls and boys seemed to endure less hostility and suspicion than German-American children. The Italian-American experience was unique in its own way, however. Italian Americans were among the most recent of America's newcomers; their ties to the homeland were strong, their feelings ran deep. In the summers during the war, Judy Milano Berman, a homefront girl, visited her grandfather at the Jersey Shore. "Grampa Milano was Italian," she wrote, "and owned a little cat called Mussolini. I know it was hard for him to realize that Italy was our enemy." Another homefront girl wrote that her Sicilian-born father was relieved when Italy surrendered. And the father of Gay Talese, the writer, had two brothers who were serving in the Italian army. "Almost every night after I went to bed," wrote Talese, who lived in Ocean City, New Jersey, "I could overhear my father's whispered prayers...." During the day, his father's behavior was volatile, "his moods abruptly shifting between resignation and peevishness, tenderness and aloofness, openness and secrecy. On this flag-waving island where my father wished to be perceived publicly as a patriotic citizen, I instinctively sympathized with his plight as a kind of emotional double agent."[62]

German Americans and Italian Americans who spoke with accents were suspected of being enemy spies. So too were other ethnic Americans, as well as pacifists and "bohemians" whose lives departed from the mainstream, such as the bearded painter on Cape Cod. Ironically, even Jewish Americans, whose relatives were dying in Nazi death camps, were suspected of being foreign agents. A couple of Jewish-American homefront boys recalled fearing that they might be taken for spies if their behavior appeared conspicuous, such as "hanging around" an area with no clear purpose. Moreover, some Americans accused Jews of having started the war, and of having done so strictly for selfish economic gain.[63]

Overhearing the suspicions uttered by their parents, the homefront children speculated that certain of their neighbors were spies. Fueling their fantasies were the spy shows on radio and at the movies, as well as the spy games that children played during the war. One homefront boy suspected that a German-American family that owned a shortwave radio was broadcasting secrets to the enemy, but he "was frightened to death & never told anyone...." Children in Charleston, West Virginia, heard that on their daily radio show a local country music group, "Dot & Smokey," was broadcasting "secret messages to Germany through their guitars." Rumors sometimes started when the FBI paid a visit to a family in the neighborhood; and on the East and West coasts, teachers and parents urged the children to report any suspicious persons or activity. A homefront girl who lived in the southern Appalachians recalled that her mother had lost a baby in childbirth; the midwife was of German descent, "so she was mistrusted to a degree...." Since the baby was born dead, the midwife "concentrated on saving the mother ... until the doctor arrived...." But then the rumors began that she had "deliberately killed the baby so there would be one less male to fight Germans one day. To her dying day, my mother believed this!"[64]

America's religious pacifists were sometimes associated in the public's mind with the enemy. After all, some Mennonites spoke German; and when they exercised their conscientious objections to all wars and thus refused to serve in the armed forces, other Americans sometimes viewed them as Nazi sympathizers. What sort of Americans, they wondered, would fail to come to their country's defense? Homefront children who were pacifists also encountered suspicion, ostracism, and even violence. One homefront girl reported that her teacher had forced her to stand up in class and publicly refuse to buy war stamps. But the day-to-day challenges confronting pacifist children often transcended the pain of other people's suspicions and hostilities. For one thing, these Quaker, Mennonite, and Church of the Brethren children had to deal with troublesome moral dilemmas. Ruby Hershberger, a Mennonite and Indiana farm girl, wrestled with issues of war and peace. She believed that it was not right to kill anyone, "not even our enemies." She prayed for peace, not victory.[65]

Born in 1935, Paul Boyer, the historian, was in grade school in Dayton, Ohio, during the war. His family belonged to the Brethren in Christ Church, a small pacifist denomination; his parents "dressed plain" and were deeply "non-resistant," or pacifist. One issue with which Paul grappled was whether he should donate his little stub of a lead pencil to the class drive. He placed it in the collection box, but "later," he recalled, "I started feeling very guilty about this. I had actually done something that was going to kill people," because the graphite stubs could be recycled as lubricant for military vehicles and weapons. "Could I have that pencil back that I put in the box?" he asked his teacher. But she responded by turning to the entire class and asking, "What are you, an Indian giver?" Paul's pacifism also embroiled him in a dispute with another nine-year-old-boy, Raymond Bland. What triggered the incident was Paul's conviction that "as Christians we had to love everyone. That included Adolf Hitler." Raymond was "really upset about this. . . . He pushed me around. . . . He called me a traitor." Later the boys settled on a compromise that "I would agree to hate Hitler's body if Raymond would love his soul."[66]

Today, Paul Boyer describes himself as "not a doctrinaire pacifist exactly," but he is strongly opposed to military spending. This outlook has shaped his historical interest in the Cold War at home; he is the author of *By the Bomb's Early Light: American Thought and Culture at the Dawn of the Atomic Age* (1985). The experience of being a pacifist child on the homefront has influenced him in another way as well, giving him "a sense of being an outsider." Paul lamented the "super patriotism" of the 1980s and 1990s, which "gets to me at a very deep level," he said, "because I associate it with my own early experiences when I was very much an outsider."[67]

One irony of wartime was that not only were German Americans and Italian Americans perceived to be of questionable loyalty to the United States, so too were Jewish Americans. Anti-Semitism was rampant on the homefront, but much of this hostility preceded Pearl Harbor. In his study of ethnic conflict in New York City, for example, Ronald Bayor concluded that the primary targets were Jews. Some cities were legendary for their anti-Semitism, and Boston apparently topped the list. In 1942 Gordon Allport and colleagues in Harvard's psychology department conducted surveys of war rumors in Boston. The rumors about Jewish Americans were telling: "The Jews brought on this war as part of an international plot." They "blackmailed Roosevelt into getting us into this war." Jewish doctors were giving Jewish young men either sugar to bring on symptoms of diabetes or capsules which "slow down the beat of their heart so that they are rejected from the draft." If all else fails, "seven of the nine medical doctors at the

Causeway Induction Center are Jews." Some of these rumors accused Jews of profiteering; one claimed that "plans to cut off the cuffs on trousers is part of a Jewish plot to profiteer on the accumulation of extra goods."[68]

Where anti-Semitic attitudes prevailed on the homefront, so too verbal and physical attacks were widespread. Again, Boston is an example. Nat Hentoff and other Jewish teenagers living in South Dorchester and Chelsea encountered hate first-hand from youthful gangs of "Irishers." In Boston and elsewhere during the 1930s, radio listeners heard the anti-Semitic diatribes of Father Charles Coughlin, broadcasting from Royal Oak, Michigan. In 1940, when Coughlin praised the Nazis for restoring moral purity in Germany, he found himself dropped from the airways. He continued to foment hate through the pages of *Social Justice*, his newspaper; but in 1942 the government barred it from the mails. Nevertheless, Coughlin had many followers not only in Boston but in New York City, Detroit, Chicago, and elsewhere.[69]

Proud of Jewish-American contributions to the war effort at home and abroad, Jewish children lauded the heroics of Barney Ross, former lightweight and welterweight world boxing champion, who won the Silver Star at Guadalcanal, and Congressional Medal of Honor winners Sergeant Isadore Jachman and Lieutenant Raymond Zussman. As a result, when incidents of anti-Semitism erupted on the homefront, Jewish children wondered why. In the summers during the war, novelist Philip Roth's family vacationed at Bradley Beach, New Jersey. Most shore communities were closed to Jews; only a few, including Bradley Beach, were "open." There, the Roths and "hundreds more lower-middle-class Jews . . . rented rooms or shared small bungalows. . . . It was paradise for me," he recalled, "even though we lived three in a room. . . ." But the summer of 1943 when Roth was ten years old, "gangs of *lumpen* kids . . . swarmed out of Neptune, a ramshackle little town on the Jersey Shore, and stampeded along the boardwalk into Bradley Beach, hollering 'Kikes! Dirty Jews!' and beating up whoever hadn't run for cover." "Race riots," Roth and the other children called these "hostile nighttime attacks by the boys from Neptune; violence directed against the Jews by youngsters who . . . could only have learned their hatred from what they heard at home." Roth at the time thought that these attacks on Jewish children were not only frightening but peculiar, "since we were all supposed to be pulling together to beat the Axis Powers. . . ."[70]

Jewish children remembered their vulnerability to abuse, with school-age children feeling especially vulnerable. Joyce Spurgeon, a homefront girl in Las Vegas, was walking home one day with her friends from school when a new girl, who had joined their group, suddenly turned and said, "I know you're a Jew and nobody should talk to you, [because] your people started the war." "We did not," Joyce shot back. "You're German and your crazy Hitler started the war." And with that, she pushed the other girl down. Another homefront girl, Carol Helfond, was five and in kindergarten in 1942 when her father, a physician, joined the Army. While he was overseas, his wife and five children stayed in their house in the predominantly German-American neighborhood of Glendale in Queens, New York. Carol and her brothers were the only Jewish students enrolled in the local grade school. "There was much overt anti-Semitism," including the swastikas that were painted on the family's house during the war. But even worse, Carol remembered, was the day she was walking home from school and "a big kid ... called me a 'dirty Jew' and spit on me." Jewish children who lived in largely Christian neighborhoods, or in contiguous neighborhoods, were subject to taunts; "dirty Jew" and "dirty kike" were the most frequent. "Christ Killer" was another. Norma Rajeck, a homefront girl in Bethlehem, Pennsylvania, said that she "grew up Jewish ... in a Protestant and Catholic working-class neighborhood where little boys rubbed thumb and fingers as I walked by and sneered, 'Money-money-money-money.'"[71]

On the other hand, numerous gentile children rejected anti-Semitism. Indeed, for them this was the most important lesson of the war. Remembering the overwhelming horror and sadness that swept over them when they learned of the Holocaust, these homefront children also recalled the refugees who moved into their neighborhoods. "I can still remember the Jewish people who became part of our ... lives in the late thirties as they came out of Europe," wrote a homefront girl who lived in Buffalo. "I can also remember the softening of the prejudice in our town against the Jews when the holocaust was uncovered." Some homefront children who migrated to new locations during the war described the loving friendships with their new neighbors and schoolmates who were Jewish. One of these children recalled that she was seven when a Jewish family moved to Corpus Christi, Texas, and opened the town's first delicatessen. Once, the girl asked Sam Levy, the proprietor, which nation he had fought for during the war. "I was young ... and he was gentle and he showed me the concentration number tattooed on his arm."[72]

Many Jewish-American children harbored an understandable hatred of their own: Germany. To a degree, many also disliked German Americans. Jewish-American children remembered the arrival of refugees in their temples and homes, and they learned that their own relatives were perishing in the death camps. Joseph Lubell, a homefront boy in the Brownsville section of Brooklyn, suffered the loss in Poland of his grandparents, eight aunts and uncles, and thirty-three cousins. "Both of my parents," he wrote, "swore at Hitler and the Germans in Yiddish, Polish, French, and Russian." For Lubell, "The only good German was a dead German. Just like the Indian pictures I saw at the Loew's Pitkin...." For him, "the real enemy was Hitler's Germany," not Japan. In fact, he said, his hatred for Germany was "unlimited."[73]

Jewish boys and girls learned of the Holocaust first-hand when they met survivors. Miriam Goldberg remembered the refugee children who enrolled in her grade school in Minneapolis. During the war Jewish-American families with relatives in Europe prayed for their safety and paid regular visits to the Hebrew Immigrant Aid Society, eager for news. And America's Jewish children felt both anguish at the enormity of the suffering and guilt at their own survival. The "war I grew up in," recalled Maurice Sendak, the illustrator and author of children's books, "was World War II, and you think about what happened to the children then, what happened to children who were my age; what happened to them, and not to me, because they lived over there, and I lived over here. When I had my bar mitzvah, they were dead. And yet they should have had their bar mitzvahs just like me. And why was I having one, and why were they not?" Or Cynthia Ozick, the writer, who in an interview said that she "was having the life that Anne Frank would have had simultaneously." She could never think of her school years, Ozick explained, "without realizing how normal they were, and how, at that very moment, the chimneys were roaring away."[74]

Still, even the horror of the Holocaust has not erased the anti-Semitic views of a minority of the homefront children. In 1990 a homefront girl who had lived in Vermont wrote that "we didn't personally know Jews much as *They* had always been *SO* self-contained (and still are in ways). If we knew any Jews they were store keepers who were rich & overcharged. . . . Even *now* when I read of the Jews and how *awful* it was—[how] they lost their furs and jewels—I can't identify with them too much."[75]

"Educators Decry Teaching of Hate," read a *New York Times* headline in January 1943. While it was appropriate for the military to motivate its combat soldiers "by hatred and revenge," stated the Educational Policies Commission of the National Education Association, the nation's schools should refrain from exposing the homefront boys and girls to "malignant indictments of entire nations and races. . . ." During the war, a host of educational, religious, entertainment, business, labor, and political leaders, not to mention numerous mothers and fathers, denounced hatred on the homefront. And they worried about its persistence in the postwar era. "In wartime," observed *The School Review* in 1944, "the aggressive feelings of people tend to be directed outward toward an external enemy. In the postwar period, when it will no longer be fashionable to hate Japanese and Germans, there is danger that Americans will fall to hating one another." Indeed, the magazine added, this was already evident in the race riots that had exploded in 1943 and in attacks on Jews in Boston.[76]

Although prejudice was virulent on the homefront, some people expressed hope that, since the war was a crusade against Nazi racism, victory would redound to the benefit of multicultural harmony. "One of our wartime gains," stated the director of the Russell Sage Foundation, "may be a new toleration and respect for other races." One will never know to what extent tolerance spread during the war—sadly, tolerance is more difficult to document than hateful behavior—but it is clear that many teachers, social workers, ministers, and others worked to defuse racial and cultural hostility and to promote intergroup understanding. In addition, the media made sizable efforts toward this goal during the war. Radio paid special tribute to African Americans; radio historian J. Fred MacDonald has written that "the persistent appearance of black cultural leaders" such as Paul Robeson, Marian Anderson, and Langston Hughes "signified that wartime respect for blacks would not be mere tokenism." Two soap operas, *Our Gal Sunday* and *The Romance of Helen Trent*, had positive black characters, one a serviceman who spoke movingly—and frequently—of his loyalty to the United States, the other a doctor who saved Helen Trent's life. Compared with radio programming prior to the war, the portrayal of African Americans during the war was "significantly progressive," with blacks portrayed as "substantial heroes, even red-blooded and All-American citizens." Moreover, white radio entertainers called for tolerance. In early 1945, for example, Kate Smith spoke on the show *We, the People*: "Race hatreds—social prejudices—religious bigotry—they are all the diseases that eat away the fibres of peace. Unless they are exterminated it's inevitable that we will have another war. . . . Of what use will it be," she asked, "if the lights go on again all over the world—if they don't go on . . . in our hearts?"[77]

Filmmakers as well as the authors of children's books also campaigned against racial and religious prejudice. One example was *The House I Live In,* a short film on tolerance, which won a special Academy Award for 1945. Directed by Mervyn LeRoy, the film starred Frank Sinatra, who sang the title song. Ministers, priests, and rabbis pitched in as well. In 1943 the Reverend Adam Clayton Powell, Jr., of Harlem's Abyssinian Baptist Church, invited A. Ritchie Low, a white minister from Johnson, Vermont, to preach to the African-American congregation. The next summer, eighty-one black children between the ages of nine and twelve left New York City to spend a month living with white families in Vermont. "It's going to be quite a worthwhile project for us church folk," Low said, who have been "content to give money and to say prayers so long as neither bring these people too close. . . ." Religious organizations also joined the war against prejudice; an example was Brotherhood Week 1944, organized by the National Conference of Christians and Jews around the theme: "Brotherhood or Chaos—History Shall Not Repeat Itself."[78]

Most of all, reformers looked to the schools to lead the way. "To construct a common American culture and to lessen hatred and group prejudice in America," stated *The School Review*, "we have to rely chiefly on the schools. Indeed, with few exceptions, those who struggled to revitalize the American ideals of freedom, justice, and equality did so by asking the schools to teach a common national culture, one that stressed the ideals contained in the Declaration of Independence and the Constitution." Their goal was to homogenize society by eliminating cultural differences. As historian Philip Gleason has written, "For a whole generation, the question, 'What does it mean to be an American?' was answered primarily by reference to 'the values America stands for': democracy, freedom, equality, respect for individual dignity, and so on. Since these values were abstract and universal," Gleason added, "American identity could not be linked exclusively" with any single ethnic group. "Persons of any race, color, religion, or background could be ... Americans."[79]

Only a handful of the homefront educators called for the strengthening of diversity. One exception was a New York City teacher, whose class of ten-year-old African-American students put on a play about an African chieftain's son, Kintu, who set out alone to conquer his fear of the jungle. The costumes, the scenery, the drums, the spears—all were African. Next, these black students invited a group of white students from a neighboring school to see the play. One white girl, saying "I don't want to have anything to do with those people," refused to go. On the street in front of the black school, a white boy pulled a knife out of his pocket and said to his teacher, "Nobody's going to mug me." Predictably, the story had a happy ending. "Why, they're nice ...," exclaimed a white student. "That play was swell."[80]

Usually, however, the wartime emphasis on patriotism and national unity, as well as on democracy and tolerance, encouraged the children to de-emphasize their distinctive ethnic, racial, and religious identities. "Blend in," was the wartime imperative: "Hush, speak English." We are not German, or Italian, or Japanese, or Jewish: "We are AMERICAN." Don Tamaki, a Japanese American, remembered that "the message passed to children" both then and after the war was: "'Be quiet.' 'Do not do anything to attract attention to yourself.' 'Anything connected with ethnicity can get you in trouble.'" And the artist Roger Shimomura, who turned three soon after he was interned at the Minidoka camp in Idaho, recalled that "we were supposed to be good and loyal Americans. Our parents' generation was doing just that, and Japanese-Americans were trying hard to be white." Self-hatred could result, said Shimomura. German- and Italian-American children also tended to mute their identity, as did children who were pacifists or who were Jewish.[81]

The problem is that most acts of self-denial in order to gain acceptance are unrewarding, not to mention psychically damaging. Writer Jerome Badanes remembered his father listening to the radio for news from Poland during the war; and he remembered when his fourteen-year-old cousin, just liberated from Auschwitz, came to stay with his family in Brooklyn. Badanes, who was a leader in the civil rights and antiwar movements at the University of Michigan, spent his younger years running away from his Jewish identity. "I wanted to be an American," he said. But the events of the 1960s and 1970s "kept connecting me back.... Part of the meaning of being Jewish is social responsibility." And his religious and cultural heritage and identity have become very important to him.[82]

Behind the façade of national unity, homefront girls and boys who were black or ethnic suffered insults from neighbors, classmates, and teachers, as well as from the cruel stereotypes that portrayed them in popular culture. In *The Nature of Prejudice*, Gordon Allport described the psychic damage done when other children began to use "linguistic tags"—"nigger," "Jap," "kike"—which were the "symbols of power and rejection." But those children who imbibed the hate of wartime also suffered. The case of Nancy Berner, who was a homefront girl in Indianapolis, was not unusual. She had learned her lessons well at home, in school, watching the newsreels at the local theater, and reading *Life* magazine. "We were taught pure hate from all these . . . ," she wrote. "The fear, hate and prejudice I learned took years to overcome."[83]

CHAPTER 11

Children's Health and Welfare

IN JULY 1944 *Life* magazine documented the tragedy befalling scores of children in North Carolina. The cause was infantile paralysis, or polio. Afflicted children brought in from small towns and backwoods in the state were overflowing the isolation ward in Charlotte Memorial Hospital. Perhaps the most evocative photograph in the *Life* article showed an older boy carrying his sick brother away from the hospital. The little boy had polio, but the hospital had space "only for serious cases."[1]

More beds were essential, so the Charlotte health department converted a fresh air camp for underprivileged boys on nearby Lake Hickory into an emergency facility. Pitching in was the National Foundation for Infantile Paralysis, which mobilized local and national resources to set up a forty-bed polio hospital. Within fifty-four hours, the hospital was in operation with a patient in every bed. But children continued arriving; still more beds were needed. The foundation dispatched doctors from Philadelphia, New York, and Chicago; the Red Cross recruited nurses; the Army supplied cots, tents, and labor; and under the watchful eye of armed guards, convicts dug ditches for the hospital's water mains. Meanwhile, local people pitched in by working as carpenters and donating blankets, linens, comic books, and toys. In less than eight weeks, the hospital had expanded to 170 beds.[2]

Still, *Life*'s photographs of Hickory's hospital wards were very sad, showing cots layed out side by side, barracks-style. In one picture, a boy lay on his side as a doctor collected spinal fluid from a puncture made in the lower back; in another, nurses placed hot packs on a boy's chest to ease pain and relieve muscle spasms. Also photographed were children in iron lungs, seven of which had been rushed to Hickory; the machines looked like coffins, except that each was filled with a living but totally immobilized person.[3]

Like the Second World War itself, the battle for the homefront children's health and welfare produced both victories and defeats. On the one hand, various epidemics afflicted children during the war. On the other hand, the government made unprecedented efforts to protect children's health, including substantially reducing maternal and infant mortality by directly paying the costs of childbirth. This chapter explores children's victories and defeats on the health front.

As frightening as any epidemic could be, nothing compared with the absolute dread which children and parents had of poliomyelitis. It was terrifying, wrote a homefront girl; a "fear that gripped my life," stated another. Parents shared this dread, for polio was a family affair; when a child contracted the illness, the entire family found itself under extraordinary stress. "Parents were as frightened of polio as they were of the Germans and Japanese," recalled a homefront boy, "probably more so." Polio was "that terrible disease," and articles in magazines of all kinds sounded the alarm. "Mothers dread infantile paralysis as a crippler of little children," wrote Dr. Herman N. Bundesen, president of the Chicago Board of Health. "When epidemics rise during the summer, 'polio panic' often spreads quickly." "Are You Afraid of Polio?" asked Dr. Howard A. Howe, head of the Poliomyelitis Research Center at Johns Hopkins University. The answer, of course, was yes, intensely afraid. The "hardest part for parents," remembered a homefront girl, "was the feeling of helplessness to protect their children. It was . . . just a case of 'luck' if you avoided the disease." "There is no disease," reported *Good Housekeeping*, "that frightens parents so much as infantile paralysis."[4]

For America's homefront boys and girls, polio was a nightmare, worse even than other bad dreams the children had—for example, of enemy airplanes swooping overhead, dropping bombs on the children's homes and strafing their playgrounds. In the latter scenario, children could fantasize heroics for themselves; they could scramble into their imaginary airplanes and be the pilots who, in their P-38s or P-40s, shot down the evil marauders. But polio was a different story, and children knew its plot as well: polio could be a lifelong affliction, leaving paralysis or, worse, years in an iron lung. Throughout the summer months of 1943, 1944, and 1945, magazines and newspapers were filled with stories about the nation's polio epidemics and with advice on how to reduce the risks to children. One thing to avoid was summertime surgery. For over thirty years doctors, suspecting there was a causal relationship between tonsillectomy and polio, had advised that children should not have their tonsils removed during the summer months. In 1942 doctors reported that in one family in Akron, Ohio, five of the six children had had their tonsils removed on the same summer day. Within forty-eight hours, all five came down with polio; three later died. Tonsillectomy was "the precipitating factor," concluded the doctors, who explained that nerves injured by surgery were more susceptible to the polio virus.[5]

This advice had not permeated the medical community of New Haven, Connecticut, even by the summer of 1943. Joan Riley, four years old at that time, had migrated with her family to New Haven when her father had taken a job in a local defense plant. Joan had had several attacks of tonsillitis the previous winter, and the doctor had recommended a tonsillectomy; the attacks persisted, but she was not well enough to have the surgery until the next summer. Joan's maternal grandmother warned Joan's mother that "she should not have it done because it was not good to take a child's tonsils out in the summer." Paying no heed, Joan's mother authorized the operation. Two weeks later, the girl's legs began to ache; fever swept her body; she had polio.[6]

Many girls and boys had first-hand knowledge of polio's effects. They saw schoolmates and neighbor children walking on crutches, their legs lifeless while their arm muscles bulged, swinging their bodies as they moved. In some families, it was a brother or sister who contracted the disease. One homefront girl remembered returning from her summer vacation to read in the local newspaper that "a schoolmate had died of polio." The children knew the disease struck at random, and they feared they might be in the wrong place when it did. Fay S. Copellman, a therapist in the orthopedic clinic of New Haven Hospital during the 1943 epidemic, wrote that there were "certain to be effects on a child who has had a serious, greatly feared disease like poliomyelitis, involving isolation in a hospital, separation from parents, strange treatment, and a long period of convalescence with the possibility of chronic paralysis as a result." The polio victim's unhappiness was evident in his or her refusal to eat as well as in nightmares, restlessness at night, chewing fingernails, enuresis, and temper tantrums. Some of the older children with polio became angry, lashing out at people and defying their parents. But "the most common manifestation of all was a tendency to cry."[7]

Beginning in 1942 with a rate of 3.1 polio cases per 100,000 population, the disease tripled in 1943 to 9.3. By mid-1943 forecasters saw "signs that an epidemic of infantile paralysis may be brewing for this summer," and within days the United States Public Health Service reported epidemics in California, Texas, Oklahoma, and Kansas. Trying to prohibit the disease's transmission through social contact, the Army declared the city of Houston to be off-limits to its aviation cadets; motion-picture theaters in Oklahoma City closed their doors to children under twelve; and Dallas responded by shutting down its thirty-four swimming pools. Polio epidemics next struck Chicago and New Haven, where officials postponed the opening of the schools. Still other cities joined the list, and during 1943 twelve states declared epidemics. In August the National Foundation for Infantile Paralysis warned that "1943 may go down in the records as one of the major epidemic years for infantile paralysis." Parents restricted their children's summertime activities, forbidding movies and public swimming pools, ordering naps "during the heat of the day," mandating that they play only in their own yards, constantly monitoring them for "chills," and issuing warnings to be "suspicious of anyone who was not part of our established playgroup" and who might "bring into our safe environment the polio germ." And if polio did strike a boy or girl, then his or her family was placed under quarantine. Still, the totals kept climbing, and by the end of the year, 12,401 Americans had contracted the disease. It was the worst epidemic in twelve years.[8]

If polio was rampant in 1943, it was even worse in 1944 and the first half of 1945. In 1944 health officials reported 19,272 cases; at 14.3 cases per 100,000 population, the polio rate had jumped by 54 percent over 1943 and 361 percent over 1942. And with a staggering 27,363 cases, polio's incidence in 1944 was the highest since 1916.[9]

Leading the battle against polio was National Foundation for Infantile Paralysis, which conducted the March of Dimes campaign and supplied equipment and personnel to stricken parts of the country. Its biggest fund-raiser was President Franklin D. Roosevelt, the nation's best-known polio victim. Americans saw newsreel clips and photographs of Roosevelt in Warm Springs, Georgia, his polio treatment facility and personal retreat. There he frolicked in the waters with girls and boys who also suffered from the disease, laughing and splashing each other without regard to who was the President. For Roosevelt, polio's eradication became a war goal. Speaking over nationwide radio on his birthday in 1944, the President stated that in Germany and Japan, "those who are handicapped . . . are regarded as unnecessary burdens to the state." In the United States, however, "if any become handicapped from any cause, we are determined that they shall be properly cared for and guided to full and useful lives." Polio, he said, was the nation's enemy every bit as much as Germany and Japan were. "The dread disease that we battle at home, like the enemy we oppose abroad, shows no concern, no pity, for the young. It strikes—with its most frequent and devastating force— against children." And that was why, Roosevelt intoned, the dimes given by citizens "are the victory bonds that buy the ammunition for this fight against disease, just as the war bonds you purchase help to finance the fight against tyranny."[10]

The March of Dimes was fortunate to have Roosevelt as its spokesperson, because its need for funds was dire during the epidemic of 1944. By August, while new polio cases subsided in North Carolina, epidemics appeared in other states, not this time in the South or the West, but in the Middle Atlantic, particularly in New York. The outbreak was moderate in New York City, but extreme in Erie County, including Buffalo with its mammoth steel factories and booming wartime population. Three children died in Erie County in July, and that month health officials in the town of Lackawanna, location of the world's largest steel plant, closed the playgrounds and motion-picture theaters. Parents took action as well, forbidding their children to swim in the public pools. But the next week, two more children died in the county. In late August, Buffalo's health commissioner announced that the scheduled reopening of public and parochial schools on September 5 would be delayed until at least the first week of October. Still, the death toll burgeoned, and by mid-September sixty-three people in Erie County had died of polio. At the same time, polio attacked other states that had been relatively immune to the disease, such as Ohio, Indiana, and Michigan. There were alarming rises in industrial cities, among them Pittsburgh and Detroit, and Milwaukee officials placed all children under twelve under "house quarantine." Before the year was out, Kentucky, Pennsylvania, Virginia, and other states had declared epidemics.[11]

In 1945 Eleanor Roosevelt read her husband's birthday message to the radio audience, expressing his thanks to the American people for the "stream of dimes" that sustained the programs of the National Foundation for Infantile Paralysis. Again, the President's metaphor was war, and his words proclaimed that "we will never tolerate a force that destroys the life, the happiness, the free future of our children, any more than we will tolerate the continuance on earth of the brutalities and barbarities of the Nazis or of the Japanese war lords." Many people pitched in to raise money, including 17,000 who attended the Pageant on Wheels in Madison Square Garden in February. Sponsored by the Roller Skate Rink Operators of the United States, the event featured a cast of 900 skaters. Another fund-raiser took place in 14,000 movie theaters and featured a film entitled *The Miracle of Hickory*, narrated by the actress Greer Garson. During 1945 the March of Dimes raised a record-setting $16,589,874.[12]

Fund-raising spectaculars were necessary in 1945, for that year's polio season gave every indication of exceeding the 1944 toll. The figure as of mid-June was higher than for either 1943 or 1944, and in late July the Public Health Service stated: "There's no question that it's epidemic again." Especially hard hit were Massachusetts, New York, New Jersey, Pennsylvania, South Carolina, Texas, Utah, and California. As in earlier years, officials responded by closing swimming pools, bathing beaches, and playgrounds. In Trenton, New Jersey, children were also forbidden to enter theaters, stores, and churches. Neighboring communities enforced similar quarantines. Such steps, however, did not halt the disease's advance in New Jersey; by mid-September, there had been 653 polio cases and 60 polio deaths. Then the number of new cases began to subside, however, and no new cases were reported in Trenton during the third week of September, when city officials lifted all the restrictions. From that point on, the downward trend was evident elsewhere. Although twenty-six states reported higher totals in 1945 than in 1944, the national rate dropped 28 percent.[13]

During the war polio was as mysterious as it was frightening, for no medical consensus existed on its etiology. Although the polio virus is actually transmitted in water droplets, theories of its origin varied widely at that time. One doctor wrote that it "may be spread from neighborhood to neighborhood . . . by human travel"; but another doctor suggested that rats transmitted the disease. A group of physicians at the University of Pennsylvania School of Medicine linked polio to vitamin B_1, while researchers at the University of California Medical School said that salt loss was a "possible cause."[14]

In articles published in 1941 and 1942, scientists reported that flies spread the polio virus, which came "directly from human alimentary dejecta or secondarily from sewage." Whether acquired from feces, open sewage pits, or garbage cans, the polio virus contaminated the flies, and they, in turn, contaminated children's food and water. "The ease with which poliomyelitis can thus be isolated from flies . . . suggests that they may play an important role in transmission of the virus and may perhaps be responsible for the special seasonal incidence of the disease," wrote Doctors Albert B. Sabin and Robert Ward of the University of Cincinnati. Putting this thesis to the test in 1945, the Army's Epidemiological Board and Air Surgeon's Office ordered a B-25 to spray DDT, the potent insecticide, on Rockford, Illinois. The site of an epidemic, the city had reported 147 cases and 17 deaths between July 1 and mid-August. First, the Army declared Rockford off-limits to the personnel of nearby Camp Grant. Then to kill off flies, it mounted a spraying unit on top of an Army truck and sprayed DDT along the streets and highways and in gutters and on embankments. Finally, the B-25 made two flights over the city; on each pass, executed at 150 feet and at 200 miles per hour, a 550-gallon tank of DDT was emptied. In 1945, other cities such as New Haven, Pittsburgh, and Savannah launched programs to spray all garbage cans with DDT; and in Pampa, Texas, German prisoners of war cut weeds, cleaned alleys, and sprayed. Yet, even though 1943, 1944, and 1945 were years of diligent polio research, the researchers were unsuccessful in understanding the nature of the disease—its origin, prevention, or treatment.[15]

With so many alleged sources of contagion, polio's threat seemed omnipresent to the homefront children. As an expert wrote in 1942: "In the light of present knowledge, poliomyelitis may be spread by contact, droplet infection, feces, polluted drinking water, polluted beaches or swimming pools, milk, food, and by insects." Yet, even though doctors had not yet developed a scientific explanation of the disease's etiology, there was general agreement on how to protect the children. "Incomplete as it is," wrote Dr. Bundesen, "this knowledge outlines the precautionary steps to be taken during an outbreak of polio: Banish flies. Take extra care to avoid contamination of foods. Guard against overtiring and chilling. Rule out swimming in ... stagnant pools." Said *Time* in August 1943: "Best advice for all: keep away from crowds, keep clean, keep rested." But what doubtless made polio as dreadful as it was was that it seemed as if children could take every possible precaution against the disease, and, still, the awful virus could silently and randomly enter their bodies.[16]

But there was hope for treatment, exemplified by Sister Elizabeth Kenny of Australia who in 1940 introduced her therapy for polio victims to the United States. With the financial support of the National Foundation for Infantile Paralysis, the Kenny treatment was quickly put to the test in America's wartime polio epidemics. Its first applications were in Arkansas and Tennessee in 1942. Kenny's therapy was simple. In fighting an epidemic thirty years earlier in an aboriginal village, she had worked out a system of wrapping patients's aching backs and limbs in woolen cloths wrung out in hot water. When pain or muscle spasm occurred, the patient was covered with these hot packs. Kenny also urged exercise; she rotated and flexed the paralyzed legs, and she taught the boys and girls to exercise their muscles by themselves—for example, by pressing their feet against boards placed at the end of their beds. The purpose was to resist the spasms, the "tight" muscles that hampered free movement. And because her treatment stressed movement, "the discomfort of being held rigidly by frames and splints [became] ... a thing of the past." Rather than frames and splints, the Kenny method required other supplies, which were promptly authorized by the War Production Board. "A quick computation," the WPB announced in September 1944, "revealed that 150,000 yards of wool, 500,000 yards of muslim binder, 150,000 yards of oil silk, 100,000 blankets, and 25,000 dozen safety pins would be needed, together with hot pack units." Three days later, the manufacturer had shipped the first hot pack units to a hospital in Washington, D.C., and the rest of the supplies soon found their way to polio treatment facilities across the country. The wartime government had accorded Sister Kenny's method a high priority. In addition, the National Foundation for Infantile Paralysis funded her training program at the University of Minnesota, which by the outbreak of the 1944 polio epidemic had graduated 1300 physicians, nurses, and physical therapists.[17]

Polio's threat did not disappear with the end of the war; in fact, its most dreadful years lay ahead. The polio rate during the war peaked in 1944 at 14.3 cases per 100,000 population. But eclipsing this rate were those established between 1949 and 1954; the 1952 rate (37.2) was more than double that of 1944. The worst year ever was 1952, with 57,900 cases. Finally, in April 1955, the federal government approved the "killed virus" anti-polio vaccine developed by Jonas E. Salk, then a researcher at the University of Pittsburgh. In the preceding forty years polio had killed or crippled more than 357,000 people in the United States alone. As Robert Locke, Associated Press science writer, put it: "Every summer for years, the polio virus raced about the country, leaving behind shriveled arms and legs, shiny steel braces, and the metal cylinders called iron lungs. And the dead. Most of its victims were children." But in the spring of 1955, American children everywhere lined up in schools and clinics, even on streetcorners, to be vaccinated. In 1962 the live-virus vaccine developed by Albert B. Sabin—an oral vaccine which induced a very small polio infection, causing the body to fight off the disease and build up a natural immunity—was licensed and replaced the Salk vaccine. Between 1969 and 1979 the United States counted only 179 new polio cases, mostly reactions to the oral vaccine.[18]

There is, however, a painful addendum to polio's history: post-poliomyelitis muscular atrophy, or postpolio syndrome. Thirty to forty years later, about one-fourth of the 300,000 Americans who contracted polio during the epidemics of the 1940s and 1950s began to feel the original symptoms returning. Beset by debilitating joint pain and by muscle weakness and spasm reminiscent of their ordeal as children, these women and men also suffered great fatigue, breathing difficulties, and intolerance of cold. "It's very frustrating and demoralizing," said one homefront girl as she pondered the need to rely again on braces or a wheelchair. Interestingly, the nation's leader during the Second World War, Franklin D. Roosevelt, might well have suffered from postpolio syndrome, or at least so his son James has written, based on his awareness of his father's fatigue, breathing difficulties, and extreme sensitivity to cold: "As I reflect on my father's later years and his fight to overcome the effects of polio, I wonder if he, too, suffered from these late effects."[19]

During the Second World War most parents' health concerns were not that their children would be the victims of enemy attack, such as aerial bombardment. They worried instead that their children would fall victim to polio or other epidemics, or that due to wartime dislocations they might suffer from poor nutrition or inadequate medical care. In all of these instances, they saw war as a threat to their children's health. The experts shared the parents' concern. All agreed, wrote Niles Carpenter, dean of the University of Buffalo's school of social work, that "whatever is lost from our present level of living—and much will be lost—the health and welfare of children must be maintained. We say this because we realize that whatever hope lies ahead for this land is in the hands of our children." Moreover, he added, in protecting America's children, "we must not think in terms of physical health alone. . . . We must also think of emotional health." The reason was obvious—Germany's interwar history, which "bears dreadful witness to the fate overtaking a generation, any considerable portion of whose members have grown up frustrated, fearful, and confused."[20]

In fact, most of America's children suffered only minor ailments during the war—sometimes nothing more than colds and sniffles, cuts and bruises, and perhaps a communicable skin disease such as ringworm or impetigo. And, presumably, many wartime illnesses would have struck children whether there had been a war on or not. Nevertheless, large numbers of boys and girls did fall victim to serious illnesses. Epidemics of measles, scarlet fever, and meningitis afflicted wartime America. The incidence of meningococcus meningitis rose throughout 1942, becoming a full-fledged epidemic by the end of the year. Influenza and pneumonia epidemics also hit parts of the United States, and in several cities epidemics of diarrhea among the newborn reversed the trend toward declining infant mortality. In San Francisco, for example, deaths from diarrhea in 1943 were twice that of the year before. Dr. J. C. Geiger, the city's director of public health, listed the reasons for infant health problems: "Lack of adequate prenatal care . . . ; crowded and unsuitable living conditions; mothers at work in war industries during too long a period of their pregnancies; traveling under adverse conditions to this community from distant areas with young infants; over-taxed hospital facilities and limited stay of mothers in hospitals."[21]

Numerous cities in addition to San Francisco reported outbreaks of epidemic diarrhea among babies, the symptoms of which included weight loss, apathy, and, in the final stages, collapse of the circulatory system. In late 1942 various Ohio cities confirmed that diarrhea was epidemic, that it afflicted babies up to thirty days old, lasted two to fourteen days, and had a 50 percent mortality rate. Respecting no class boundaries, the epidemic struck the nursery of St. Luke's Hospital in Shaker Heights, Ohio, an affluent suburb of Cleveland. After several weeks, 19 of the 34 babies stricken were dead, including the niece of Cleveland's Mayor Frank J. Lausche.[22]

At the same time that some diseases were virulent, others such as typhoid fever, smallpox, and tuberculosis reached all-time lows on their road to virtual extinction in the United States. Inoculations—for smallpox and diphtheria, for example—were partly responsible. So too were early-warning testing programs. In 1942, for example, New York City's health department offered free chest X-rays to personnel working in 400 nursery schools in the city. With an increase in tuberculosis reported from wartime England, health officials wanted to avoid a similar situation in their city. Conscious that wartime was a breeding time for disease, doctors were determined to avoid epidemics. "The past two years," boasted Dr. Irvin Abell, chairman of the board of regents of the American College of Surgeons, in 1944, "have seen hospitals and doctors score a major triumph over disease and death under circumstances that in previous wars would have meant epidemics and rising mortality rates."[23]

Even while there were health victories as well as defeats on the homefront, accidents remained a major cause of children's deaths and injuries during the Second World War. For years accidents had been a leading childhood health problem. In fact, in 1900 accidents were the leading cause of death (31.9 percent) among school-age children, more lethal even than pneumonia and influenza (8.3 percent), heart disease (6.5 percent), and appendicitis (5 percent). Infant mortality dropped dramatically over the next forty years, but in 1943 accidents still caused a third of all deaths among children five to fourteen. Homefront children received constant urgings from parents and teachers to "Be careful." "Men prepare themselves to meet the danger of war," stated *The American Home* magazine, "—that's commando training. We, at home, can and should be sensible in training ourselves to avoid accidents and other dangers that are both unnecessary and wasteful." Children should not yield to the temptation to climb trees, poles, and ladders, but "if you must . . . use your head as well as your arms and legs." Be careful in and around the water, and be careful at night "unless you know where the clothesline is that may clip your Adam's apple. . . ." Finally, "the most important home commando order is: Be smart, but not 'too smart.'" Do not, for example, pull chairs out from under people, or "push the head of someone drinking from a faucet; broken teeth . . . are likely results." A special fear was of blasting caps, which children found while playing around mines, barns, and abandoned quarries. "Parents should warn their children about the danger of playing with blasting caps," urged the Bureau of Mines in a 1945 pamphlet that described accidents in which children, unaware of the danger, hit blasting caps with hammers or tried to drive nails through them. In the consequent explosions, some children lost their vision as well as fingers and hands.[24]

While accidents were a worry throughout the year, epidemics were usually seasonal. One of the most feared was rheumatic fever. Even though it was declining in frequency due to penicillin and improved standards of living, rheumatic fever was still "riding high" during the war, noted one expert. "It causes more child deaths than the next four children's diseases. In an average year it kills five times as many youngsters as infantile paralysis, and handicaps a great many more." A streptococcal infection, rheumatic fever was labeled "the occupational hazard of schoolchildren" because its victims fell in the age bracket between five and fifteen. Dr. Martha M. Eliot, associate chief of the Children's Bureau, called it "the great scourge of the school-age period." First came a sore throat, then fever; breathing quickened, then the child's joints became extremely painful. For its survivors, rheumatic fever bequeathed lasting scars to the heart valves, leaving children with "the red plague" of rheumatic heart disease. "Conservatively estimated," wrote Dr. Morris Fishbein, editor of the *Journal of the American Medical Association*, in 1944, "perhaps 500,000 children and perhaps an additional 500,000 adults in the United States ... have crippled hearts that represent a serious infection with rheumatic fever." Medical experts predicted that children affected by rheumatic fever would die before they reached thirty years of age. Until that time, most would live as invalids.[25]

Faced with this prognosis and the absence of any cure, preventive medicine was the only possible protection against rheumatic fever. Since late winter and early spring were the "danger months from childhood's new No. 1 enemy," wrote Gladys Denny Shultz, specialist on children for *Better Homes & Gardens*, that was the time "to protect your family." Just as July, August, and September were the polio months, so February, March, and April were "the worst of all the year for rheumatic fever." So parents kept the children warm and dry during these rainy months; made them "wear rubbers or galoshes where there is slush or cold rain. If feet are wet, get them into warm water quickly, or by a fire." Otherwise, they prayed that the fever would strike elsewhere.[26]

The health problems that beset America's children during wartime had their origins not only in the rapid social change that marked the years between 1939 and 1945 but also in the nation's social structure. Even in peacetime, boys and girls from poorer classes, races, and regions were far more likely to be sick—and to die young—than children from privileged groups. Especially vulnerable were children of color; rural children, especially those living in the South; and poor children from all parts of the country. Assessing the variability of infant and maternal mortality rates, Dr. Thomas Parran, Surgeon General of the United States, decried "the shocking disparity between geographic as well as economic groups." And the participants at the White House Conference on Children in a Democracy in 1940 premised their deliberations on a frank recognition that there were enormous variations in American children's life circumstances and prospects for the future. The investigator of child welfare in America could factor in every conceivable developmental influence in children's lives, and still, the White House conferees concluded, there was no denying that most "important for the children of America are the variations that arise out of the national economy:

Differences between economic strata—the relatively secure and those living on the edge of economic self-maintenance; differences between urban and rural populations, between white and Negro." Although these differences expressed themselves in numerous ways, they were most evident in education, housing, and health.[27]

Educational expenditures at this time varied widely among the states, and in the South the disparity between funds spent for white children's schooling and the piddling amounts used for black children's education was a disgrace. Equally so were other educational distinctions based on race, such as the availability of a school library, the length of the school year, and the average number of days attended per pupil. Differentials of class, race, and region also determined the quality of children's housing. But in some ways the most direct relationship was that between socioeconomic status and race, on the one hand, and children's health, on the other. And nowhere during the Second World War was this relationship more apparent than in the rates for infant and maternal mortality.[28]

In 1941 infant mortality—death in the first year of life—was at its lowest in the states of Utah and Connecticut, each with a death rate of about 30 infants per 1000 live births. Outside these exemplary states, however, the rate soared, exceeding 55 in Alabama, the District of Columbia, Georgia, Kentucky, Louisiana, Mississippi, North Carolina, Texas, Virginia, and West Virginia. South Carolina's infant mortality rate was 75 deaths per 1000 live births. The states of the South were thus among the most backward of the states in this regard. But the worst rates of all prevailed in the Southwest—in Arizona (90.9) and New Mexico (97.5), with devastating mortality among Native-American and Hispanic-American children. With the return of prosperity, many states improved their rates in 1942—some of them dramatically—while New Mexico's rate got still worse (97.9).[29]

As always, racism and poverty helped to explain the South's health statistics. The entire South was well below the national average in such vital requirements as hospital births and births attended by a physician, and southern black women and children ranked at the very bottom. In fact, in all twenty-four states in which African Americans constituted a significant portion of the population, the infant mortality rate was higher for blacks than for whites. Most black families had little money either for a physician's care or for hospitalization for childbirth. In 1942 the proportion of American births occurring in hospitals ranged from 21.5 in Mississippi upward to 95.9 percent in Connecticut. In Mississippi, most babies—about 55 percent of total births in the state—were black; and the vast majority of these were delivered at home, sometimes by a midwife, but other times unattended except for relatives and neighbors. In Jackson County, Mississippi, where 14 of the 16 midwives were blacks, the birth data for 1943 were: Births with midwife—black 119, white 11; births with physician—white 604, black 37.[30]

By the midpoint of the war, the infant mortality rate had declined for both white and black babies, but the racial disparities were still shocking. In 1943 the risk of death for white babies in America was 37.5 per 1000 live births; for black babies, however, the figure was 62.5 deaths per 1000 live births. Equally tragic, while the risk of childbearing among white mothers was 21 per 10,000 live births, the rate for black mothers was 51, most of whom died from preventable infection, toxemia, hemorrhage, trauma, and shock.[31]

In 1943 Connecticut again had the country's lowest infant mortality rate, with 29.8 deaths per 1000 live births. The National Commission on Children in Wartime calculated the implications of this statistic for the rest of the country. If the state of Mississippi, for example, had a record equal to Connecticut's, 1,017 babies would have been "saved." For Texas, the total was even higher: 3,552. Likewise in 1943, Minnesota had the country's lowest maternal mortality rate (14.4 deaths per 10,000 live births). If New York, North Carolina, Texas, Georgia, and Pennsylvania had achieved the same standard, the numbers of mothers saved would have been 163, 170, 183, 194, and 206, respectively. Adequate prenatal care was lacking in large portions of the country. As the White House conferees learned: "Out-patient clinics serving persons unable to pay for medical care exist in all the largest cities and in about half of the cities of 50,000 population but in only 2 percent of the cities with less than 10,000 population. Yet more than half of America's children live in small towns and on farms.[32]

Some homefront children suffered grievously during the war, but for reasons that had nothing to do with either the war or variables such as class, race, religion, or region. The damage was not evident on the surface. The scars borne by these homefront children, who were victims of wartime physical and sexual abuse, have been largely internal. There is no way to know how widespread child abuse was at this time. As historian Linda Gordon has observed in *Heroes of Their Own Lives*, her study of family violence, the decades of the 1940s and 1950s "represented the low point in public awareness of family-violence problems. . . ." Complete and accurate data do not exist for the homefront years—far from it. In fact, it was not until 1961 that a group of pediatricians headed by Dr. C. Henry Kempe of the University of Colorado School of Medicine coined the term "battered child syndrome" to account for unexplained fractures or other severe manifestations of physical abuse. Clearly, child abuse occurred during the homefront years, but it evoked avoidance and denial among family members as well as neighbors, doctors, and teachers who tended to explain the burns, bruises, and broken bones as accidents. Nevertheless, asserted *Survey Midmonthly* in 1944, there were "Skeletons in the Closet."[33]

Abused homefront children have lived with their battered feelings and their memories, and only years later, if then, could they discuss these things. Some of these experiences were direct results of the war. For example, reports appeared of defense workers whipping their children, who had become the "family scapegoats" for a host of stressful problems. In addition, several homefront children whose parents worked wrote of being sexually abused at child-care centers or by babysitters. Others remembered encounters with servicemen who were eager to fondle little girls whom they encouraged to sit on their laps; some homefront girls also recalled seeing exhibitionists. Other children's experiences reflected a kind of institutional timelessness—children's lives in orphanages, for example. The sexual abuse in orphanages which several homefront girls remembered experiencing usually had little to do with wartime forces; rather, it had a "Dickensian" quality that was as suggestive of nineteenth-century England as it was of the United States during the war. Still, for these homefront children, sexual abuse, which disgusted and shamed them, constituted painful wartime memories.[34]

Marilyn Van Derbur, born in 1937, gained fame twenty years later when she was crowned Miss America. What her public did not know was that from 1942 until 1955, her father had often entered her bedroom at night and sexually abused her. Not until she was twenty-four could she voice this secret to those close to her. In 1984 she suffered a physical and emotional breakdown. After additional therapy, she told her mother, and in May 1991 she made a public statement, telling a Denver audience of "the greatest accomplishment of my life—surviving incest."[35]

Although most victims of incest were daughters, a few were sons. One of the homefront boys, born in 1937, wrote that his mother performed fellatio on him when he was three, and she derived pleasure from whipping him with a lilac branch and abusing him with enemas. She simulated intercourse with him, and she often entered the bathroom when he was taking a bath and fondled him. And when he showed his shame, she ridiculed him. Although his father apparently did not directly abuse him, both parents permitted other men to abuse him in the darkness of his bedroom. As the youth reached adulthood, his life, he wrote, was "pretty grim . . . anxious and depressed, suicidal, full of rage, unsociable, aloof, full of terrors, addicted." The therapists he consulted did not help. Finally, he found one who did, and through treatment he found his "voice, autonomy, a self." Moreover, he discovered that he could confront his abused childhood by writing poetry about it. He entitled this poem "Riddle":

> Night brings it.
>
> Don't.
> Don't breathe.
> Don't fight it.
> Don't feel it.
> Don't think.
>
> Fog hides it.
>
> It wasn't, was it?
>
> What was it? (1990)[36]

Infant Death Rates by Race: United States, 1933–45
(Deaths Under One Year Per 1000 Live Births)

Year	Total	White	Nonwhite
1933	58.1	52.8	91.3
1934	60.1	54.5	94.4
1935	55.7	51.9	83.2
1936	57.1	52.9	87.6
1937	54.4	50.3	83.2
1938	51.0	47.1	79.1
1939	48.0	44.3	74.2
1940	47.0	43.2	73.8
1941	45.3	41.2	74.8
1942	40.4	37.3	64.6
1943	40.4	37.5	62.5
1944	39.8	36.9	60.3
1945	38.3	35.6	57.0

Source: U.S. Children's Bureau, Statistical Series Number 4, *Further Progress in Reducing Maternal and Infant Mortality. Record of 1945 and 1946* (Washington: GPO, 1949), 19.[38]

Unquestionably, poverty and race consigned numerous homefront children to premature debility and death. Undoubtedly, too, there were girls and boys whose physical and sexual abuse never became part of the official record. But there were also optimistic data reported during this period of social change; nationwide, maternal and infant mortality declined markedly both during the New Deal and throughout the Second World War (see table). Between 1933 and the end of the war, American medical care registered its most dramatic improvements to that time—due to such factors as wartime prosperity, rural-to-urban migration, and the development of penicillin, antibiotics, and the sulfonamide "wonder drugs." One of these drugs introduced in the early 1940s, sulfadiazine, cut children's deaths from meningitis, pneumonia, and dysentery, while another, sulfanilamide, helped to neutralize the threat posed by scarlet fever and measles.[37]

A final reason, however, for the dramatic decline in maternal and infant mortality was the government's assumption of health care costs for growing numbers of women and children. With passage of the Social Security Act in 1935, the federal government authorized grants to the states not only for homeless, crippled, dependent, and delinquent children but also to provide maternity and infant care. Initially, this was a small program, but its success indicated that the direct application of public money could reduce the incidence of death among the nation's children and could mitigate suffering from disease and illness. An incidental but important benefit was that in administering this program the women officials of the Children's Bureau, most notably Katharine F. Lenroot and Martha

M. Eliot, learned important lessons that would help them to confront the new and bigger health challenges of the war. Between 1933 and 1943 the maternal and infant mortality rates slumped to all-time lows. Significantly, the maternal mortality rate fell from a 1933 rate of 61.9 deaths (per 10,000 live births) directly due to pregnancy and childbirth to 24.5 such deaths ten years later. The infant mortality rate declined from 58.1 per 1000 live births in 1933 to 40.4 in 1943.[39]

After the attack on Pearl Harbor, the national defense migration swelled and millions of Americans swarmed into overcrowded communities, leading medical personnel, government officials, and other children's experts to argue that funding for the health of mothers and their babies needed to be increased. "Maternal and infant mortality rates for 1940 were the lowest in our history," boasted a writer in *Parents' Magazine*. Moreover, these record low figures were just half those "of the first year of this country's participation in World War I," when 15,000 mothers and 300,000 infants died from "causes that were preventable." But "that is the toll which Americans must prevent from happening again."[40]

The government accorded a special priority to the health and welfare needs of military families. Because of the magnitude of America's military involvement, it was evident that the Army would eventually have to conscript fathers with young children. To supplement the incomes of young military families, Congress in 1942 passed the Servicemen's Dependents Allowance Act, which authorized the payment of monthly allowances to the wives and children of soldiers, sailors, Marines, and members of the Coast Guard. Initially, wives received $28 per month, with increments of $12 for the first child and $10 for each additional child. Dependent parents, grandparents, and siblings also were entitled to allowances. Like Social Security, the plan's financing consisted of a sum deducted from the soldier or sailor's paycheck matched by a contribution from the government.[41]

Under the Servicemen's Dependents Allowance Bill, the government's scheduled first mailing of allotment checks would be November 1, 1942, meaning that they would arrive in the mail just a day or two before the congressional elections. Republicans denounced this as dirty politics, contending that the payments should begin at once. But the manipulation of benefit-check payments to garner votes was what these politicians had come to expect of the New Deal. Even some Democrats criticized the timing, but the family-allowance legislation was not only popular but necessary, and it swept to easy passage. Congress, in addition to voting to commence the payments on September 1, not November 1, later enlarged the monthly payments from $12 to $30 for the first child, and from $10 to $20 for subsequent children.[42]

The Army administered the Servicemen's Dependent Allowance Act through its Office of Dependency Benefits (ODB), located in an unfinished skyscraper in Newark, New Jersey. There, 10,000 women clerks working in two shifts handled an average of 130,000 pieces of mail a day, exclusive of the allowance checks. About 3.5 million checks were mailed out monthly. By June 1943 the ODB had issued eleven million individual checks for family-allowance payments; and the figures kept mounting. A year later, the ODB announced that it had added allowances for 250,000 babies born to military wives since September 1942.[43]

But for numerous service families living in urban areas, the allowances were inadequate. A study done in Detroit concluded that "the family allowance system does not provide an adequate income for a mother and her children unless this is supplemented by private means," usually from war-production work. Some young mothers did not make it. In February 1944, Elinor Stefanski, twenty-one years old, was arrested by Brooklyn police for child desertion. The wife of a soldier stationed in England, she had abandoned her eleven-month-old daughter Barbara on the steps of the Brooklyn Blood Donor Bank. The family's monthly allotment check was $80, recently raised from $62. "I used the money to care for the baby and myself," Elinor Stefanski told police, "but as the months went by and the baby grew and the cost of living went up, I found that the allotment money was too small. I just wasn't able to make ends meet. I decided to look for a job, but I had one worry to over come—that was to get some one to care for my baby while I worked." By that time, she was deeply in debt to her landlord and grocer, her gas had been shut off, and Brooklyn charity organizations had not responded to her appeals for help. And so she abandoned Barbara. Several hours later, however, when she saw Barbara's photograph in the newspaper, she turned herself in to the police. The judge paroled Elinor Stefanski and placed her in her mother's custody, but Barbara remained at the foundling home.[44]

At times, the ODB had to make tough decisions, but it was sensitive to cultural differences. According to Gypsy tradition, for example, an uncle's orphaned nieces and nephews became his responsibility. But would the Army pay for their support if an uncle was drafted? The answer was yes. Moreover, for many poor families, the Army offered employment and a pay check. One recent inductee, a husband and father of a very large family from Pennsylvania, had worked for the WPA for $87.50 a month. "To the surprise and delight of all concerned," reported writer J. C. Furnas, "he found that allowances would better than double the family income the moment he was inducted—one wife and eleven children equaled one hundred and sixty-two dollars a month, not to mention that papa was also getting free lodging, clothes and food and twenty-eight dollars a month cash besides."[45]

The ODB's program was enormous. "In the 15 months since September 1, 1942," wrote ODB director Brigadier General Harold N. Gilbert, "we have sent a total of 37,384,217 separate allowance and allotment checks to almost 9,000,000 wives, children, and dependent relatives of soldiers." There was no denying either that for many families government allowances made the difference between survival on the one hand and unmitigated suffering on the other.[46]

As significant as family allowances were, another government innovation with the initials EMIC touched even more directly the lives of America's war babies and their mothers. In March 1943 the government initiated a massive health care program called Emergency Maternity and Infant Care for the dependents of enlisted men. Dr. Martha Eliot recalled a wartime experience that attested to the need for a program that would pay the doctor's bills and hospitalization expenses for servicemen's wives and their infants. In April 1942 Eliot met Ruth Jensen, an Army wife who was eighteen years old and six months pregnant. Ruth was living in Tacoma, Washington, "in one room in a down-at-the-heel rooming house converted from an old fourth-rate hotel." Her husband was a soldier at nearby Fort Lewis, and her acquaintances in the rooming house were also soldiers' wives. Like her, most were natives of farming communities in the Dakotas, Montana, or Wyoming, and they all had babies or were expecting soon. Ruth Jensen told Eliot that while the $30 a month her husband gave her paid the rent, she was "short on food and couldn't buy milk." "No wonder," Eliot observed, that "Mrs. Jensen welcomed my friend the nurse, for she brought with her news that the plan for Mrs. Jensen's care at the hospital for her delivery had been completed and the cost would be borne by maternal and child health funds of the State Health Department." As Eliot rose to leave, she asked Jensen why she had not returned to the family farm in North Dakota. "And leave Carl while he waits to go heaven-knows-where?" was Ruth Jensen's immediate response.[47]

Washington was the first state to finance military dependents' maternity and infant care, but by March 1943—the month in which Congress established EMIC—twenty-seven other states had authorized plans. Funded jointly by the Social Security Administration (under Title V of the Social Security Act of 1935) and the states, these plans were jointly administered by the Children's Bureau and the state health departments. Although the Children's Bureau had long advocated medical payments for military dependents, additional justifications arose. Some experts pointed, for example, to American men's high rejection rate for military service and concluded that the nation's men were not as fit as most Americans had thought. Indeed, the IV-F "pool" grew throughout the war; the final tally was 17,955,000 men examined and 6,420,000 rejected. And as the Selective Service System's chief medical officer explained: "This has continued despite the drastic lowering of standards," including "the induction of limited numbers of men with uncomplicated venereal disease, with hernia and illiteracy, and large numbers of men with dental defects." Concerned that the next generation of American young men should be physically and emotionally fit for future military service, some members of Congress listened receptively to the arguments for a government program of maternity and infant care.[48]

The more immediate issue, however, was the morale of military personnel and their families. Since not all states had federally funded programs, tens of thousands of service wives with monthly incomes under $50 unhappily faced the medical expense of giving birth and providing care for themselves and their newborn. The question was obvious: How could the country expect America's soldiers to do their best job if fraught with worry over their families' health and welfare? Katherine F. Lenroot, the Children's Bureau director, put it bluntly:

"There is one casualty which no responsible nation should ask a fighting man to face. That casualty is the preventable injury of his wife or child back home." The challenge was to provide federal funding for military wives' prenatal and child-birth expenses, along with pediatric care for the infant's first year of life. And this is what Congress authorized in the bill which President Roosevelt signed on March 18, 1943.[49]

Martha Eliot was elated, as were others who worked to improve children's health and welfare. "The expansion of this program made possible today by act of Congress," Eliot wrote, "indicates a wide recognition of the responsibility of the nation for the protection of the lives and well-being of mothers and children." This "marks a red-letter day for the United States...." But others were not so thrilled, particularly political conservatives and a vocal portion of America's obstetricians and pediatricians, who denounced the program as "socialized med-icine," if not outright communism. One conclusion was inescapable: by extend-ing free medical care to this vast new constituency of Americans, EMIC repre-sented a radical departure in the relationship among the federal government, the medical establishment, and the American people. Here was a group, the govern-ment said, which was entitled—as a right, not as welfare — to free medical care.[50]

"These [EMIC] babies constitute the prize exhibit of the biggest public health experiment ever conducted in this country," stated *Collier's* magazine. A year after its inception, the program was functioning in all forty-eight states, Hawaii, Alaska, Puerto Rico, and the District of Columbia. During the short period of three years, from March 1943 until July 1946, Congress appropriated $130,500,000 for EMIC. The program's sweep was broad, and it made its benef-icence felt in countless communities. "For the first time," wrote Martha Eliot, "state and local health departments are carrying out programs of maternity and infant care that reach out into every county and municipality, that are as wide-spread geographically as the draft of enlisted men under the Selective Training and Service Act of 1940." Under EMIC, wives had freedom of choice provided that the doctors, clinics, and hospitals they selected had the approval of the state boards of health. EMIC set minimum requirements: running water, screened win-dows in the hospital or clinic, the presence of a trained nurse, and a separate bed for each patient. Pregnant women rushed to apply, and EMIC responded by pay-ing for the medical care of every eligible child; this held true even when, as one doctor explained, "the baby's father and mother aren't married yet."[51]

By July 1944 EMIC had authorized 402,112 maternity cases and 38,591 infant care cases; by the end of EMIC's life in late 1946, these totals had climbed to 1,163,571 and 189,740, respectively. New York State alone accounted for 100,407 maternity cases and California for 92,424. Payments varied by state, ranging downward from $135 per maternal case to $45. Doctors generally received $40 to $50 per delivery. At its peak in mid-1944, EMIC funded about one out of every six births in the United States and, in the process, facilitated the trend toward in-hospital births. States with an already high percentage of births in hospitals, such as Massachusetts, New York, and California, could not markedly improve their records. But other states could. In Nebraska, for example, inhome deliveries dropped from 28.6 percent in March 1944 to 5.1 percent in June 1946. Nationwide, the percentage of in-home births dropped by half during the same 15 months.[52]

During the Second World War governmental funding was instrumental in reducing childhood illness and mortality, but it was not the only factor. Wartime prosperity was another, as was the migration of millions of rural Americans to urban areas where they had access to public health facilities. In 1941 the ten chief causes of death among children were premature birth, pneumonia, accidents, congenital malformations, diarrhea and enteritis, injury at birth, tuberculosis, influenza, diseases of the heart, and appendicitis. Of the 180,787 children who died in that year, 63 percent died before their first birthday and 76 percent before age two. By the end of the war, deaths before age one from illnesses such as pneumonia, diarrhea and enteritis, influenza, and tuberculosis had plummeted, although deaths from the other causes remained as high as before.[53]

The mortality rate for children of color declined during the war, as it did for whites. Nevertheless, the racial differential—that is, the mortality ratio, nonwhite to white—changed only marginally; mortality was 70 percent higher for African Americans in 1939, and 60 percent higher in 1945. Children of color suffered particularly from pneumonia, influenza, accidents, acute infections, and tuberculosis. And the rates were highest in states where blacks, American Indians, and Hispanic Americans lived in large numbers, such as New Mexico, Arizona, Texas, Virginia, and Tennessee. Rates for whites were also high in these states. But in other states, black infants were almost twice as likely to die as white infants. In Kentucky, the rate for white children was 38.2, high in its own right, but not in comparison with the 75.3 rate for black children. Indiana's gap was between 30.5 and 57.3, and pronounced disparities existed in several other states.[54]

While childhood epidemics explain some of these statistics, their root cause was inequality. But medical access was opening up, and the ten-year period between Social Security's passage in 1935 and the end of the Second World War heralded the entry into the modern medical and social welfare world of hundreds of thousands, perhaps millions, of American children from all racial and ethnic groups and from every part of the country. Naturally, some children benefited more than others. The reasons for the gaps were obvious: money and commitment. "More than half the child welfare workers of public agencies," Katharine Lenroot explained in 1945, "are working in New York, Massachusetts, Minnesota, Indiana, Washington, Illinois, Ohio, and Connecticut, or only eight States." At the other end of the scale, certain states had only one welfare office per fifteen or twenty counties.[55]

An important homefront lesson from the Second World War is that federal funds, if targeted to children's special needs, can significantly enhance American girls' and boys' quality of life. Health was the area in which improvement was most evident. The question many parents, politicians, and medical people at this time began to ask was: If through EMIC the country could save pain and death, why not extend these benefits to civilian children and mothers—and why not renew these benefits for peacetime America as well? By equalizing infants' life prospects, national access to maternal and infant care would represent a significant step in the direction of a democratic society. Looking toward the country's postwar health prospects, one writer concluded: "The field for expansion is big. The difficulties are many. But the stakes are high because one out of every four U.S. citizens is a child under 15 years of age." The challenge was there: "If all these boys and girls can be reared to their fullest potential strength, what a boost they will give to America's future."[56]

How serious was the United States about instituting publicly funded health care? Was it likely that Congress and the new President, Harry S Truman, would vote to extend EMIC into peacetime and expand its clientele? Truman was sympathetic; he endorsed a peacetime public health program for children and announced: "We should resolve now that the health of this Nation is a national concern ... [and] that the health of all its citizens deserves the help of all the Nation." EMIC's longtime advocate, Martha Eliot, was in total agreement. EMIC was a beginning, "a small beginning, to be sure, but," Eliot wrote, "its significance is great. It marks a major break from our present system under which some mothers in the United States get the best care that is known anywhere in the world while others get along with no skilled assistance of any sort." Educators also supported the continuance of EMIC. "Although initiated as an emergency measure and limited in its service function to the families of men in the armed forces, the program may reasonably be expected to set the pattern for broadened social legislation of like import in the post-war world," stated the *Elementary School Journal*. One of the most enthusiastic of EMIC's proponents was Vice Admiral Ross T. McIntire, surgeon general of the Navy, who called it that "splendid program." "In my opinion," McIntire told a subcommittee of the Senate Committee on Education and Labor, "it should be expanded and made a very permanent affair throughout the entire nation."[57]

In 1945 Senator Claude Pepper introduced legislation to make EMIC a peacetime program for civilians as well as military dependents. His bill would provide $100 million for maternal and child health. If the bill were renewed annually, the senator said, within ten years all America's women and children could have the same inclusive treatment, at government expense, as servicemen's families enjoyed under EMIC. Pepper's program also would be broader than EMIC, including psychological counseling and dental care. EMIC should be extended, wrote the *Washington Post*, "so that good medical care can be made available to mothers and children wherever they live and whatever their economic circumstances. As a Nation, we could make no wiser investment."[58]

But the medical establishment disagreed. Numerous doctors had accepted EMIC as a wartime measure only, and others had resisted even that. Even with the war going on, doctors in Cleveland, Detroit, and elsewhere had refused to accept the government's fee for prenatal care; they had insisted on higher payments and referred maternity patients to public clinics. Claiming that their methods were the only ones that would "preserve the patient-physician relationship," many doctors, including the house of delegates of the American Medical Association (AMA), wanted the federal payments to be made directly to the servicemen's wives, who would pay the bills, rather than to the physicians and hospitals. The American Academy of Pediatrics also announced its opposition to EMIC as administered by the Children's Bureau. Contending in August 1944 that the program should be run instead by the Public Health Service (a close ally of the conservative AMA), the academy's executive board accused the Children's Bureau of having "wandered rather far afield" from its original mission. "The function and purposes of the Children's Bureau have been abruptly changed so that it is now an active factor in the practice of medicine throughout the United States, dictatorially regulating fees and conditions of practice on a Federal basis." In short, stated the academy, the Children's Bureau was trying to implement socialism. It was apparent, charged the academy, that a "free to all service with full-time salaried physicians, paid for directly from general taxes and controlled and directed by a Federal bureau is the plan of the Children's Bureau." Also at this time, the medical establishment flexed its political muscle and had legislation introduced to transfer EMIC to the Public Health Service, but the bill failed in Congress.[59]

Speaking for the Children's Bureau, Dr. Eliot denied the doctors' charges, but a bitter disagreement had begun. An angry retort to the American Academy of Pediatrics came from Dr. Grover F. Powers, professor at the Yale Medical School. It might seem, Powers wrote, that the doctors were "more interested in preserving certain professional mores than in promoting the welfare of mothers and babies." And as for the Children's Bureau, he applauded its orientation that "a Federal children's bureau should not stand for health alone, but rather for an integration of all aspects of child life." At the same time, the American Academy of Pediatrics was correct in accusing the Children's Bureau, Senator Pepper, and their allies of wanting to make medical care available to every child and adult in the United States. As Pepper told reporters, he hoped his legislation would serve as prelude to a "total medical-care plan designed to lift the levels of health and medical care."[60]

Pepper's bill failed to pass, and EMIC became history in November 1946. The medical establishment was a potent lobby, and not until 1965 did Congress enact federal health insurance under Social Security: Medical Care for the Aged and Medicaid for the needy and disabled. Even then, America's relative ranking in the world in infant mortality slipped from sixth at the end of the war to seventeenth forty years later.[61]

There are no easy, inexpensive answers, but attention to the history of the United States' welfare and health programs during the Second World War—particularly to the history of family allowances and the EMIC, as well as to Extended

School Services—should illuminate not only the specific public programs but also the political commitment essential to eliminating longstanding barriers to equality and democracy in America.

There has been some progress in postwar America. Spurred by prosperity and the civil rights movement, for example, black infant mortality dropped by half from 1960 to 1980. In the 1980s, however, it began rising again. "More people are below the poverty level," explained Dr. David Allen, an epidemiologist at the Center for Disease Control, "and poverty means more medical problems. . . ." In fact, as the United States considers its health care and welfare needs in the 1990s, economic and racial inequality is becoming more pronounced every year. And it is the children who suffer the most because of it.[62]

CHAPTER 12

"Daddy's Coming Home!"

"IN THE LATE afternoon, all the factory whistles went off, sirens sounded & everyone in town was honking their car horns. . . . So much excitement!!" It was V-J Day, August 14, 1945. That evening this home-front girl and her townsfolk in Waukesha, Wisconsin, "gathered at the park & sang patriotic songs." And so it was across the country. The war was over; the Japanese had surrendered. Americans celebrated. A Massachusetts boy's family was staying at a beachside resort when two women from a nearby cottage, Docky and Eddie, appeared bearing tumblers of whiskey, a large Chinese gong, a trumpet, and news of America's victory. "In all the hubbub," the boy remembered, "punctuated by Eddie's gong, a parade was organized." Beating pots and pans and shouting "THE WAR IS OVER," the parade stopped along its route, and "at every house, kids and their parents joined us. . . ." At the end, "the zaniness waned" and seriousness prevailed as the marchers sang "God Bless America."[1]

People in communities of all sizes were jubilant, and they danced and shouted and marched in spontaneous parades. But the nature of the V-J Day celebrations varied depending upon whether they took place in cities or towns, on farms, or in the resort areas where many Americans were vacationing that August. Cities and towns indulged in raucous celebrations which the homefront children remembered well. "The most exciting time of my childhood was the morning we received news that the war was over," wrote a girl who lived in Brooklyn. "Everyone ran outside in their pajamas, dancing & hugging each other. . . . Screaming with joy . . . & then came the block parties. . . . I thought the world was going crazy but it was great." Jean Beydler, who had just turned twelve, went to downtown Los Angeles with her parents. It was "going crazy," she wrote, "—cars were not even able to move in the streets and people were everywhere." The joyous soldiers and sailors were "so glad to see a child that I was the belle of the ball. . . . So many sailors were hugging and kissing me and crying at the same time. . . ." Jean came home tired but happy and wearing a sailor's cap. "What a wonderful day—the war was over—I'll never forget it."[2]

Whether in Wausau, Wisconsin, or Talequah, Oklahoma, Brooklyn, Chicago, or Los Angeles, Americans erupted with joy. At the courthouse in Ironton, Ohio, "people were wildly throwing confetti." Davenport, Iowa, was also "in a state of pandemonium," with thousands of people jamming the business district. In amazement, children watched the frenzy of their parents and other adults. In numerous celebrations, people imbibed too deeply and reeled in their drunkenness. One girl returned home to find her mother "being put to bed—drunk! Her friends took her shoes so she wouldn't go out again." For a few boys and girls, the celebrations were loud and upsetting. Neva Jean Kmen, an eight-year-old in Wausau, went downtown with her family for the victory parade. "What a celebration! Confetti and shredded paper showered over us. . . ." Then, in the parade came a pick-up truck attached to the bed of which was a wooden gallows; from it hung an effigy of Japanese war minister Hideki Tojo. People cheered wildly, Neva wrote, but "it frightened me."[3]

Church bells summoned Americans throughout the country to prayer. In Glenshaw, Pennsylvania, just north of Pittsburgh, Roberta Post and her teachers and schoolmates walked the quarter-mile from their school to the Lutheran church "for a time of thanksgiving and celebration." Connie Amaro, a seven-year-old in New Braunfels, Texas, heard the bells of her local Catholic parish and "saw people in our neighborhood running out of their houses to the church to Thank God. . . ."[4]

Celebrations in rural America were different from those in the towns and cities. Marge Marshall, a Kansas farm girl of eleven, heard her mother shouting the news from the front porch and immediately ran barefoot down the driveway, across the road, and into the freshly plowed field where her father was driving his tractor. "The war's over," she shouted over the roar of the engine. Another Kansas farm girl wrote that one of her "happiest childhood memories" was V-J Day, which she celebrated with her grandmother. Sitting on the porch swing, they hugged and clapped their hands. Although rural children missed out on the tumultuous urban festivities, they nevertheless experienced the joy of the war's end. For some, the victory treat was a watermelon; for others, a large bowl of ice cream. A Wisconsin farm girl remembered that on V-J Day her father drove to town and brought back a gallon of ice cream, fulfilling his frequent wartime promise that "when the war was over he would buy us more ice cream than we could eat."[5]

Vacationing children celebrated V-J Day in resorts and family hideaways across the country, from Atlantic City, New Jersey, to Bodega Bay in Northern California. Twelve-year-old Loralee Jenz and her family were traveling by train on their vacation to California and had a ten-hour layover in Kansas City's Union Station. "There were conga lines all through the depot and people were cheering, laughing and crying." While great fun, such an emotional outburst was no substitute for being at home on this historic day. The same was true in Bodega Bay, where the vacationers built a huge bonfire and "laughed and sang till the late hours." A girl who was there enjoyed the festivities, but remembered "wishing that we were home . . . where we could have celebrated . . . with our own friends."[6]

That evening, exhausted from the festivities, children in various places opened the pages of their diaries. "Dear Diary," wrote a Detroit girl, "*THE WAR IS OVER!!* Japs have quit! . . . I still can't believe it's all over!" And in her diary, Janet Sollett, a homefront girl in Elkhart, Indiana, recorded everything that happened on V-J Day and then reflected on the meaning of the atomic bomb:

> The war is over! Completely! Today is V-J Day. . . . Just think, nine days ago people were prepared for a long long war but that was before they discovered the Atom Bomb. It is a bomb about the size of a baseball [,] some say. . . . One of those bombs would completely destroy Elkhart. At least most of it. . . . Japan has unconditionally surrendered. . . . Daddy, Pat, Mother, Grandma, & I went downtown & everybody was going stark, raving, hysterically mad. The street was *jammed*. People were yelling, honking horns . . . Everybody was happy . . . except the loved ones of those 152 Elkhart boys who sacraficed [*sic*] all they had. . . . Tom won't be home. We laugh and cry, we honk horns, yell and throw things, and we go to church and pray. The war is over. All wars are over [,] we hope.[7]

Not surprisingly, the factors of age and gender filtered and refracted the experiences and emotions which the homefront boys and girls had on that glorious day. Of course, the babies and infants were largely oblivious to V-J Day's import, as, indeed, were most preschoolers. But for those school-age children whose political consciousness had awakened with the war, the opposite was true. "I will never forget V-J Day . . . ," wrote a homefront girl. "I was almost eleven years old and the war had been part of my life for most of my conscious memory."[8]

School-age boys enthusiastically welcomed the war's end, as did the girls their age; but there was a significant gender difference. Preadolescent boys shouted, smacked each other on the back, and banged pots and pans, but kissing seemed out of place. For girls nearing adolescence, however, biology and popular culture had conspired to stir romantic day dreams. And it was clear whom many of the homefront girls wanted to kiss on that victorious day: young men in uniform. An eleven-year-old girl in Shelton, Connecticut, hurried downtown to join the crowds on V-J Day. "What excitement! . . . There was singing and shouting, kissing and hugging. I can still remember my disappointment that I was not old enough to be grabbed and kissed." Another girl, a thirteen-year-old celebrating in Chicago's Loop, saw a sailor coming toward her. He was "kissing every female along the way. With alarm, I realized he was looking at me." Although she wanted to be viewed as old enough to be kissed on the lips, she panicked and ducked into the crowd. He followed: "'Hello, Baby.' That was all he said. I could not answer a sound. He gave me a big buss on the mouth and lurched on. . . . It was my first grown-up kiss."[9]

Not all Americans joined in their neighbors' jubilation. After all, more than

400,000 Americans had died in the war. Henrietta Bingham's two older brothers, both Air Corps gunners on bombers, had been killed while still teenagers. Her parent's sadness had greatly aged them, and Henrietta, a thirteen-year-old, was trying to deal with the enormity of her own losses. On V-J Day, she remembered, "my parents just sat & listened to the excitement in the small [Montana] town & the sirens going off & did not even leave the house." Her parents, who never did recover, "lived there until their death." Eight-year-old Vicki Lacount was playing in a park near her home in Minot, North Dakota. "There were shouts of joy, dancing in the streets, clanging of pots and pans. I was not rejoicing. The war never ended for me because my Dad never came home." He had been killed in Europe and was buried in Luxembourg.[10]

Most fathers and older brothers did return, and for their families the end of the war had special meaning, promising to make the family whole again. Anita McCune's father's homecoming was the fulfillment of her dream. She was only four when her father had joined the Marines; in 1945, he was fighting on Iwo Jima, and it seemed to Anita, then seven years old, that he had been gone a long time. "I couldn't remember when he had left home to go to war," Anita wrote, "so it seemed like he had always been there." Still, she never doubted that he would return. "I just figured that someday someone would call and say that he was coming home, and that's exactly what happened. From that day on," she remembered, "I daydreamed about how proud I'd be if I could take my Dad by the hand and walk him through my neighborhood." And then he *was* home, looking "so tall and so wonderful." Wearing his Marine dress blues and spit-shined black shoes, he asked his daughter: "Girl, are you ready for that walk?" Anita had not told him of her daydream, but she quickly took his hand, and together they started down the block. It was Saturday; everyone was at home; and from all the houses came the neighbors "to shake my Dad's hand, and welcome him home." People smiled and grinned "and pumped my Dad's arm.... The kids gathered around us," but they were "quiet and shy for once, and just stared and grinned at us. I grinned back and hung onto my Dad's hand.... It was my very own moment...."[11]

Happily, Anita McCune's dream came true. But if the homefront children's letters are an index of experiences, the dreams of numerous others did not. While the homefront girls and boys remember their fathers' return and its effect on their families and on their own lives, their recollections often are of the negative effects. The problems of reunion were widespread, for nearly one-fifth of the country's families had suffered wartime separation from fathers, sons, and husbands for periods ranging from one to four years. An eight-year-old in Pueblo, Colorado, Judy Fisher shared in the jubilation of V-J Day. "Daddy's coming home! Daddy's coming home! Daddy's coming home!" she shouted as she stood on the sidewalk and stopped everyone who passed. Both her father and grandfather had been construction workers on Wake Island in December 1941, when they were captured by the Japanese and held prisoners under gruesome conditions for thirty-four months. Judy's excitement on V-J Day reflected her fervent wish "to be a part of a family. With ... parents who cared and real birthday parties. ... I knew that when Daddy came home, I would have what I wanted. (I never did.)"[12]

Scholars have investigated myriad aspects of father absence, ranging from studies of Norwegian sailors' families to inquiries into father absence's effects on the development of a child's sexual identity. Some psychologists who have studied America's homefront children have hypothesized that father-absent boys tended to develop cross-sex identification with their mothers. During the war years, there was a fear of growing maternal influence in the lives of children, especially boys. Philip Wylie's *Generation of Vipers*, published in 1942, asserted that "megaloid momworship has gotten completely out of hand" and warned that "Mom" was a monster who emotionally infantilized her children so they would never leave her. The first order of business, wrote Wylie, must be "the conquest of momism, which grew up from male default." Obviously father absence is a crucial developmental factor in a child's life, but with few exceptions, scholarly studies dealing with the wartime father's absence do so almost to the exclusion of the other half of the equation: that is, his return.[13]

For America's veterans, life in 1945 was fundamentally different from what it had been before Pearl Harbor. Hollywood recognized this; the Academy Award–winning best film of 1946 was *The Best Years of Our Lives*, the painful story of the readjustments of three veterans and their families and friends. In addition to those who died in the war, 670,846 Americans suffered war wounds, many of them permanently disabling, while uncounted others bore severe emotional scars. Eighty-three thousand were placed in Veterans Administration hospitals. Yet there was much that was unspoken—in the military, the medical profession, and inside and outside of the family—about the veterans' nightmares and flashbacks to the battlefield; about their alienation and loneliness, depression and withdrawal; their restlessness, agoraphobia, and alcoholism; and about the anger from which these veterans suffered. Medical labels in use at the time resulted not only in misdiagnoses but in mistreatment as well. And such labeling stigmatized the veterans. During the First World War, psychiatrists classified combat soldiers' emotional suffering as a "neurosis" the origins of which preceded the battlefield trauma. Similarly during the Second World War, doctors tended to label nervous breakdowns on the battlefield as "combat fatigue," suggesting that this condition called only for rest and recuperation, not therapeutic intervention.[14]

Even in 1945, however, there was graphic evidence of the trauma of warfare. That year, film director John Huston made *Let There Be Light*, a documentary showing treatment given in an Army hospital "to battle casualties ... of a neuro-psychiatric nature." The War Department, Huston recalled, "told me they wanted a film to show industry that nervous and emotional casualties were not lunatics; because at the time these men weren't getting jobs." Huston used concealed cameras; in recording several men's initial interviews with Army psychiatrists, the film focuses on facial expressions and shows tics, spasms, and uncontrollable weeping. "Every man has his breaking point . . . ," explains the narrator. The veterans' symptoms include speechlessness and stuttering, amnesia, loss of the use of limbs; treatment consists of hypnosis, narco-synthesis, and group therapy. It is gruelling to watch. Down a corridor, a nurse and an orderly help a man who cannot walk; his feet hang limp. "Now what's the trouble?" asks the doctor. "Nervousness," he replies; it had come on suddenly, with "crying spells . . . and I felt something funny in my shoulders here, and my back bothered me, and I started finding things wrong with my legs and arms." Next he was paralyzed, not from a spinal lesion or any organic cause, but from post-traumatic stress. Although in the end some of the other patients "recover," the War Department restricted the film to psychiatrists, saying that it did not want to raise "false hopes for incurable cases." It "was too strong medicine," Huston believed, bringing the public too close to the human toll of war.[15]

It was not until 1980, in response to Vietnam veterans' experiences, that the American Psychiatric Association officially diagnosed the serious psychological disorder afflicting combat veterans—a disorder with numerous secondary physical and emotional symptoms. The psychiatrists called it Post-Traumatic Stress Disorder (PTSD), the essential feature of which is "the development of characteristic symptoms" such as "intense fear, terror, and helplessness" following "a psychologically distressing event that is outside the range of usual human experience. . . ." Among these events are automobile and airplane crashes; violent assaults, including rape; natural disasters such as earthquakes and floods; and combat situations—bombings, firefights, torture—that "involve either a serious threat to one's life or physical integrity . . . or seeing another person who has recently been, or is being, seriously injured or killed as the result of an accident or physical violence. . . ." Finally, these symptoms "involve reexperiencing the traumatic event, avoidance of stimuli associated with the event or numbing of general responsiveness, and increased arousal." From corroboration in the homefront children's letters, it is evident that many of America's returning veterans suffered from PTSD.[16]

In order to understand the centrality of the Second World War to the postwar lives of many of America's homefront girls and boys, it is necessary to re-examine families' experiences in light of the burden of PTSD, which America's veterans brought home with them. At first glance, the frequency of homefront children's letters that detailed troubled reunions when fathers returned was surprising. In fact, these letters outnumber by four to one those that describe happy reunions with fathers, or tell of postwar bonding and warm friendship.

Naturally, the children worried that their fathers might not accept them, or even recognize them. These children had matured, and the oldest of the girls and boys would soon become adolescents. "I worried how he would react when he came back," recalled a homefront girl, who was thirteen in 1945 and had not seen her father in over three years, "because the long, thick braids I had had for years were now cut off and I was a developing young woman—very unsure of myself and afraid he would not accept me with the changes...."[17]

On the other hand, the younger children worried that it would be they who would not recognize their fathers or older brothers. One little girl thought her father was an imposter and expressed amazement that her mother could be so deceived by him. Another girl, seven years old, answered a knock at her front door. There stood a young soldier; he said hello and they "looked at each other.... He kind of reminded me of someone but I could not think of who." Not until her mother ran to the door, threw it open, and began kissing and hugging him did the girl realize that this was her twenty-one-year-old brother, who had enlisted when she was just three years old. Likewise, Cheryl Kolb's father had left when she was three and returned when she was seven. At the end of the war, Cheryl, her sister Val, who was three years older, and her mother traveled to New York City to await the arrival of her father's troop ship. Soon after they had checked into their hotel, they received a call from the lobby; it was her father saying that he was already there and would be right up on the elevator. But when the elevator door opened, "Oh, oh, here was a whole group of men, all in uniform, all with aviator glasses on—I couldn't recognize my own father. I never told him that in all these years!" Val did recognize her father, however, so Cheryl "just followed suit as she gave him hugs & kisses."[18]

At other times, the returning father, though recognizable, was different—and, in the eyes of his children, diminished—from the person he had been upon entering the service two or three years earlier. Joyce Rose remembered that she had just turned four when her father, "a fat man," had left for the Army. Two years later, she and her family went to the railroad station to meet him. "This little white haired man," Joyce wrote, "got off the train and cried in Mom's arms...." Some veterans had lost limbs or suffered other serious injuries, and children awaiting these fathers were apprehensive. Most disturbing, though, were the behavioral changes. "My father came back with ... battle fatigue and flashbacks," wrote a homefront girl; more than once, thinking that he was bailing out of a burning airplane, he tried to jump from his second-story bedroom window. Like veterans of earlier wars, some slept on the floor for months afterward. Children described their fathers as "solemn," "intolerant," "frustrated," "abusive," "very nervous." Especially devastated were those who had been prisoners of war. Judy Fisher's father, who had been a Japanese prisoner for nearly three years, had "no expression and showed no emotion" when they met him at the train station. He suffered from flashbacks, night terrors, alcoholism. "Always distant," he even "seemed afraid to touch us."[19]

Alcoholism was rampant among the returning fathers; indeed, alcohol abuse throughout the postwar population was a significant—though largely unacknowledged—feature of American family life. Born in 1944, Ruth Behrens was a war baby; but "for so much of my life [after the war] all I could remember was his bad nightmares—him screaming [sic] in the night or drunk. Either way," she concluded, "we had no bonding, no love as a girl should have with her father." The homefront children could never be sure that their fathers' alcoholism began during the war, but they suspected it had. "Dad came home a different man," recalled a homefront girl, "he didn't laugh as much and [he] drank a lot. . . ." Other alcoholic fathers abused their families, mentally and physically. "It started with me," wrote the homefront daughter of an alcoholic, "I was never good enough. . . . Punishments became standing in a corner at military attention," in shoes nailed to the floor. The recollection of a homefront boy also was bitter: "My father was a boozer, wife & kid abuser. . . ."[20]

For many veterans the abuse of alcohol showed no surcease over the years. "My dad began to spend his free time at the American Legion Hall with others who shared his experiences. Unfortunately they drank as much as they talked, and this became a problem," leading to her parents' divorce in 1949. For some veterans it never stopped. Forty years after the war, Dr. Arthur Arnold, a psychiatrist with the Veterans Administration, talked of a medic who was on a Pacific island where he and his comrades were assaulted by Japanese troops for three days: "Nearly everybody was killed. He diligently sat there, trying to put little pieces of bodies together to be evacuated. He spent the next forty years drinking his life away, obsessed by this, and being treated as merely an alcoholic. He stayed drunk in order to survive." For some veterans, PTSD's symptoms never disappeared.[21]

But PTSD was not the sole cause of alcoholism in postwar America. Sue Herron, born in 1939, had a different interpretation of the alcohol abuse which she saw in her Rocky River, Ohio, neighborhood. "I remember," she wrote, "[that] the adults worried me. They drank a lot—my parents, my friends' parents, my aunts and uncles. They drank together for fun, and sometimes they drank alone to get drunk. It seemed as if we kids . . . spent a lot of time together so we didn't have to be with adults. And this seemed to be okay with the adults—they were just as happy not having to bother with kids." Lying in bed at night, Sue heard the conversation grow boisterous "after dinner and many drinks. I would hear them talk about who did what in the war, and who weaseled or 'goldbricked' his way out." (Bitterness against civilians who profited financially or professionally during the war was a hallmark of veterans' behavior.) Although her father, a dentist, had served in the Army Air Corps, he had been stationed in Illinois throughout the war, and his family had lived with him in base housing. While her parents had drunk heavily at the officers' club, the war itself had not been traumatic for the Herron family. "The *real* effects of the war, however, could be felt in the decade from 1946–1956." The "pervasive attitude," as she discerned it, went "something like this: 'We've been through the Depression; we've been though the war. . . . We *deserve* a good life now. . . . Let's make some money and enjoy ourselves.'"[22]

If the homefront children's letters can serve as a guide to the postwar history of veterans' families, alcoholism was a major cause of the upward spiral of wartime and postwar divorces. In 1942, the first year of American involvement in the war, 321,000 couples were divorced. By 1945, that total had reached 485,000; and in 1946, the first full year of peace, it soared to 610,000. By 1950, a million veterans had been divorced. Of the score of homefront girls and boys who wrote of their parents' divorces, one half ascribed the breakup to the chronic drinking of one or both parents. A somewhat smaller group—40 percent—attributed divorce to adultery, but many also indicted heavy alcohol consumption as coterminous. And it is clear that extramarital affairs, or suspicions of such, demoralized both wives and husbands and destroyed their marriages. Trust dissolved and depression often took its place. That the age-old double standard prevailed was not surprising. "Some Wives Hurt Soldiers' Morale," stated an article in the *New York Times*, which, while condemning the unfaithful wives, did not comment on how wives should deal with their husbands' infidelities. One homefront girl reported that when her mother learned of her father's affairs while in the service, it "triggered a long and deep depression. . . . My mother's escape was sleep and an ever deepening agoraphobia. . . ." The war devastated families. A homefront boy recalled that when his father joined the Air Corps and his mother moved to New York for a defense job, he stayed with his grandmother in West Virginia. The boy saw his father only twice in four years, while his mother, "a country girl basically," had "a number of male acquaintances, mostly military. At the end of the war my parents divorced, and I became a pioneer among the 'Latchkey Kids.'"[23]

Father's return unsettled the homefront children in a variety of ways. Some children feared their fathers would not stay. "I wasn't sure he wouldn't leave again," a homefront girl recalled, "and I didn't want to become too attached." Some were bitter that their fathers had left in the first place. But others, especially preschoolers, hid in fear; they were frightened of these strangers. A four-year-old watched "the stranger with the big white teeth" come toward her and as he did, she "ran upstairs in terror & hid under a bed."[24]

In addition to fright and fear, some homefront boys and girls resented their fathers for disrupting their lives. "I have often wished," wrote a homefront girl born in 1943, "that my life had not been touched by World War II." For her first three years, she lived with her mother and maternal grandparents: "My time before my father returned was very peaceful and protected. . . ." She heard that, before the war, her father had been "kind, sensitive and gentle"; but when he returned, he was "troubled and violent." Another homefront girl remembered feeling "particularly connected" to her mother during her father's absence: "I felt her great sadness and worried about her when she cried." Also in her father's absence, the girl's older brothers were very protective of her. "We were a tight, tough unit," she recalled. "In some ways, I resented my father's return to the family. Everything changed!"[25]

Jealousy ran both ways, and homefront children resented fathers for displacing them. Oedipal problems surfaced when children were evicted from mother's bed to make room for father. Billy was three and a half when his father left for the service. According to a psychiatrist who observed Billy's family, the boy "expressed freely in play and in fantasy his unconscious gratification over the relief that the father, his rival, had disappeared." His father served overseas for three years, and during this period Billy's fantasies subsided. The "father did not exist again in his reality until the emotions of the family were stirred up by the anticipation of his return. The little boy awaited the father's return with great anxiety," as though he were coming back from the dead. The boy, who was nearly seven when his father returned, withdrew from him. Billy was "an active little boy, whose masculine, assertive tendencies had had occasion to be stimulated; he found outlets in school and in the yard, but mostly in the yards of neighbors, since he avoided his father shyly."[26]

The homefront girls who awaited their fathers' homecoming did so with different sets of fantasies. A girl who was four when her father had left fantasized that he when he did come home, he would "return to her alone." It was "a deep disappointment," she remembered, "that her father, upon his return, treated her as a child, that he had really come home to the wife and mother." Not all the homefront daughters embraced their returning fathers, however; girls also expressed resentment at the strangers who had re-entered their lives to change them forever. "The man I could not remember but who had been referred to as 'Dad' was coming home," wrote a homefront girl who was six when he returned. "And, I, displaced from my mother's bedroom by this stranger, hid in a closet."[27]

Little children were not the only ones facing adjustment challenges; school-age children too—those homefront girls and boys born before Pearl Harbor—also suffered as a result of fathers' return. George R. Bach, a psychologist who studied children from six to ten years old, found that the homefront child whose soldier father was away constructed an idealized conception of "Daddy," due, in part, to the mother's glowing accounts of the uniformed man whose photograph adorned the mantle. Yet, as Bach understood, "The stereotyped, idealistic fantasy picture . . . may initially be a handicap in the re-establishment of a realistic father-child relationship." Not only would there be disappointment, but also resentment and anger, as "the father's resumption of domination and authority would certainly come in conflict with the child's idealistic expectations."[28]

Studies of father's perceptions of their familial roles suggest that many returning fathers believed that their two most important functions, in addition to being the breadwinner, were to discipline the children and to enforce what they considered to be appropriate sex-role behavior. For fathers coming home after prolonged absence in a military environment in which they either gave orders, or promptly obeyed them, there seemed little doubt that their sons and daughters should respond to them as buck privates had to first sergeants. Often there was a problem, however. For example, homefront children who had lived with their grandparents during the war enjoyed being pampered. "We lived in my grandparents' house," wrote a homefront girl who was two years old when her father enlisted in the Navy, "& life went on without him—I think I forgot him." In a variation on a family drama that played itself out frequently during the war, Pam Geyer, born in January 1944, reported seeing her father "only once before he left—for one day—and returned in October 1945." Meanwhile, Pam, her mother, and two aunts moved back home, to the family farm near Beverly, Kansas. "For me," Pam, wrote, "that was a happy time, because.... I was the first grandchild and got lots of attention. I was their hope in a desperate world." Then her father returned, determined to fulfill his paternal role and impose discipline, and "it turned my world upside down.... My dad and I were never close when I grew up."[29]

No doubt, the homefront children challenged the authority of returning fathers. Sandy Newton, born in 1943, was only three weeks old when her father, a fighter pilot, left for overseas. Shot down and captured during the invasion of Sicily in 1943, her father escaped from a prison train and reached asylum in Switzerland. Meanwhile, Sandy and her mother moved in with her paternal grandparents. But when her father returned in 1945, "he was totally appalled at my lack of discipline.... Suddenly, I was no longer a princessly 'half orphan' but a spoiled brat! He scolded, I cried, Mom and Grandma stood up for me.... There were numerous tears and quite a power play.... Many years later, Mom told me what a rocky time that was for all concerned." Jacqueline Dowd Hall, the historian, also was two years old when her father returned. Convinced that "I was a spoiled brat," her father tried to force her to eat her food and not to cry. "It quickly became—and remained—a test of wills." He just could not understand," Jackie recalled, "that little children don't understand. He thought you could just tell them to behave.... There would always be some crisis over me. I wouldn't eat all my food or something. He would leave. I would be crying. My mother would be crying. We would lie down on the bed together and cry."[30]

Some of these children never really got to know their fathers, especially if after the war another baby was born with whom the father developed a genuine friendship. A clinical psychologist, Arthur L. Rautman, has written that in such wartime cases, "the family circle will then consist of the father, the mother, and the new baby, with no place for the war baby." The war child would become "an extra-familial child." Such children, Rautman warned, were in danger of becoming "strangers in their own families, often overworked as kitchen maids or choreboys, called upon to raise ... younger brothers and sisters."[31]

Readjustment was obviously difficult for both the homefront children and their fathers, many of whom had trouble acknowledging the extent to which their families had coped in their absence. Exemplifying these problems were America's "war-born children," born while their fathers were overseas. Anticipating postwar difficulties, *Parents' Magazine* sent free issues to fathers serving abroad who had never seen their young children. In these families, psychologist Lois Meek Stolz explained, there was greater familial calm while the father was gone than there was after he had returned. According to Stolz, these children "showed less tendency to aggression than the children whose fathers were at home," since home life had been quieter and less punitive with mother in charge of decorum and discipline.[32]

In addition to the fractious relationships provoked by fathers' unrelenting need to impose discipline, another adjustment challenge was posed by the presence of an illegitimate child, or one thought to be illegitimate. Illegitimacy jumped during the war, although one could not ascertain this from governmental vital statistics, which cited 103,000 illegitimate births in 1940 and 128,000 in 1945; unofficially, the total was much higher. Fathers returned to children of whom they had been totally unaware; sometimes, these children did not resemble their older siblings. In one case, a young child was much stockier than his small-boned older siblings. In another family, the child had facial features (for example, his cheekbones) that differed noticeably from his sister and brother born before the war. Even if the child were legitimate, there could be a problem. Arthur Rautman, who worked with veterans, reported that he found that a "recurring difficulty . . . when a veteran returned to his war-bride . . . was a tormenting doubt that the child he found with her was indeed his own offspring."[33]

Moral judgments about illegitimacy were harsh in the 1940s. So too were the accompanying terms of disparagement for both the mother and her "bastard" child. But as the family history of the Second World War repeatedly makes clear, nothing was quite so simple as it appeared. War was an ambivalent time, filled with unprecedented life-or-death exigencies as well as with moral conflict. Margaret Sams's *Forbidden Family: A Wartime Memoir of the Philippines, 1941–1945* tells her story of internal conflict. In 1941 Margaret, her husband Bob Sherk, and their three-year-old son David lived in Manila. Not unexpectedly, the attack on Pearl Harbor changed their lives. Bob joined the United States Army, but was captured by the Japanese; he survived the Bataan Death March only to be sent to Cabanatuan, a "death camp" known for its atrocities. Meanwhile, Margaret and David, along with 5000 other people (mostly American and British) were interned in Santo Tomas, a Japanese camp for civilians in Manila.[34]

In the Santo Tomas prison camp, Margaret met Jerry Sams, another American. She was twenty-six, he was thirty-one. Inventive and courageous, Jerry accomplished feats that became myths in the camp; outwitting his captors, Margaret wrote, he did "EVERYTHING that the Japanese had strictly forbidden." The first time she saw Jerry, she thought, "Whew—what a good looking man!" Jerry was kind to Margaret and her son, and he did "wonderful things" for them, such as making David a little tin pail with his name painted on the side, for carrying his food from the mess line. Although Margaret and Jerry started eating meals together, she did not permit herself to think of romance. Brought up "to think of divorce as a work of the devil," she considered herself a "model wife." Still, she could not deny her feelings, and after he had given her a surprise birthday party—again defying camp rules—she realized that she had fallen "completely, irrevocably in love." At just this juncture, the Japanese announced that Jerry and

800 other men were to be transferred to another camp, Los Banos. With that news, Margaret wrote, "My philosophy had become: TODAY." They made love, and then he was gone. When she learned that she was pregnant, she decided that she would have the baby. "Believe me," she wrote, "we who have decided to have our 'love children' give up the world." Margaret chose to risk her reputation, her ties with her family, and, being pregnant in an unhealthy prison camp, even her life. And in January 1944 she gave birth to her daughter Gerry Ann.[35]

Margaret earlier had written Bob and told him about both Jerry and her pregnancy. She did not hear from Bob for months, but when she did, it was a love letter. He said that he would help to raise the child as if it were his own and that he did not want a divorce. The Japanese made the next decision in Margaret's life; they asked for volunteers to go to Los Banos. Gerry Ann was just three months old, but Margaret knew that she and the children had to go. When Margaret and the children arrived by truck at their new camp, she did not see Jerry at first, "when suddenly, she looked down and there he was at the side of the truck, grinning and holding up his arms. . . ." Finally, in February 1945, American troops liberated the prison camp. Soon afterward, Margaret saw a man who had also been a prisoner of war at Santo Tomas; he took her aside and told her that Bob had been killed on a Japanese transport ship, which—with 1600 Americans aboard—had been sunk by bombs from American airplanes. "A prisoner of war for three years," Margaret wrote, "and then to be killed by our own Americans just a month before he would have been liberated. How could there be any justice in the world. . . . That I had given him a death blow the year before will always be my cross to bear. . . . My conscience will always grill me, no matter how devotedly I love Jerry."[36]

On January 26, 1946, Margaret and Jerry were married in Virginia; but the anniversary they celebrate is September 13, the day they met in 1942. Two other children were born after the war, one in 1946, the other in 1954. The war had turned Margaret and Jerry's lives upside down, but they offered "no apologies or excuses for our actions." Instead, Margaret believed that it was important to tell their story. "First, we think it is a love story deserving the name," she wrote. "Second, it may help our children judge us a little more dispassionately when the time comes."[37]

The fathers who returned from battlefields around the globe worried about the future, and many arrived home not knowing where they would fit into the familial scheme. "Domestic life, so longed for in foxholes," wrote two child development experts in 1946, "may seem in the first difficult days to be a woman-dominated, intricate array of trivialities in which a man has neither the wish nor the power to find a real place. The pattern of life for his wife and children seems to be complete without him. . . ."[38]

Some fathers, however, were determined to make a place for themselves, and this they did by becoming the master disciplinarians. The children would obey, they thundered, or else. Naturally, some of these boys and girls were fearful of their fathers, while others were also jealous of and angry at these strangers and intruders. One would think that these hazards alone would challenge any father-child relationship. But there was still one more obstacle—one exemplified by the homecoming father who railed that his son had become a "sissy," surrounded as he had been by women and in the absence of a strong male role model.[39]

Fathers, possibly fearing homosexuality in themselves as well as in their sons, did not want their boys to cross the boundaries of gender and identify too closely with their mothers' "feminine qualities." And fathers were not alone in expressing this concern. In a 1942 article, for example, *Hygeia*, a magazine on health care, warned its readers of "The Menace of the Maternal Father."[40]

Moreover, in the soldier-father's absence, cross-sex identification was a possibility for the boys; in fact, it often made sense. Consider the following paradigmatic story, from an observational study done by a sociologist, of a girl and a boy playing house in a front yard:

> The little girl was very busy sweeping up the play area, rearranging furniture, moving dishes about, and caring for baby dolls. The boy, on the other hand, would leave the play area on his tricycle, disappear to the back of the ... house, remain for a brief while, reappear in the play area, and lie down in a feigned sleep. The little girl had a rather extensive knowledge of the mother role, but, for the boy, a father was one who disappeared, reappeared, and slept, *ad infinitum!*[41]

Now if the homefront boy realized what he was missing, he would decide to park his tricycle, and come inside the house to join his mother, enjoy her company, help out with the household tasks, and possibly be rewarded with milk and cookies. In wartime, other females were likely to be there as well, including sisters and grandmothers. During the war numerous homefront boys lived in "families of women," and in a little boy's life especially, this situation could lead to cross-sex identification. And there is evidence from the homefront that it did so.

Some of this evidence exists in the field of developmental psychology. At the outset, it is important to note that these studies focus only on the homefront boys. Conducted fifteen years after the war, this research analyzed scores on Scholastic Aptitude Tests taken by the father-absent boys. In one article, psychologist Lyn Carlsmith studied men of the Harvard classes of 1963 and 1964, who had been born between 1941 and 1945. Her father-absent group consisted of students whose fathers had gone away before their sons were six months old and had stayed away for at least two years. A matched sample consisted of students whose fathers were civilians and had remained at home.[42]

Carlsmith cited studies of math and verbal aptitudes which demonstrated that, on standardized tests, "females are generally superior to males in Verbal areas, while males are superior to females in quantitative pursuits, particularly numerical reasoning." These studies suggested that there were two styles of conceptualization: "an 'analytical approach' which is characterized by clear discrimination between stimuli, a direct pursuit of solutions, and a disregard for extraneous material; a 'global approach,' characterized by less clear discrimination of stimuli and a greater influence from extraneous material. The first approach," Carlsmith wrote, "is more typically used by boys while the second is more typical of girls."[43]

More recently, psychologists have focused again on gender differences in analytical styles. Notably, Nancy Chodorow, a sociologist and analyst, and Carol Gilligan, a cognitive psychologist, have argued that while girls derive their basic sense of identity from experiencing themselves as like their mothers, boys develop a male-gender identity that not only depends upon seeing themselves as different from their mothers, but also requires separation from them. Throughout childhood, then, girls tend to emulate their mothers' behavior and identify with the same-sex parent who was their first love object. But since the boy separates from his mother while the girl experiences herself relationally, explains Chodorow, "The basic feminine sense of self is connected to the world, the basic masculine sense of self is separate."[44]

Viewed through these lenses, Carlsmith's findings take on fresh meaning. Accordingly, it is noteworthy that Carlsmith found that the father-present boys achieved predictably higher math than verbal SAT scores, but that the father-absent boys achieved the opposite result. Their scores were similar to those scored by girls. Moreover, Carlsmith found that "the relative superiority of Verbal to Math aptitude increases steadily the longer the father is absent and the younger the child is when the father left." Finally, the boys' career goals were different. The father-absent boys were ten times more likely than the father-present boys to aspire to a career in social work, and twice as likely to want to become a psychologist. The father-present boys, on the other hand, were three to four times as likely to express interest in careers in mathematics and science, and twice as likely to want to become doctors.[45]

Carlsmith's research goal was to "provide evidence relevant to any general theory of identification by showing certain strong effects of father absence at various ages." She argued that these theories "usually agree on two points: for the boy to identify successfully with the father, the father must be present during at least some portion of the boy's childhood; development of an appropriate masculine identity or self-concept is predicated upon the success of this early identification with the father." But in the absence of the male role model provided by the father, the boy identifies with the dominant role model in the family, that is, the person who merits the highest status: the mother.[46]

One finding on which the psychologists, not to mention the psychoanalysts, agree is that the age or stage of development is the crucial factor mediating the effects of father absence. Ross Parke, the psychologist, has described these findings: "Boys who had been separated from their fathers before the age of 5 were more dependent on their peers and less assertive. They played fewer rough physical contact sports. . . . Instead they chose reading, drawing, or working on puzzles. . . . In contrast, if the father was available until 6 years of age, his later departure did not have this effect." These boys, Parke explained, "behaved in these areas the same as boys raised in intact homes."[47]

Given recurrent lamentations about the "erosion of masculine authority" in American society, it is not surprising that the topic of father absence provokes speculation about the origins of homosexuality.[48] Does cross-sex identification result in homosexuality? No one really knows. The answer might reside not in behavioral development, but in biology.[49]

Homefront children who are gay usually ascribe their homosexuality neither to father absence nor the war, but to an agonizing process of self-discovery spurred by the gay rights movement. While it is possible—and indeed likely—that a substantial minority of the homefront children are homosexuals, father absence is not necessarily the explanation. Perhaps the answer is that more men and women from the homefront cohort than from earlier generations have openly acknowledged their sexual orientation. Perhaps the answer has less to do with father absence than with political awakening, inspired by events such as the Stonewall riot of 1969. In the 1950s and 1960s, homefront children who came of age as homosexuals feared that disclosure would cause them to lose not only their jobs but even their families and friends. But in 1969, with the homefront children then in their twenties and thirties, that began to change. In New York City, a riot erupted between police and the patrons of the Stonewall Inn, a gay bar in Greenwich Village. Police who raided the inn expecting little resistance were unprepared for the volley of beer bottles that greeted them. Rioting continued well into the night, and graffiti calling for "Gay Power" appeared. As historian John D'Emilio has written: "Stonewall thus marked a critical divide in the politics and consciousness of homosexuals and lesbians. A small, thinly spread reform effort suddenly grew into a large, grass-roots movement for liberation. The quality of gay life in America was permanently altered as a furtive subculture moved aggressively into the open."[50]

Having considered the homefront boys, one wonders about the homefront girls. Having discussed the little boy on the tricycle who decides to come inside and join his mother, one wonders about his sister who was already in the kitchen with her mother. How did father absence affect daughters? Did the father-absent daughter engage in cross-sex identification with an idealized father? If boys identified with the "feminine" side of mother, did the father-absent girls tend to identify more with her allegedly "masculine" side: independent, self-possessed, and capably running the family on her own? No one has definitive answers.

For guidance perhaps there are only the homefront girl's individual experiences. Father absence proved liberating for one homefront girl, Patricia Wide.

First, she wrote, there were "no male role models" to shape her expectations and behavior. Second, "I probably would not have been allowed so much personal freedom to roam the library, the city, the fields, neighborhoods, woods, and have a paper route . . . if my father had been home." Looking back, she concluded that her homefront years had prepared her "even better for life. . . . Now I know there is not a whole lot I can't do, even as a woman."[51]

This tidbit is helpful, but there is little else to go on. There is, for example, only one contemporary study involving the homefront girls. Done in early 1945, this research focused on doll play as an indicator of aggression. The children were three, four, and five years old, and all were enrolled in child-care centers. Half were girls, half boys; half of the children lived in father-present households, the other half in father-absent families, usually with the father serving in the armed services. "Boys from father-absent homes," the study concluded, "portrayed much less fantasy aggression than boys from father-present homes." But the reverse was true for girls, with those from father-absent homes portraying slightly "more aggression than girls from father-present homes."[52]

While postwar research done on daughters offers confusing findings about father absence, it does provide retrospective insights into another issue: the war's impact on the daughters of working mothers.[53] A 1982 study, for example, concluded that "girls whose mothers work (contrary to the feminine sex-role stereotype) develop less traditionally feminine sex roles themselves."[54] Was this also true for the daughters of "Rosie the Riveter"? Did some—or even many—become activists in the rebirth of feminism in the 1960s and 1970s? We now know that many homefront girls, as well as boys, believed their mothers were the real wartime heroes. Mothers' wartime competency both in the home and on the job inspired homefront girls to be independent in their own lives. "My mother had become a very independent woman" and she remained that way, wrote one girl; "If there had been no War, perhaps she would not have worked. . . ." Another girl expressed pride in her mother's wartime contributions, but sadness in her retreat back into the domestic sphere afterwards. "She was the sole provider & decision maker at that time, but had to take a back seat to my Dad when he came home." As the historian Susan Hartmann has observed, this veteran's wife was not alone. Whatever difficulties attended the reunion of a veteran and his spouse, the experts all assigned to women "the crucial responsibility for solving this major postwar problem," and "to do so in terms of traditional female roles." As an article in the *Ladies' Home Journal* asked: "Has Your Husband Come Home to the Right Woman?"[55]

In 1970, it seemed that historian Gerda Lerner had the homefront girls in mind when she wrote: "It was left to the college-age daughters born of the World War II generation to furnish the womanpower for the new feminist revolution. . . . They felt personally cheated by the unfulfilled promises of legal and economic equality." Many were baby boomers, but many too were born in the 1930s and during the war years. For them, homefront events involving father's absence and mother's working first awakened the questioning of gender roles.[56]

But for other girls, there was a downside. One homefront girl, in assessing the effects of her fathers' death in the war, described later difficulties in maintaining relationships with men. Another wrote that she had searched for "father figures" and now realized "how during most of my life I have sought the approval of men to make up for the loss of my father." She endured a philandering husband for several years "before I could admit that I could live without a man." In some ways, even greater problems existed for girls and young women whose fathers returned home only to treat them with disdain, contempt, and anger, sometimes contributing to divorce. Studies show that while the daughters of widowed mothers tend to idealize their fathers, the daughters of divorced mothers have negative views of them. Several homefront daughters confided that to escape their fathers they had left home at a relatively early age and married men who were twelve to fifteen years older. "I got pregnant," one wrote, "then married a man 12 years older than myself—right out of high school—1960." Separated in 1970, they divorced two years later, and since then, she wrote, she had been "in and out of several relationships." In psychotherapy, she learned, "The biggest problem [was] my relationship (or lack thereof) with my father. . . ." Another woman's experience had similar origins, but a different ending. Her father suffered from alcoholism, "flashbacks," and other symptoms of Post-Traumatic Stress Disorder. She left home and "married a war Hero 15 years my senior & already the father of 4 children. After 32 years & 2 more children, we are still married."[57]

For most homefront girls whose fathers did not make it home, there were only photographs and perhaps dim memories. An exception was Marion Daly Cipolara, who remembered her father well. He was a seaman, and she was twelve when his ship was sunk by a German submarine in the North Atlantic. From then on, Marion's mission was to succeed. She worked for three years after high school to earn enough money for college; after raising four children, she has earned a master of arts degree and is writing her doctoral dissertation. ". . . I'm still thinking of making Dad proud of me," she wrote.[58]

The Second World War thus shaped America's homefront girls and boys in varied ways. And even afterward, its influence continued to be evident in individuals' lives, whether the developmental issue was sexual orientation or sex roles; whether the personal issue was marriage versus a career; or whether the occupational issue was to become a psychotherapist instead of a surgeon.

There were indirect political outcomes from the war as well—for example, its democratic message spurred the civil rights movement for African Americans, which in turn inspired other struggles for equality. At the same time, the war retarded change. It reinforced sex-typing not only in children's war games but also in their radio shows, comic books, and Saturday matinees. Adventure, patriotism, and defined gender roles went hand in hand.

Father absence and the tragedy of a father's death had to have been the most significant influences in a homefront child's life, but even these were different experiences for girls than they were for boys. Indeed, in ways both subtle and obvious, gender neatly divided the homefront children, and, in so doing, it provided boys with wartime experiences often substantively different from those which girls had. But most children paid little attention. "In those days," reminded a homefront girl, "'Boys were boys' and 'girls were girls.'"[59]

CHAPTER 13

Confronting War's Enormity,

Praising Its Glory

T HE SECOND WORLD WAR transformed not only the United States but the entire world. In the United States, many of the homefront children contended with the war's enormity. First, by war's end, six million Jews had been forced into concentration camps and worked to death or systematically killed by firing squads, in gas chambers, or in medical experiments. The Nazis also exterminated 250,000 Gypsies, and 60,000 men who had been convicted of homosexuality ended up in concentration camps where they too perished. The second enormity was the atomic bomb, which the United States dropped on Hiroshima on August 6, 1945. It deepened the psychic numbing that accompanied people's knowledge that governments and their citizens were capable of previously unimagined brutality. The Hiroshima bomb immediately leveled the central city; soon after, radiation fell from the sky in oily black rain. Seventy thousand people died that day, and thousands more died from radiation sickness in the following weeks. The toll by 1950 was 200,000. "One day, one bomb," the writer Jim Miller noted. "And three days later it happened all over again—this time in Nagasaki."[1]

There is no way to measure the psychological impact of these events on the homefront children, but it had to have been considerable. "Probably," wrote Norman Mailer, "we will never be able to determine the psychic havoc of the concentration camps and the atom bomb upon the unconscious mind of almost everyone alive in these years. For the first time in civilized history . . . ," he wrote, "we have been forced to live with the suppressed knowledge that the smallest facets of our personality . . . could mean . . . that we might . . . die as a cipher in some vast statistical operation in which our teeth would be counted, and our hair would be saved, but our death itself would be unknown, unhonored, and unremarked . . . in a gas chamber or a radioactive city. . . ." More than that, the "Second World War presented a mirror to the human condition which blinded anyone who looked into it." For, Mailer continued, if death camps and massively destructive weapons comprised the policies of governments "one was then obliged to see that no matter how crippled and perverted an image of man was the society he had created, it was nonetheless his creation . . . and if society was so murderous, then who could ignore the most hideous of questions about his own nature?"[2]

The Holocaust introduced America's homefront children to the depths of evil. In 1945, sitting in a darkened theater, a homefront girl saw a newsreel about the death camps: "I left the theater totally stunned as to how human beings could do such things to other human beings." Another homefront girl wrote that "those images have haunted me my whole life. Sometimes you would see stacks of bodies, and I would look at the newsreel through parted fingers, not wanting to see, but unable to look away." Magazines and newspapers also printed photographs of the Holocaust. One girl remembered that these pictures "horrified and terrorized" her, and that afterward she did not want to eat and could not sleep. Indeed, numerous children were unable to rid their minds of the horrible images.[3]

The responses of gentiles to the Holocaust ranged from total ignorance to profound anguish. "We were never told about the Jews," wrote a homefront girl, "and my father never believed the Holocaust existed!" In addition to anti-Semitism, there was parental protectiveness; "I guess Mom wanted to protect us . . . ," recalled a homefront girl. Another remembered that it was not only her family that failed to tell her of the Holocaust, but also her neighbors, teachers, and minister.[4]

Many other gentile children felt deep guilt. For a girl whose parents were Austrian immigrants, the Holocaust was a personal burden: "When I think of the Holocaust I feel ashamed and want to cry." Freda Jones, born in Chicago in 1936, was in the fourth grade when her teacher told her that the Germans had "ignored what was taking place around them. . . ." As an African American, Freda Jones "knew the German people hated Black people," but it was the Holocaust that seared her conscience. She could not sit through the newsreels then, and cannot even sit through that topic "in the movies of today. . . . I feel as if I was in those concentration camps." Freda Jones added a postscript to her letter: "Maybe that is why I feel so comfortable working at a Jewish Agency [the Jewish Vocational Service in Kansas City, Missouri] for the past 10 years—it's like I'm home."[5]

As for the atomic bomb, certainly few American children expressed opposition to dropping the bomb. As far as they were concerned, it had to be done to win the war. "The bomb saved my brother's life . . . ," wrote a homefront girl. Another child's father, a doctor stationed in the Pacific, told her that "many more people would have died if we had not done that. . . ."[6]

But while America's homefront children accepted the necessity of using the atomic bomb, they nonetheless felt terror when they contemplated their own deaths in a nuclear war. "Packaged Devastation," a November 1945 article in *Boys' Life* called it. A homefront boy recalled that "fear, bewilderment, and confusion swept the minds of grade schoolers when they heard the news." Moreover, once they saw newsreels of the bomb's explosion, a new horror entered their lives. "Every newsreel," wrote a homefront girl, "showed the mushroom cloud over and over," and *Life*'s photos were also "imprinted in my mind's eye forever." For her the message was clear: "THERE IS NO PLACE TO HIDE."[7]

The war's effects did not cease with the Japanese surrender. "For years afterwards," wrote a homefront girl, "the nightmare" of the Holocaust "persisted in news stories," while the air raid drills continued in her school. Children remembered lying in bed, worrying about the bomb. "The Bomb" was also a theme in films and popular music; "Atom bomb gonna' get you," was a line from one postwar song.[8]

Psychiatrist Lenore Terr was nine years old in 1945. One Saturday she saw a newsreel of the United States Army entering Hiroshima. "They were wearing space suits not unlike what the actors wore on *Green Lantern* and other Saturday serials. . . . What got to me was a shadow." On a foot bridge at ground zero, which otherwise had been bleached of all color, lay the shadow of a man who had been walking there. The atomic bomb had "vaporized" him, explained the announcer. "I took it all in. And I understood what I saw. . . . The shadow still lives today in my mind," Terr has written in the preface to *Too Scared to Cry*, her book about psychic trauma in childhood. Even at age nine, she remembered, "I recognized a psychological symptom when I had one. . . ." From then on, whenever she was awakened by a sudden noise, or by a light going on in the middle of the night, her heart would begin pounding and she would say to herself, "This is it. The bomb." And she would wait to be vaporized. *Too Scared to Cry* is Tarr's study of the twenty-six Chowchilla, California, children who in 1976 were kidnapped from their school bus and buried alive in a hole; all survived physically, but they have suffered degrees of PTSD. As a psychiatrist treating children, Terr was naturally drawn to this subject. But, she added, she also "was impelled to the book as a fourth grader at a Saturday afternoon movie matinee. Some books take a while in the writing. This is one of these books."[9]

The Holocaust and the atomic bomb were historical watersheds. A watershed of a different kind resulted from the nearly total mobilization of all Americans and their institutions during the Second World War. The homefront children had united with adults behind the war effort, planting victory gardens and collecting scrap iron and old newspapers. But more than unity and enthusiasm were required to win the war. Essential to victory was the mobilization of all sectors of society and the economy—industry, finance, agriculture, and labor. The federal government had the monumental task of coordinating these several elements, as well as new ones: science and technology.

The United States underwent profound change during the war, which was a time of economic growth and consolidation. The country's big businesses got even bigger, while hundreds of thousands of neighborhood enterprises ceased to function. Similarly, the average farm grew in size as agricultural equipment replaced the millions of people migrating to urban areas. Membership in labor unions increased dramatically, as did the size of the budget and bureaucracy of the federal government. In the next few years, government agencies that had been conceived as temporary would become permanent and grow in size and influence—the Department of Defense (consolidating the War and Navy departments), for example, and the Central Intelligence Agency (succeeding the Office of Strategic Services). Moreover, the seeds of the military-industrial complex were sown during the war. On the other hand, the mobilization provided opportunities to Americans for employment, education, and training, and, most of all, for elevation to the middle classes. For better or worse, then—and clearly there were elements of both—the Second World War was a turning point for America and its homefront children.

The homefront children are of many minds about the war's impact on their lives. One group, which includes those children whose fathers stayed at home, views the war years as personally uneventful and, in that sense, "normal." "I'm not sure," conceded one homefront girl, "that . . . my life was much different than it would have been without the war." And a homefront girl of six asked after the Pearl Harbor attack: "Is this a war? It doesn't feel like a war. It's like peace to me. Nobody's doing anything to me."[10]

A second group, which was larger, remembers the war as a somber time, a period of worry, of sadness, and of "scary days." A Jewish-American boy wrote that he was constantly fearful of enemy bombs or even an invasion. Being aware of the Nazis' special hatred, he added, "certainly compounded whatever paranoia a child of the times could be expected to have. . . ." He felt "cheated out of whatever security a peacetime childhood affords." Other homefront children agreed that they too had grown up too fast, having lost "years that should have been happy and carefree." Moreover, in some families suffering downward mobility, the children remembered being hungry and cold during the war. A Tennessee farm girl, whose father was a private in the Army, recalled "a hard cruel life of poverty, of two frightened women," her mother and grandmother, "trying to care for three hungry children with no money to do it, of going to bed hungry. . . ."[11]

Largest by far, however, is a third group composed of homefront girls and boys who remembered the war as a time of rewarding adventures. "Terribly poor. Terribly frightened. Terribly happy," wrote a homefront girl. Another wrote that even though her father was a Marine wounded at Okinawa, she had happy homefront memories of playing war games, chanting "jump rope ditties" about Hitler and Mussolini, and singing the popular songs of the day. Similarly, a girl whose father had not gone to war recalled that her war years were "a wonderful, warm time to grow up in." Just as enthusiastic was another girl: "I feel guilty saying this, but those years were probably the best years of my life."[12]

What the homefront children warmly recalled was the patriotism and national unity evident in the pursuit of victory. Repeatedly, they commended the homefront Americans for their "cohesiveness," their "pulling together in body, mind, and spirit." "We had a goal—victory," wrote a homefront boy, "[and] we worked towards it." "My personal feeling," explained another homefront boy, "about the way Americans joined together after Dec. 7 and put their combined efforts into protecting American soil is one of pride."[13]

For numerous children, the war's most important offshoot was full employment. After a decade of depression, with a jobless rate that as late as 1940 still exceeded 20 percent, a job was a welcome commodity. But during the war, employment abounded; in some families both parents worked, and they worked overtime as well. "Our family had money," wrote a homefront boy who, born a "Depression baby" in Cleveland in 1933, remembered years of poverty before the war. Sudden prosperity in the early 1940s changed the course of American family history. Personal savings in the United States sextupled during the war boom, jumping from $6.3 billion in 1940 to over $37 billion in 1945. For one thing, there were no new houses to buy, since none was being constructed. The same was true of automobiles. "My parents saved a lot of money during the war . . . ," wrote a homefront girl, "—some of it because we couldn't find a decent place to live. . . ." Once the war was over, American families looked forward to using their savings to embrace their dreams. Houses and cars topped their lists. But their most fervent hope for peacetime was that their upward mobility would become permanent.[14]

As the homefront girls and boys entered peacetime, age remained a significant variable, dividing the younger children from the older ones. The war babies, who were still in early childhood in 1945, have few clear memories of wartime. On the other hand, the Depression-born children were grade-schoolers in 1945, and most of their childhood memories are of the homefront years. Born in 1934, a homefront girl wrote that it was "difficult to remember a time when we were not at war." She was seven when the war began at Pearl Harbor, eleven when it ended: "It seemed that it had always been with us and would always be with us. . . . The War was the eternal backdrop of the routine of our lives." Likewise, for children born later in the 1930s, the war also subsumed all else. "WWII was my childhood," wrote a homefront boy born in 1937. "There was talk of little else, then," remembered a homefront girl born in 1939.[15]

In myriad ways, the homefront girls and boys have had to confront the enormity of the Second World War. The shadow cast by the war did not lift with victory in 1945, and its imprint is still visible even fifty years after Pearl Harbor. The war bestowed both special strengths and special vulnerabilities on the homefront children. And as their lives have evolved over the next five decades—from adolescence though adulthood and into middle age—they have struggled to balance the verities taught by the war against the realities of a changing world.

CHAPTER 14

Age, Culture, and History

UMAN DEVELOPMENT, as well as social change, often proceeds in fits and starts. Sometimes it is linear, sometimes orderly. Frequently, however, its movement reflects the chaos of its major elements—age, culture, and history—which are constantly shifting and realigning.[1] For this reason, it is clear that historians and social scientists need each other. Historians understand the nature of social change, including the significance of culture, but they generally fail to recognize that social change has differential effects based on whether children, adolescents, or adults are affected. Psychologists fall into the other trap; they value age as the most significant variable, while discounting the other variables.

Psychologists, however, have begun to question their priorities. In 1985 Erich W. Labouvie and John R. Nesselroade criticized their discipline's *"ahistorical* paradigm," which has "resulted in a slavish reliance on chronological age as the only time parameter of importance." At the same time, other psychologists have lamented the "progressive fragmentation" of their field, "that is, looking more and more at less and less." In a discussion published in the *American Psychologist*, Urie Bronfenbrenner, Frank Kessel, William Kessen, and Sheldon White argued that psychologists generally fail to take serious account of the outside world in which children develop because that world—in its economic, social, cultural, and political realities—is simply too overwhelming to contemplate, let alone quantify. Kessen stated that "the fractionalization or modularization of developmental psychology represents ... primarily a loss of nerve. One tries to get the research domain that is small enough and in which there are few enough competitors that one can live without raging anxiety." Bronfenbrenner was also critical, arguing that developmental psychology has become "the study of variables, not the study of systems, organisms, or live things living." The result, said Bronfenbrenner, was "the almost pointed avoidance among developmental researchers across the decades of speaking to the question, What do we mean by human development?" And "that, you see," said Kessen, "runs against the notion of ... the community of scholars. It's perfectly all right for people to till their own gardens, but once in a while they are going to have to talk over the back fence."[2]

This book is my effort to talk over the back fence. Historians and developmentalists share a commitment to the study of human life, and from examining the American homefront, it is evident that the Second World War's effects upon children varied widely depending upon the child's developmental stage. As a result, certain war events had a major impact on some children and only a secondary impact on others. An example is father absence. When fathers left home to go to war, their daughters and sons felt the effects regardless of age; but the most telling developmental impact was on the "war babies," those infants and toddlers born during the war who were under the age of four when the war ended. Also deeply affected were the little boys and girls who were "camp followers," or who suffered as "extra-familial children" when fathers returned. For these reasons, two experts noted in 1946, "The youngest children, up to about six years old, knew the least about the war, yet they were the group most deeply affected by it."[3]

For school-age boys and girls, however, the situation was different. These "Depression children" born in the 1930s had known their fathers prior to the war, and having already passed through the Oedipal stage into latency, they suffered fewer long-term consequences from father absence than did their "war-born" younger siblings. In practically every other respect, however, it was the children between the age of five and six and the onset of puberty who felt directly the tumult as well as the excitement of homefront life. "Rosie the Riveter"'s children, for example, usually were not preschoolers, but school-age children who assumed the extra responsibilities that went along with being latchkey girls and boys. School-age children too became politically aware during the war and participated in the patriotic enthusiasms that had become part of the school day. Because of air raid drills in school, fear too was more a part of their everyday lives than it was for the younger children. Likewise, it was school-age children who suffered most from war-inspired nightmares and who most dreaded the polio epidemics. Finally, these boys and girls played war games, followed the radio and film adventures of "The Just Warrior and the Beautiful Soul," and experienced the homefront's pervasive sex-role stereotyping.

Age, culture, and history determine individual development and shape social change.[4] These variables do not operate independently of each other; rather, they function in varying configurations. First, there is age, chronological and developmental, beginning with earliest childhood. While not denying that people continue to change throughout the life course, few people—whether doctors or psychologists, parents, teachers, or casual observers—would disagree that the earliest years are the most formative ones. In fact, it is because early experiences have lifespan consequences that it is instructive to look at the homefront children's lives. Fifty years after Pearl Harbor, America's homefront girls and boys have entered the 1990s as middle-aged women and men. And they, in turn, have reared another generation; many of their children have since added another generation of their own, thus making grandparents of many of the homefront children.

An emerging interdisciplinary field is lifespan studies, in which there is an appreciation of the importance of, and the unique challenges represented by, all stages of the life course from infancy to old age. Historians' most popular explanation of these stages is still Erik Erikson's "Eight Stages of Man," first published in 1950 in *Childhood and Society*, and later renamed the "Eight Ages of Man." Although Erikson received his clinical training in Freudian psychology, his particular message for historians is more helpful than that conveyed in orthodox psychoanalysis. Sigmund Freud explained human development in terms of sexually oriented tasks, or challenges, and strict Freudians adhere to the psychosexual sequence of oral, anal, phallic, and latency stages in childhood. Erikson, on the other hand, focuses on the individual ego, which he illustrates with references to "personal character," "inner solidarity," "a sense of individual identity."[5] The ego is constantly engaged because each of Erikson's psychosocial stages involves a struggle for the direction of an individual's life. For example, in infancy—up to age one—the struggle is between trust and mistrust. In early childhood—in the second and third years—it is between autonomy on the one hand and shame and doubt on the other.[6]

Just as a dialectic is evident in each of Erikson's stages, so too Jean Piaget's psychology emphasizes stage development. Piaget theorized that children progress through cognitive or intellectual stages, each of which results in a radical reorientation of the child's worldview. Piaget's writings have so influenced developmental psychologists that even some committed behaviorists, who formerly posited an "age-irrelevant" view of human development, concede that cognition springs from biological maturation and proceeds in recognizable stages. And when Piaget's theory is placed alongside Erikson's, one can see that their stages begin and end at about the same times in a child's life (see table).[7]

While Erikson's first major contribution to historical understanding is his "Eight Stages of Man," his second is his suggestion of a motivational construct for human development: "ego identity." A central theme in Erikson's own life history has been his personal search for identity in a world in turmoil. In fact, it was during the Second World War that Erikson, a German-born Jew (of Danish parents) who had fled Nazism and immigrated to the United States, first proclaimed identity to be a human need that "especially in times of change in the structure of society . . . becomes as important as food, security and sexual satisfaction." It was also during the war, according to Erikson's biographer, Robert Coles, that "Erikson's work made him more than ever aware that a child can grow into a very sturdy young man, in spite of all the problems a psychoanalyst might observe in him at any given age. . . ."[8]

Women and men who missed an encounter with Erikson's "Stages of Man" in college or graduate school had another opportunity to read about developmental stages when Gail Sheehy's *Passages: Predictable Crises in Adult Life* was published

Age	Erickson's psychosocial stages	Piaget's cognitive stages	Age roles
0–1	Trust v. Mistrust	Sensorimotor (age 0–2)	Infancy
2–3	Autonomy v. Shame	Preconceptual (age 2–4)	Early Childhood
3–6	Initiative v. Guilt	Intuitive Thought (age 4–7)	Play Age
7–12	Industry v. Inferiority	Concrete Operations (age 7–11)	School Age
13–20 or so	Identity v. Identity Diffusion	Formal Operations (age 11–15)	Adolescence
20s	Intimacy v. Isolation	Consolidated Formal Operations (over age 15)	Young Adulthood
30–50	Generativity v. Self-absorption		Adulthood
50s+	Ego Integrity v. Despair		Maturity

in 1978. People who wanted to read about "passages" experienced by men could pick up *The Seasons of a Man's Life* (1978) by Daniel Levinson and others. Sociologists and historians have contributed to the life-course literature, an important example being Glen H. Elder's *Children of the Great Depression: Social Change in Life Experience* (1974). The growing scholarly recognition of stages tends to validate the thesis that if each developmental stage is significant, the most formative are those of childhood. This seems to be the case whether the developmental domain is physical, social, affective, or intellectual. Certainly, as the homefront children grew to adolescence, adulthood, and now middle age, they carried much with them from childhood, including strengths and vulnerabilities, anxieties and coping mechanisms, and fears and hopes.

After age, the second element in the developmental configuration is culture, but it too is age-specific. The list of individuals who transmit culture to children begins with mother and father, and it takes its first form as child rearing.[9] Other family factors are determinative as well, such as income and number and birth order of siblings. At the beginning, the little child's world is that of the family, and the distinctive forces in his or her life—race and ethnic culture, socioeconomic class, region—not only shape the way they are reared but do so in richly diverse ways. Soon the list of cultural transmitters includes the surrounding ecology, whether rural or urban; the church and school; peers; and popular culture. These complex acculturative processes are all part of a child's socialization.

The world outside the family is different. Entering school at the age of five or six, children move into settings where they encounter new cultures. At the same time, the schools during the Second World War sought to homogenize language and dress as well as beliefs and attitudes, partly by discouraging cultural diversity. For America's school-age children during the patriotic war years, the outside world entered their lives not only through new friends and new neighborhoods to explore but also through comic books, radio shows, and the Saturday matinee. During the first half of the 1940s, there was "the War" itself, towering over everything else. Because of the wartime influences inside and outside the family, the homefront children shared both an ideology and, as one homefront girl wrote, a sense of participating in a cause "greater than ourselves."[10]

With age and culture in place, the final element is history. Historians generally can explain why something has happened, but they have a far more difficult time measuring the impact of what it was that happened. To correct this problem, the historian might adopt an age-specific perspective in assessing social change. Glen Elder has suggested ways for scholars to do this. "The imprint of history," Elder begins, "is one of the most neglected facts in [human] development." Lives "are shaped," he writes, not only "by the settings in which they are lived" but also "by the timing of encounters with historical forces, whether depression or prosperity, peace or war."[11]

Obviously, the timing of historical encounters is important not only during childhood but beyond it as well, for timing can conspire to produce both an immediate impact and one that resonates throughout one's life. Elder has given a vivid definition of the life course as "pathways through the age differentiated life span, to social patterns in the timing, duration, spacing, and order of events; the timing of an event may be as consequential for life experience as whether the event occurs and the degree or type of change."[12]

For the homefront children's cohort born between 1932 or 1933 and 1945, it was crucial how old they were when later events changed history—for example, how old they were in 1950 when the Korean War broke out, or in 1954 when the Supreme Court announced that segregation in public education was "inherently unequal" and must end. For example, in 1950 the oldest of the homefront boys were eighteen years old and of draft age, while the youngest were only five years old. On the other hand, in 1964, when Congress passed the Gulf of Tonkin Resolution giving President Lyndon B. Johnson the "equivalent" of a declaration of war, the oldest of the homefront children were thirty-two and well beyond draft age, while the youngest were nineteen, twenty, and twenty-one.

Political movements also reflected the significance of chronological age. It mattered, for example, how old a homefront girl was in 1963, when Betty Friedan's *The Feminist Mystique* was published. Also in 1963, it mattered how old one was when Martin Luther King, Jr., declared his dream that one day his "four little children will . . . live in a nation where they will be judged not by the color of their skin but by the content of their character." Age has continued to influence people's responses to popular culture; in 1964, for example, when the Beatles visited the United States, the youngest homefront children were nineteen, the oldest were thirty-two. Similarly, the struggle against the Vietnam War involved most of all the youngest of the homefront children, many of whom were called upon to fight, and least of all the oldest, who were too old to be drafted and who had other concerns anyway, such as supporting a growing family and paying the mortgage.[13]

Another aspect of age stems from the reality that children's wartime experiences often placed them in adult situations, which reinforced both strengths and vulnerabilities. Gertrude Stein explained this in 1945: "In time of peace what children feel concerns the lives of the children as children but in time of war there is a mingling there is not children's lives and grown up lives there is just lives...." Bolstering personal growth during the war, children took on additional responsibilities. As the older siblings either left home to join the service or took war-production jobs, their younger sisters and brothers assumed household responsibilities. They did the cooking, cleaned the house, and baby-sat. When the father of an Illinois farm boy took a job at an ordnance plant forty-five miles away, the boy milked the cows and took care of livestock. Some children went into business for themselves, including a group of ten- to fourteen-year-olds in a suburb of New York City. For fees ranging from a nickel to 50 cents, the children ran errands, raked leaves, and walked dogs while the owners were engaged in war work. Moreover, if an older sibling left a part-time job at a soda fountain or grocery store, a younger brother or sister sometimes filled the slot. In fact, so many teenagers were leaving school to take full-time employment that high-school enrollments declined by 1,200,000 from 1941 to 1944. By the fall of 1944, the situation was of such concern that the U.S. Office of Education and the Children's Bureau sponsored a nationwide "Go-to-School" drive. "During the war," recalled a homefront boy who was eleven when he got his first job at a gas station, "so many young men were gone that it was an easy chore to get a delivery or menial job...." A fourteen-year-old girl's first job was as the head cook in a local restaurant; she next worked as a nurse's aid in the hospital; and as she reflected on her childhood, she wrote that that war had changed "our life from Kids to 'Mini-Adults' with much responsibility."[14]

But while some experiences strengthened the homefront children's sense of autonomy, initiative, industry, and self-worth, others were negative and exacerbated childhood vulnerabilities. Some of America's little children, for example, suffering early and prolonged separation from a father in the service, were likely to harbor fears of subsequent abandonment. Other homefront boys and girls saw their emotional security eroded by air raid drills and blackouts, and by images of the Holocaust and the atomic bomb. And there was concern about the long-term consequences of these fears. "If ... the possessing or losing of a sense of security in early childhood," stated *The Contemporary American Family*, a survey published in 1947, "contributes significantly to the emotional character of adult life, only when these war children mature will the sizableness of one of the most important penalties of World War II be disclosed." Since there were so many affected children, "their misfortunes prophesy much future suffering and instability." This prophecy proved to be far too gloomy.[15]

One goal of this book is to show that history is a study of both human nature—that is, the factors that are unique, individualistic, idiosyncratic, and, therefore, quite unpredictable—and social and cultural trends, which are quite predictable. Another goal has been to suggest the importance of the shifting configurations of age, culture, and history; for it was at the intersection of these variables that the homefront children lived their lives. There is, however, a final question to be addressed, one that also derives from the configurational perspective on individual development and social change: Do the homefront children, based on collective memory and collective identity, deserve to be called a "generation"? The best way to answer this question is to turn to the homefront children themselves.

America's homefront girls and boys who wrote letters recalling their wartime years readily identified themselves as members of a "generation," and they were proud to be so indentified. The issues for their generation, they remembered, were unambiguous, as were the moral choices they made. "Right and wrong were clearly defined . . . ," wrote Mary Maloney, a homefront girl. "I didn't realize it at the time, but life would never be that simple again for my age group, or any other for that matter." Another girl, Catherine Meyer, compared homefront Americans to "passengers on a great ship," sailing through enemy waters; the voyage constituted "an important part of our memories," particularly of "the hopes and fears we'd shared." And she articulated a fundamental faith of the homefront children: "We had witnessed the triumph of total good over total evil. . . ."[16]

Because of their collective memories, the homefront children sense a kinship with each other. The Second World War "shaped my life," wrote a homefront girl, not only inculcating patriotism but also providing a link "to every one my age." It still "imbues everything I think and do," providing a generational perspective and giving a "sense of time and place that fixes me in history. . . ." There is a self-congratulatory element in the children's recollections. "Perhaps," mused another homefront girl, "the war helped us to grow up 'well' because of that sense of participation and inclusion" in a righteous crusade. "I like us!" she exclaimed: "You could call us . . . the 'responsible generation,'" which she defined as "more optimistic, less competitive, and [less] money oriented" than either the children born before them, including their older siblings who had been marked by the Great Depression, or by their own children born during more affluent times in the 1950s and 1960s. Joan Chalmers, a homefront girl who lived in Hawaii, reflected on her childhood during the war: "I think it is only now, in our late 50's, that we realize how unique our experience was." Certainly, the war was "an enormous disruption" in children's lives, particularly for girls and boys who suffered separation, absence, and loss. But for the rest, Joan wrote, "we were innocent participants in an adventure. Our lives were disrupted but we were rarely victims. . . ." And now, "as we assemble at our 40th high school reunions and reminisce, we begin to appreciate how fortunate we were . . . and are."[17]

Most of the homefront children have positive sentiments about these years. They were patriots then—and most still are. The Second World War was "a catalyst for patriotism," observed a homefront girl, so "I suspect you will find that most of the kids who grew up in the 40s were inoculated with a pretty fierce sense of patriotism. . . ." Another homefront girl, now a music teacher, "always makes sure" that she teaches her students patriotic songs: "I'm sure I would not feel so strongly about it if I had not established a sense of pride during W.W. II." The homefront children's letters abound with such sentiments: "As a child of the war, I still get 'Goose Bumps' when I see the American Flag. . . . I feel anger when I see the disrespect that is sometimes shown [it]. . . . They have no right."[18]

There are some homefront children whose patriotism has calcified over the decades, and who measure all political behavior by their childhood standards from the Second World War. Seeing dissent as treason, they wrote that if their sons had refused to serve in Vietnam, they "would have disowned them." "I say 'Love It or Leave It,'" wrote a homefront girl. "If you can't afford to leave it, I'll get enough of my American Loving People to buy you a 1-way ticket to another country." But do America a favor, she implored: "stay there . . . we don't need rebels without a cause here."[19]

Considering the patriotic wartime atmosphere, it is unsurprising that the homefront children have embraced these lifelong sentiments. Nor is it surprising that they exhibit elements of a collective identity—relevant to which are glorious notions about the destiny of the United States that infused the children's school curricula and their popular culture. The homefront children's sentiments are heartfelt, and should not be minimized.

Nevertheless, other memories have left far more lasting imprints. Many of the letter-writers, after alluding to the patriotism of their childhoods, then turned to topics of greater personal importance. First, they remembered their unhappiness, and when some of these homefront children reached to open the doors to their childhoods, sadness engulfed them. Some could not finish their letters. One homefront girl apologized that as she reflected on her father's departure for the service and then on her mother's leaving to join him at a stateside post, "suddenly as if I was paralyzed I could go no further." Another homefront girl confessed that she had "tapped into a lot more" than she had bargained for; she called it "reminiscence therapy." A homefront girl whose two older brothers had been killed in the Army Air Corps waited a month before writing. "Finally, I just decided to write this down. I'm glad I did."[20]

It has been understandably difficult for homefront children to relive the pain of father absence or the death of a close relative. The pain, which is still there, has been profound. When Vicki Lacount came home from school one day in late 1944, she learned that her father had been killed in the Battle of the Bulge; thus began her decades-long "grieving process." Another homefront girl wrote of her beloved older brother Bill, a pilot whose bomber was shot down over Germany: "Not a day goes by that I don't think of him." In fact, "nothing has ever affected my life and my way of thinking as much as those years of loving and losing Bill."[21]

For almost four years, the homefront children knew of the war's suffering and bloodshed—even if from afar. They listened to the news on the radio, read about battle casualties in the daily newspapers, and saw death depicted in news photographs and at the movies. They knew that children got killed along with adults, because they saw those images too: newsreels showed dead bodies of children, sometimes strewn in the road after being strafed by fighter planes, sometimes stacked with adult bodies in trenches. Visual images also depicted children who were refugees jamming roads in Europe and Asia. And in America's schools and on its playgrounds, homefront children talked with each other about the gore and heartbreak they were witnessing. Still, they dealt with this intense emotional input largely on their own, unable for one reason or another to share their fears and concerns with their parents or teachers, many of whom never thought to ask how they were doing.

"How do I describe how those years affected me?" asked a homefront girl, Bernadine Ratliff. "Well, there was horror that our country would be invaded." Her Uncle Bob Jack had been killed in France, so "we had ... seen death in our family. I saw pictures of the Jews found in the labor camps. I guess we thought we would live a short life and maybe die a frightful death." Jean Johnson compared her childhood years with the vicissitudes of childhood today. "It seems," she wrote in 1990, "that I [have] lived all of my life under the fear of war and I am 51 years old. When people say young people have it worse today, I do not agree." Jean recalled seeing neighborhood "men badly crippled from the war," some in wheelchairs, and there were "also young people crippled from polio." She saw families "suffering in silence" as they tried to welcome home emotionally distressed fathers. Her fears did not subside after the war. Next, there was "the fear of nuclear war," and then there were other wars in Korea and Indochina— "they all have been very much a part of my life."[22]

On the other hand, some of the homefront children still revel in the Second World War. Usually, it is the men who have become collectors of Nazi memorabilia, and who display Japanese Zero flags and samurai swords on the walls of their homes. A letter arrived from a homefront girl who wrote not about herself, but about her husband, born in 1936. "Although he refuses to write you, I believe his life has been tremendously affected by the war." The only books he reads, and only movies he sees, are about the Second World War; and he is "extremely patriotic." "Airplanes fascinate him. Believe it or not, he has hundreds of 2 to 3 inch thick notebooks of detailed scale drawings, illustrations and facts regarding planes of that period, compiled over our almost thirty years of marriage." The notebooks seem to have "no purpose," since he "plans to write no books—he just seems to be obsessed with learning every minute detail about the planes used in the war." He enlisted in the Navy during the Korean War, "and to this day regrets getting out...." Sometimes he "fights battles in his sleep," and it is sad because he is "reliving a battle he never really fought." Now he is about to retire, and when he does, she wrote, "I have a great fear that working on his notebooks is how he'll spend his remaining years! And, if it has no purpose, when he's dead and gone we'll pitch years and years of work."[23]

Another negative personal consequence recalled by the homefront children has a universality which the above example does not. "We are known as the Brain-Washed generation," wrote a homefront girl. Another girl, Mary Ann Cunningham, wrote that she has long resented being "propagandized by our government and taught to hate at an early age...." As a result, she became skeptical rather than patriotic "regarding government policies and the information on which they are based." In 1951, when Mary Ann was eighteen, she joined a Roman Catholic religious order; during the 1960s she became "involved in anti-Vietnam War activities [and] disarmament protests including a number of arrests and brief (8-days) jail terms." Another homefront girl, remembering that children "were pretty heavily indoctrinated," confided that "the fear, hate and prejudice I learned took years to overcome." She had learned hatred from her reactionary and bigoted grandfather as well as from the newsreels and the pages of *Life*. Her parents lied to her too; it was not until high school that she learned that the Holocaust had "actually happened ... and that what my folks had told me was not necessarily the truth." When the Vietnam War began, she was a college student and a mother, and she "tried very hard," she recalled, "not to give my daughters the hate that I was given at the same age they were."[24]

"I do not know what the war did to others of my generation," wrote Jerry Wade, a homefront boy who had lived in Jefferson, Iowa, a town of 5000 people, but it has had a lasting impact on him. Wade, like most homefront children, considered himself to be a member of a "generation." Never defined quantitatively, this generation was a product of shared memories and collective sentiments. And now, as these women and men gather to celebrate class reunions, they talk about the patriotism that motivated their sacrifices and participation during "the good war." Assuredly, they speak with pride of their country's history during the Second World War.[25]

There is nothing unique in an older generation boasting of its past, sometimes while bemoaning the collective shortcomings of succeeding generations. Parents and grandparents before them have always been convinced that "the older generation"—that is, theirs—"had it a lot rougher than the kids of today." At the same time, there is no denying that the past fifty years have borne monumental changes, many of them highly disillusioning and demoralizing. The world of the 1990s is a far different place from the one they knew on the homefront. Today they see not sacrifice and moral purpose, but selfishness and moral decline. "The people of today don't stick together...," lamented a homefront boy. "The thinking of the . . . protesters would not have won the war. They are 'me' orientated," he wrote. "I think I am more of a team player. . . ." "ME, ME, ME is the order of the day," added a homefront girl. "No one seems . . . willing to make any sacrifices toward a united effort."[26]

From the perspective of middle age, the homefront children still see themselves as members of a generation. The Second World War was the distinctive experience they shared and which separated them from those who went before and those who came after. Moreover, they perceive the war years to have been an ennobling time for children; and they tend to believe that their childhood years were blessed compared with the desperate years of the Great Depression, when children and youth suffered great privation, or blessed even when compared with childhood during the postwar years of child-centered prosperity and "permissiveness," which marked the baby boomers born in the 1940s and 1950s and which, some critics have contended, encouraged narcissism and rebelliousness. Moreover, many homefront children believe that their own offspring have been shortchanged, coming of age as they have in an era of tarnished ideals, economic uncertainties, and environmental collapse.

But whether the population composed of the homefront children actually is a generation depends upon one's definition. The Bible tells of three generations occurring in a century, and some demographers define a generation as roughly thirty years, while others say it is about twenty years. Classically, a generation is conceived of in terms of parents and children, of lineage relations, the father and mother comprising one generation, their children the next. Novelists and poets have pictured generations in terms of conflict, as F. Scott Fitzgerald did when he wrote of a generation as "that reaction against the fathers that seems to occur about three times in a century."[27]

Yet, when one looks carefully at distinctive generational groupings in twentieth-century United States history, a couple of things are apparent. First, rather than being thirty years in duration, these generations are about half that long. Second, for each two generations, there seems to be an "in-between" generation. Perhaps, as the writer Frank Conroy, a homefront boy, has observed in assessing his generation, "life has accelerated to the point where [these] changeovers occur much more rapidly. . . ." For another thing, the conception of what constitutes a generation has moved away from lineage relations toward a greater concern with a period's "cohort effects," such as war or depression. In this vein, two sociologists have written that they "treat *cohort effect* and *generation*" as functionally synonymous, since both have the "implication of long-lasting, if not permanent change in the cohort," including "the still further implication of ideological distinctiveness." In terms of cohort effects, then, the homefront children qualify as a generation.[28]

But how does a generation function as a force in history? According to Peter Loewenberg, an exponent of the use of psychoanalysis in studying history, the scholar should turn to the famous 1928 essay by Karl Mannheim entitled "The Problem of Generations." "Here," Loewenberg explains, "Mannheim speaks of the human mind as 'stratified' or layered, with the earliest experiences being the basis, and all subsequent experience building on this primary foundation or reacting against it."[29] From this perspective, Loewenberg has studied German children during the First World War who suffered prolonged parental absence as well as extreme hunger, and whose fathers returned defeated "and unable to protect them from further sorrows." The insight Loewenberg gained from reading Mannheim is that "those of a generation who experienced the same event, such as a world war, may respond to it differently. They were all decisively influenced by it but not in the same way." Among Germany's homefront children from the First World War, some became pacifists, others turned to revolutionary Communism, and still others espoused a return to the monarchy. And, finally, there were those who followed Adolf Hitler, seeking "personal and national solutions in a violence-oriented movement subservient to the will of a total leader."[30]

When the homefront children are looked at, in Mannheim's words, as having had "a common location in the social and historical process" during a period when events were novel and change was fast, they again seem to constitute a "generation."[31] In truth, the issue is far more complicated. Just as very little in life is perfect, so there is no modern generation that is "uncontaminated" by contradictory factors. There is always some overlap in generational characteristics, as one generation bleeds into the next. Even though numerous baby boomers grew up amidst prosperity, for example, their parents, never having outgrown their "depression psychosis," harped that their sons and daughters should not become too accustomed to their good fortune because it could vanish overnight. The parents' behavior makes vivid the insight provided by José Ortega y Gasset that "at every moment of history there exists not one generation but three: the young, the mature, the old. This means," he continued, "that every historical actuality, every 'today' involves ... three different actualities, three different 'todays.' Or to put it another way, the present is rich in three great vital dimensions which live in it together, linked together whether they like it or not...."[32]

Some observers argue, however, that rather than comprising a distinctive generation, the homefront children are actually split between two generations— between the *Zeitgeist* of the Great Depression, with its small birth cohort, and that of the "Affluent Society" with its baby boom. Lyle E. Schaller, a minister and social surveyor, has written that a generational split occurred in 1942. Schaller sees American generations as comprising those born between 1928 and 1942, 1943 and 1955, 1956 and 1970, and 1970 to the present. At work in Schaller's demarcations are significant historical factors, particularly the economic cycles, wars, and epochal demographic and cultural changes. The smallest of Schaller's generations is that for 1928 to 1942, during which only 38 million babies were born. By contrast, there were 47 million born between 1943 and 1955 and 57.5

million between 1956 and 1970. "In other words," Schaller has noted, "the members of the generation covering [this] fifteen-year period" are much more numerous than "the generation born in the fifteen years of 1928–42 inclusive."[33]

Along with some other observers, Schaller disputes that the years of the baby boom are either from 1946 to 1961, or from 1946 to 1964. He points out that while the peak birth years of the baby boom were 1956 to 1962, those baby boomers born in 1946 had little in common with those born ten or fifteen years later. Furthermore, a book on the "Sixties Generation" sees a first wave of baby boomers being born between 1944 and 1949, followed by a second wave from 1950 to 1957. In *Great Expectations*, Landon Jones has observed that those babies born during the war and in the years up to 1957 "experienced far more turbulence in their life than those born" after 1957. "The earlier group," Jones explained, "living on the boom's cutting edge, suffered the growing pains of a world constantly straining and expanding to make room for it. . . . At the same time, they have always enjoyed an acute sense of their own power. The later group, however, entered a society already overexpanded by their older brothers and sisters. The biggest battles had already been fought. Events like the assassination of President Kennedy mean something altogether different to them." A man born in 1955 as a second-half baby boomer expressed this dilemma: "Lost in Time: Too Young for the '60s, Too Old for the '80s."[34]

Finally, it is important to remember that the parents of the first wave of baby boomers probably came from a different generation than the parents of the second wave. John Womack, the historian, observed that his students' attitudes were strongly influenced by their parents' politically formative experiences. "The students of the '60s," Womack has said, "to a great extent, were born during the Second World War or right after. . . . It wasn't so much important what *they* went through but what their *parents* had been through . . . the New Deal and the Second World War, two great democratic, left-leaning episodes in their parents' lives. . . . The children who have come to school in the '70s are mostly the children of the Korean War generation" and of "a period of deepening conservatism in the United States."[35]

Further confounding the generational issue is the presence of overlapping political, cultural, and economic cycles. The Project on the Vietnam Generation, for example, in identifying the years of birth of "the 60 million men and women who came of age during that time," bracketed the Vietnam generation between the birth years of 1937 and 1954. While birth cohorts are the most logical way to delineate these groupings, there are conflicting cycles. Technological innovations, such as television, have introduced their own cycles. There is a "television generation"; in fact, there are probably several such generations, including since 1981 the most recent one, that is, children reared on the music-video channel MTV.[36] Also superimposed on everyday life are varying economic cycles.[37] Still other notable historical cycles have been triggered by medical epidemics. To top it all, these cycles may interact with each other, gaining energy through "synchronicity," a concept developed by C. G. Jung to indicate "a *meaningful coinci-dence* of two or more events, where something other than the probability of chance is involved."[38]

Thus, there are cycles at every hand—geological cycles, meteorological cycles, lunar cycles. In addition, biofeedback instrumentation measures cycles known as biorhythms. Likewise, historians have explained political movements in cyclical terms. Arthur M. Schlesinger once identified eleven separate swings in United States history between conservatism and liberalism. Averaging 16.5 years in length, America's twentieth-century cycles include the Progressive era, 1901–19; "the Republican restoration," 1919–31; the New Deal era, 1931–47; and a "recession from liberalism," lasting from 1947 to 1962. In 1949 Schlesinger predicted that liberalism would return in 1962. "On this basis," Schlesinger again predicted with accuracy, "the next conservative epoch will commence around 1978."[39]

All of these cycles notwithstanding, the homefront girls and boys still see themselves as members of a single generation. But the evidence—including the homefront children's own recollections—suggests otherwise.

The homefront children do comprise a cohort, the standard definition of which is that it is that "aggregate of individuals (within some population definition) who experienced the same event within the same time interval." Still, while this appropriately defines the homefront children, it conceals the vast differences in outlook that existed between the oldest members of this cohort and the youngest. In fact, the homefront children themselves insist that their cohort should actually be subdivided into three subcohorts. Although these divisions undermine hypotheses about collective identity, they are nevertheless significant to the homefront children's self-identification. They have divided themselves into the following age groups.[40]

The *first subcohort* consists of those homefront children born between about *1932 and 1935*. While they were still preadolescents during wartime, their early childhood years had been deeply etched by the economics—and psychology—of the depression. Another source of this subcohort's traditionalism is its rural roots. This group, having been reared largely before the massive migration of the Second World War took place, was more likely to have spent its early childhood years on farms and in small towns. Moreover, being on the verge of adolescence during the war, to these children sexuality and military service seemed not so far away. Many of these girls and boys carried heavy burdens with them; they worried about the future, especially about making a living. "Boys of the Depression generation," wrote journalist Russell Baker, who was one of them, "were expected to have their hearts set on moneymaking work. . . . Boys who hadn't yet decided on a specific career usually replied that their ambition was 'to be a success.' That was all right. The Depression had made materialists of us all. . . ." Glen Elder, in his study of children who were eight and nine years old when the Great Depression began, concluded that "the depression's main legacy took the form of shaping values and attitudes. Men and women who had grown up in depression-marked homes were most likely to anchor their lives around family and children, perhaps reflecting the notion of home as a refuge in an unpredictable world." It is therefore not surprising that numerous parents of the baby boom came from the first subcohort.[41]

The *second subcohort* consists of boys and girls born between about *1936 and 1941*. This age group sees itself as having functioned as a bridge between the Depression children and the war babies. This group's self-definition is that it is mediative because it sits astride two generations whose great divide occurs in 1941–42. True, a good many men and women of the second subcohort still consider themselves children of the Great Depression. Along with year of birth, social class was a key factor in determining whether a Depression-born child from the second subcohort identified with scarcity and uncertainty or with comfort and security. Those girls and boys born into continuing economic uncertainty between 1936 and 1941—those whose parents were unemployed and who were hungry—tended to identify with the first subcohort, with its serious concerns. Other members of the second subcohort, however, identified with the baby boomers and later joined movements for social change. In fact, some of the radical leaders and countercultural innovators of the 1960s, when the nation's culture and politics changed, came from this group. A decade later, two politicians from this subcohort, Jerry Brown (b. 1938) and Gary Hart (b. 1936) tried politically to mobilize the baby boom. As Eli Evans (b. 1936), another member of this age group, has observed: "By the 1970s and 1980s, Hart and Brown emerged as the two most successful practitioners of 'generational politics.' But," befitting this in-between group, "neither was of the generation they were said to lead."[42]

It is evident from the homefront children's letters that the subcohort born between about 1936 and 1941 has been pivotal in defining the entire cohort. "I was a member," a homefront boy wrote, of the group "that fell between the sharp pincers of the Depression on one side and World War II on the other." Born in 1936, another homefront boy wrote: "We had been born in the first flickers of recovery from the Depression: we were symbols of hope for our parents." He believed that his birth cohort was "the last . . . to grow up and share in a deeply rooted, purely American sense of security and invincibility," and "the first to have had that security removed at an early age"—on August 6, 1945—by the atomic bomb dropped on Hiroshima. "This loss of innocence was a pivotal psychological event. It separated us not only from the generation that landed on the beaches of Normandy, won World War II and continued to believe in America's superiority and invincibility, but also from the generation that was born following the dawn of the nuclear age."[43]

"Born in 1939," a homefront girl observed, "I was both a child of the depression and the war. I recall hearing about both almost all my life." Another homefront girl, born in 1937, became a nurse and encouraged her patients "to tell the stories of their lives." She found that the stories related by people older than herself were "those of great pain," and younger than herself were those of "great rejoicing." "In many ways," she observed, "I think that those of us falling in that time period between being depression babies/baby boomers have struggled to fully experience the ordinariness of our *own* lives." And she wondered whether "a high percentage of us ... became care-givers [and] teachers. ..." Others in this in-between group commented that they had made an early accommodation with the prevailing mores. And a homefront boy ventured that the real "organization men" had come from this subcohort. It was likely, he said, that the homefront emphasis on democratic cooperation, respect for other Americans, and "fitting in" had contributed to "other-directed conformist behavior" and the development of "a bureaucratic personality."[44]

Most of all, wars have defined this middle group. Born in 1936, for example, Eli Evans reflected that he was of "the generation too young for World War II and Korea and too old for Vietnam. ..." "Too young for Korea and too old for Vietnam ... ," agreed a homefront girl born in 1938. "Korea was my adolescence," observed a homefront girl born in 1937; "Vietnam found me in the throes of raising children. ..." Thus, in vital ways, the second was the in-between subcohort—in its way, it was like being the middle child in a family. As Eli Evans has observed of the transitional role played by these Americans, "Perhaps our greatest legacy will be the way we served as interpreters of the rebellious young to our elders. Of neither generation, we had ... been called on to build bridges between the generations on either side of us."[45]

Many members of the second subcohort not only believe they alone are the homefront children, they also question whether those babies born next, between 1942 and 1945, remember enough even to qualify for membership in this cohort. "My high school class, 1957," wrote Valerie Drachman, born in 1940, "was the last one in which the youngsters could remember the war." Michael Ferman, born in 1935, agreed that "the age of the war children makes a big difference in what they understood. Those born after 1940 really learned about the war" afterward, at the movies over the next fifteen years—from the *Best Years of Our Lives* (1946) and *Battleground, Twelve O'Clock High,* and *The Sands of Iwo Jima,* all released in 1949; to *Stalag 17* (1953), *The Bridge on the River Kwai* (1957), and numerous others in this genre. Arnie Isaacs, born in 1941, said that he and the other children born that year "have the quality of last survivors—like children in a flood. ... We were the last, perhaps, to inhabit the same America our schoolbooks taught us about, the last to believe that you had to obey the rules, but that if you obeyed them you could win the game."[46]

Although there is disagreement over whether those children born in 1940 or 1941 are old enough to have wartime recollections, many do. Born in 1941, Gerry Lunderville remembers well the war's terror, including the comic books that depicted "the hated Japs, with buck teeth and thick glasses. I detested these yellow people. ..." Gerry, now married to an Asian woman, is completing a degree in Asian history. "World War II," he wrote, ". . . your power has all but dissipated the haunting cobwebs of a child's first knowledge of that world gripped in the terror of the early 40s."[47]

Finally, the *third subcohort* consists of the "war babies" born between *1942 and 1945*. Significantly, it was the upsurge in births during the war years that launched the demographic bulge known as the "Baby Boom." Because this became such a large group over the next twenty years, its members developed a greater consciousness of being part of a special generation, of being what Landon Jones has called the "'pig in a python' to describe the resulting motion of the baby-boom bulge through the decades as it ages." It was the children born between 1942 and 1945 whose presence first put pressure on government to build more grade schools, then more high schools, and finally more colleges. And it was this subcohort's members who first began to worry whether Social Security, along with other pension, disability, Medicare, and trust accounts, could fund their retirements in twenty to thirty years.[48]

Perhaps most significant, from the third subcohort came many tens of thousands of soldiers, sailors, and marines who helped constitute the first wave of the massive Vietnam buildup from 1965 to 1968. It is further evidence that war has been an element of monumental significance in the lives of the children born between 1942 and 1945. Because of these children's early developmental stages during the war—because they were neonates up to age three or four—they suffered emotional damage from father separation and absence. And the sad irony is that it was they who went to Vietnam as twenty- and twenty-one-year-olds in 1965. Moreover, for some members of this subcohort, there has been a four-generational relationship with war. Their grandfathers, born during the 1890s, were children during the Spanish-American War and were of military age themselves during the First World War. Their fathers, who were born during the First World War, were in their twenties in 1942 and ripe for conscription. And the men who were born between 1942 and 1945 came of draft age themselves during the Vietnam War. Moreover, in 1991 a fourth generation—including the sons and daughters of homefront children—waged war in the Persian Gulf.

Martin Wangh, a psychoanalyst, has ventured that there is "a psychogenetic factor in the recurrence of war," namely, that every twenty-five years or so, the sons refight their fathers' war. Wangh hypothesized that young men's "predisposition" for war "is the result of the experience of the stresses of a previous war which has traumatized these people so that they unconsciously seek a revival of the traumatic situation, in consonance with the 'repetition compulsion'" described by Freud. The fallacy here is that it is old men, not young men, who declare war and order the country's sons and daughters and grandchildren to do the fighting. On the other hand, this "psychogenetic factor"—when combined, first, with the intensely patriotic popular culture that prevailed from the 1940s well into the 1960s and, second, with the prevalent sex-typing that defined warfare as a rite of passage for young men—does help to explain the ready response of millions of young people to the call to the colors not only during the Second World War, but also during the early stages of the Vietnam War. Ironically, returning fathers of the Second World War, knowing the horrors of the battle-field, prayed there would be no more wars. When one homefront girl's older brother—a medic who had won both the Silver and Bronze Star—returned in 1945, his wife was pregnant with their first child. "I don't want a little boy," he said; "I don't want him to have to serve in a war." But when war broke out in Vietnam, the boy volunteered for the Marines.[49]

Age, culture, and history are the key ingredients in individual human development. Obviously, fortuity is involved as well, including one's race, class, and gender. Was it a girl, for example, or was it a boy, born on the homefront in 1944, just twenty years before the Gulf of Tonkin resolution launched the war in Indo-china? These factors shape historical change, determining what comprises a generation.

The postwar history of the United States tells of shifting individual and societal configurations caused by wars, economic fluctuations, technological innovations, and, perhaps most of all, shifting family fortunes—marriages and births, divorces and deaths. The homefront children have experienced great change since 1945. Only they are no longer children; it is time to look at the homefront boys and girls almost fifty years later, that is, at middle age.

CHAPTER 15

The Homefront Children at Middle Age

E ACH GENERATION receives a distinctive imprinting from the social, political, and cultural events of its childhood. Shaping America's homefront girls and boys were the "cohort effects" of the Second World War; but the war's influence did not surcease in 1945. In fact, the wartime beliefs and values with which the homefront children grew up were largely the same beliefs and values that guided them later, during their adolescence and adulthood. The prevailing ideology about patriotism and America's leadership of the Free World, as well as about marriage and the family and the need to "get ahead," changed little from the early 1940s to the mid- to late 1960s.

It is easy to caricature the postwar years as America's "Good Times Generation"—an epoch of innocence, simple truths, and apathy, before the onslaughts of the 1960s and beyond. Cold War historians, liberals and conservatives alike, have praised this period, entitling their books *American High: The Years of Confidence, 1945–1960* and *The Proud Decades: America in War and Peace, 1941–1960.* A radical writer, on the other hand, has called his book *The Dark Ages: Life in the United States, 1945–1960.*[1] But whether liberal or conservative, few would disagree with Godfrey Hodgson, the British journalist, who has discerned the twin moods of postwar America: "Confident to the verge of complacency about the perfectibility of American society, anxious to the point of paranoia about the threat of communism—these were the two faces of the consensus mood. Each grew," Hodgson added, "from one aspect of the 1940s: confidence from economic success, anxiety from the fear of Stalin and the frustrations of power." By the 1960 presidential election, there were few Americans who contested this "ideology of the liberal consensus." As Hodgson noted, "Some voted for Kennedy, and some for Nixon. But, except for a small fringe of conservatives and an almost insignificant sprinkling of radicals, political differences were less important in 1960 than the underlying consensus. Most Americans then accepted an ideology of imperial liberalism whose chief tenets were simplicity itself: the American system worked at home, and America must be strong abroad."[2]

When Americans entered the postwar era, however, many wondered whether it would resemble another postwar epoch, the 1920s. In April 1946 a *New York Times* writer predicted a return of the Roaring Twenties. Not everyone agreed. Reminding readers that the 1920s had culminated in economic depression and world war, another writer responded that "there are too many people who, knowing the results which flowed from the attitudes of 1920, are going to see to it that history does not repeat itself." Indeed, most Americans expected a replay of the 1930s. After all, it was the war that had created jobs and prosperity; surely the end of war would bring both an economic slump and another "Baby Bust." As it turned out, neither prediction was correct.[3]

The United States in 1945 entered one of its longest, steadiest periods of growth and prosperity. Blending "New Deal ideology and wartime pragmatism," Robert J. Samuelson of *Newsweek* has observed, the federal government led the way in declaring that the keys to postwar happiness were increased output of and demand for consumer products. Indeed, there was a national "obsession with growth," which Americans saw as "producing higher living standards and less poverty," as well as "social and economic stability." The wartime success in mobilizing the country's resources, Samuelson explained, had "established the government's reputation as an economic manager," with the ultimate responsibility "for assuring prosperity and correcting capitalism's flaws." In the twenty-five years after 1945 the American economy grew at an average rate of 3.5 percent per year. Even with occasional recessions the gross national product seldom faltered, rising from just under $210 billion in 1946 to $285 billion in 1950, $504 billion in 1960, and close to $1 trillion in 1970.[4]

Among the chief beneficiaries of this twenty-five-year boom were the homefront children, particularly the "Baby-Bust" constituents of the first and second subcohorts. A homefront girl born in 1934 recognized her good fortune as "a member of the 'Chosen Few' Generation." The low birthrate throughout the 1930s had produced a small group of children; and yet it was these relative few who were born enough ahead of the baby boom to be able to take advantage of jobs created by the existence of the largest generation in the nation's history. For example, young men and women—born in the mid- to late 1930s who were straight out of graduate school in the second half of the 1960s—found professorships in colleges and universities, teaching the early wave of baby boomers.[5]

Even before the war's end, it was evident that a new world of comfort and convenience was emerging. "Device That Sprays Cosmetics on Skin Is Outlined by Industrial Designer," read the *New York Times* headline in June 1944. This newly designed spray container, with pressurized gas inside, would simplify women's lives. "In the future," the article explained, "women will no longer be bothered removing the caps of bottles and jars, pouring or scooping out their cosmetics, and replacing the tops again." Predictions abounded that science and technology were charting "new ways of life" for peacetime, including family airplanes and wall-sized televisions. And there was optimistic talk of a "Miracle

Bean"—"the lowly soy—eventually you may walk on it, sit on it, ride on it, and wear it."[6]

Material comfort became the goal of the postwar middle classes. Whether considered in terms of income levels or lifestyles, in the 1940s and 1950s more Americans were better off than ever before—and most counted on their good fortune to continue. An expression of postwar prosperity was the baby boom. Propelled by what Landon Jones had termed "the Procreation Ethic," from 1946 through 1964 births hit record highs, with a total of 75.9 million babies being born in the United States. And the baby boom spelled business for builders, manufacturers, and school systems. "Take the 3,548,000 babies born in 1950," wrote Sylvia F. Porter in her syndicated newspaper column. "Bundle them into a batch, bounce them all over the bountiful land that is America. What do you get?" Porter's answer: "Boom. The biggest, boomiest boom ever known in history. Just imagine how much these extra people, these new markets, will absorb—in food, clothing, in gadgets, in housing, in services. Our factories must expand just to keep pace."[7]

To numerous people, it seemed the American dream was coming true. But it was a selective dream, the exceptions to which went unnoticed by most Americans. And it was a time when the lack of equal opportunities for women was concealed by an emphasis on femininity, piety, and family togetherness. Just as the Cold War era was conservative in international affairs, so too conservatism prevailed in the areas of courtship, marriage, sexual orientation, and, underlying all else, gender roles. As the demographer Richard A. Easterlin has noted, for example, America's older homefront girls born in the early to mid-1930s were "in their teens or preteens during World War II, and ... married and became mothers in the 1950s and 1960s." Moreover, these homefront girls, as another demographer has observed, "married at a younger age than any other age group in American history.... Demographers expect 96 percent of the women born in the 1930s will marry compared to 87 percent of those born in the 1950s." Likewise, women born in the 1930s gave birth to their first child at a young age—"the last generation of parents to not postpone the birth of the first child by many years"—and they had few children after their thirtieth birthdays. "Thirty percent of the wives born in the 1930s had at least two children by their third wedding anniversary.... Only 7 percent of the women born in the 1930s are childless, the lowest proportion of childless women to any generation in American history." At the same time, this would be "the first generation in which divorce was more likely than death to terminate the first marriage."[8]

Between the women of this generation and those in the next, there is a cultural divide. Women born in the 1930s, Katherine S. Newman has observed in a study of downward mobility in America, "instantly recognize the values and expectations they hold in common and feel a kinship to their age-mates...." But women born in the 1940s and "marked by the sixties generation," are "products of an entirely different formative experience." When divorce occurred, both groups of women had to deal with "shattered expectations for their own lives and those of their children. But the nature of those expectations, and the ways in which they come to grips with their destruction, is a matter of generational culture."[9]

While dating had been popular before the war, "going steady" became popular afterward. Historian Beth L. Bailey has written that going steady became "a sort of play marriage, a mimicry of the actual marriage of their slightly older peers." Already, these youth—the homefront children five to ten years later—were looking for lifetime security. Going steady, historian John Modell has explained, "added a strong institutionalized node in the career that led from early heterosexual sociability to early marriage and extended downward into the high school years some of the emotional comforts of marriage."[10]

While going steady and marrying early helped to meet the emotional needs of the homefront children, world conditions exacerbated those needs. Elaine Tyler May, the historian, has recognized that youth growing up in the 1940s and 1950s, "amid a world of uncertainties brought about by World War II and its aftermath," took their marriage vows early, thus lowering the marriage age and quickly bringing "the birthrate to a twentieth-century high, after more than a hundred years of steady decline. . . ." The family and the home had become a refuge—a refuge from "Red Fascism," from the saber-rattling of the Cold War and the all-out policy to "contain" Soviet communism, and from the relentless threat of worldwide thermonuclear annihilation.[11]

But there was another reason for marrying early: it was the only societally approved way to be sexually active. While Alfred Kinsey reported in his volumes on male and female sexuality published in 1948 and 1953 that a wide variety of sexual conduct was practiced by Americans, it was also evident that taboos against premarital sexual activity, especially by women, were powerful. "Our sexual mores," observed writer Frank Conroy, who was in college in the mid-1950s, "were conservative in the extreme. . . . Fiercely monogamous, we frowned on playing the field and lived as if we were already married." Parents accorded the highest value to virginity, though they often did so obliquely—through object lessons. There were disappointingly few frank conversations about sexuality, its mechanisms and responsibilities; but mothers and fathers were quick to denounce the "bad girls" in the neighborhood. The worst shame was to have a baby out of wedlock, next to that was to have an abortion. In 1958, when Letty Cottin Pogrebin, the writer and feminist, was a nineteen-year-old college student, she had an "illegal abortion." "For those who didn't live through the 50s," she has written, "it's hard to believe the suffocating terror of sexual shame and the coercive power of social propriety." And for a Jewish young woman, "pregnancy out of wedlock was the ultimate disgrace," a "'*shonda*,' a scandal discussed in contemptuous whispers behind closed doors."[12]

It was apparent during the postwar era that just as the Russians were to be "contained," so too sexuality should be held in check. The burden for fighting the Cold War against heavy petting, not to mention against sexual intercourse, fell mainly to the girls. Another aspect of the double standard emerged in the

1950s, when men began to dream of the "Playboy" lifestyle: living in a bachelor apartment, with a stereo set, imported wine, and young women visitors. The "sexual containment ideology," Elaine Tyler May has stated, "was rooted in widely accepted gender roles that defined men as breadwinners and women as mothers. Many believed that a violation of these roles would cause sexual and familial chaos and weaken the country's moral fiber. The center of this fear was the preoccupation with female 'promiscuity.' ... "[13]

All the indices point to a postwar era of traditionalism, centered on sexual continence, early marriage, and large families; plus a belief in progress, upward mobility, and the American dream of "a four-bedroom ... ranch house in the suburbs with two children, a dog, and a station wagon." It included as well respect for authority in all forms, beginning with the family patriarch and including the leaders of government, education, religion, business, and labor. There was an almost sacred belief in the essential goodness of the United States and its mission in the world. The verities on which the homefront children fed during the war were essentially the same ones on which they came of age as adolescents and young adults. But it was these same truths that the 1960s would shake to their roots.[14]

In their recollections, the homefront children focused not only on their war-inspired attitudes and values but also on how subsequent events have challenged these beliefs. The civil rights movement, the assassination of President Kennedy, the Vietnam War, Watergate—these traumas have caused some of the homefront children to reformulate the meaning of the Second World War in their lives. Nancy Morris, who was seven at the time of the Pearl Harbor attack, wrote that her homefront childhood caused her to form "such a strong bonding with my country that for years I overlooked many of our failings." Over the years, her "feelings about what it means to be an American have matured. The blind patriotism is out. ... " Vietnam was instrumental in people's reassessments. One homefront girl wrote that her beliefs had "changed instantly ... with the My Lai incident. ... " Another, Marlene Larson, born in 1936, explained that she "finally 'grew up'" at about age thirty, when she "started reading between the lines" and "started listening to the young people in the streets." During the Vietnam War, her "'blind trust' in our government died." "We were taught" during the Second World War, Marlene recalled, that "to ask questions was to show lack of patriotism. What a crock!" Worse yet, this unfailing obedience persisted during the Cold War, and "we were still a bunch of sheep. ... " The result was "Vietnam, Watergate, and countless other situations. ... "[15]

The homefront children who wrote of their postwar disillusionment dated its arrival to the 1960s. The assassination of John F. Kennedy was a watershed. For homefront girl Donna Zimmerman, this event "most affected the way I feel about my country ... and made me much more cynical toward politicians. ... " It "stripped away forever" the innocence with which she had grown up. "Bad things happen in other countries," Donna had thought; now she knew the United States was not immune. Gay Hollis pointed to different reasons for her disillusionment. The pride which she had felt had been eroded by "our leaders' lack of personal integrity, business' lack of American fair-play, [and] government officials being caught in lies that we, the American public, were duped into supporting.... I'm angry!" she added. "A change of perspective? You bet." As the homefront innocence shattered in the 1960s and 1970s, so too did much of the unquestioning patriotism. During the war, remembered a homefront girl, Americans had "rallied together with fierce pride toward a common goal and this made me feel very secure and patriotic." But "the selfishness of individuals has eroded this feeling of 'oneness.'... No," she sighed, "I do not feel as proud to be an American as I once did."[16]

Some of the homefront children not only questioned the zealous patriotism of the war years, they redefined patriotism itself. Beginning with the civil rights movement, these women and men came to believe that it was the protesters who were the real patriots. Best exemplifying the noble ideals of the Second World War, a homefront girl ventured, was the Reverend Martin Luther King, Jr. The writer Nicolaus Mills, who marched at Selma, Alabama, in 1965 and would become an organizer for the United Farm Workers, has made the same connection. "Growing up in Ohio during the 1950s, I most admired those men who had served in battle during World War II.... They had seen death firsthand. They had done what was asked of them in the face of danger. They didn't have to guess about their courage." Mills was a college student in the 1960s at the height of the civil rights movement, and to him there was "an analogy between the fate of the Jews in Nazi Germany and the fate of blacks in the South." For him not to have marched in the South, knowing the evils of segregation and disfranchisement, would have been to repeat the moral cowardice of the "good German." Also like the Second World War, the civil rights movement had elements of "a Holy Crusade," and for Mills, those who risked their lives in the South gained "a rite of passage" richer even than that experienced by "the World War II vets."[17]

Sexual mores were changing too in the 1960s, partly as a result of the government's approval of the birth-control pill in 1960. Use of the pill accelerated, and by the middle of the decade, 42 percent of married women aged fifteen to twenty-nine were on the pill. Age again was a factor, and it was the homefront girls born in the 1940s who were more inclined than their older sisters to go on the pill when it became available. Americans formerly had linked sexuality to romance and reproduction; but in the 1960s people viewed sexuality as a means of self-expression and as a gauge of personal happiness. Eroticism and sensuality, magnified by drugs and music, loosened inhibitions and eroded the barriers erected for sexual containment. For increasing numbers of Americans too, living together no longer equaled living in sin. And as attitudes toward premarital sex changed, so did notions about pornography, homosexuality, and sex roles.[18]

Another factor that seemed to separate the older homefront children from their younger cohorts was the notion of "getting ahead." The culture of scarcity had resulted in early job-taking for males and traditional gender roles for females, particularly in the working classes. The postwar culture of abundance, however, eliminated some of the need for young people to make major job choices while still teenagers; increasing millions entered college before taking on life responsibilities, while the availability of the birth-control pill encouraged delays in marrying and starting a family. Thus, those reaching late adolescence in the 1960s might well decide to extend this stage into the 1970s. Finally, there were changes in the structure of work in American society; whereas children born during the 1930s tended to follow "a lockstep pattern" of school followed immediately by work, those born later were more inclined to follow "a postindustrial pattern of greater flexibility, more interspersing of school and work, and a more subtle adaptation between family and work patterns."[19]

War remained an ongoing concern for the homefront children. First, there was the Korean conflict from 1950 to 1953. Of the homefront children, it was mostly young men from the first subcohort, those born before the mid-1930s, who fought in Korea, but a number of men from the second subcohort also enlisted by falsifying their enlistment forms. Larry Stanton, born in 1936, joined the Army when he was fifteen years old. "The Korean War was in full swing," and Larry was afraid that it would end before he became of age. "So," he wrote, "I changed my birth certificate and got in. This feeling of patriotism had remained with me from W.W. II." Rather than wait to be drafted, young men enlisted at eighteen, or seventeen with a parent's signature. The patriotism from the Second World War, homefront boy Edward Kaucher has observed, "carried over to the Korean War. Although Korea has mostly been forgotten," he added, "those of us who served . . . did so with the same fervor as World War II."[20]

For the homefront girls, there was another expectation—indeed, a fear—that their husbands, or brothers, or sons would someday go to war. During Korea they dreaded that their fathers or older brothers would be recalled or drafted, and they wept when the men marched off again. As before, they lived with the sadness of their mothers. When one homefront girl's older brother was drafted and sent to Korea, her mother "cried day after day" and announced that "if anything ever happened to him, she would keep his casket in the living-room as long as she lived." For the homefront girls too, through several wars, there was the ordeal of waiting. As one remembered:

> It seems as if all my life I have been waiting for men to return from war.
> In the 40s my uncles.
> In the 50s my college friends. Some did not return.
> In the 60s I waited for my husband to return from two tours in Viet Nam.[21]

And during the Vietnam War, some older homefront girls waited for their sons. ". . . I thought I would lose my mind," wrote Ann Laramie, born in 1933; "& it was only then that I finally realized what my mother must have gone through." For many of these mothers, Vietnam presented a challenge to cherished beliefs. While still patriots, these women, as one explained, were "unwilling to sacrifice" their sons for that war. "Every mother's son whose life was lost there was a terrible waste. . . ." After visiting the Vietnam Wall in Washington, a mother of six children, four of whom served in the military, had a photograph she had taken blown up to poster-size. She hung it on her wall, but it makes her cry to see it "because it reminds me that I felt patriotic during that time also and later came to realize what a tragedy it was."[22]

America's homefront boys grew up thinking that one day they would have their war to fight, and while most ridiculed any suggestion that they might be afraid of war's violence, some had secret doubts about their own courage and feared that, under fire, they might flee. But the homefront girls harbored a different fear. "As a teen," one wrote, "I remember thinking I would never get married because I wouldn't want my husband to get drafted and I wouldn't want to have sons, because they might have to go to war." But she had two sons. "When my first son was born," she recalled, "the fear returned . . . and I prayed he would never be sent to war, but when our second son was born in 1964, I remember hearing on the radio while in the hospital about Vietnam. Then, I thought about this THING that would never go away. Please, God, no war for my sons!" Other homefront girls shared this deep apprehension, including one who wrote that "when I grew up I wondered did I really want to bring children into the world where life was so cheap that we could kill and kill and never stop killing. . . ."[23]

Now that the homefront children are in their late forties and fifties, they have begun to evaluate their lives. They are in late midlife; before long the first of them will enter old age. They now have to deal with personal losses. Their children have grown up and moved out. Their parents have become enfeebled, and many have died. For their part, the homefront children, particularly the older ones, have suffered physiological and neurological losses; and they have had to relinquish their heroic egos and accept their limitations. Some are saddened too by the world which their children are inheriting. "We no longer believe," wrote a homefront boy, "that our children's lives will be better than our own, [or] that history is a chronicle of unending progress. . . ."[24]

According to Erik Erikson's "Eight Stages of Man," the younger of the homefront children are in stage seven: generativity versus stagnation. Erikson has defined generativity as the virtue of caring and reaching out to others, and as "the interest in establishing and guiding the next generation. . . ." Contrarily, stagnation is a self-indulgent "regression from generativity" resulting in "a pervading sense . . . of individual stagnation and interpersonal impoverishment." By the term "generativity," Erikson said later, "I mean everything that is generated from generation to generation: children, products, ideas, *and* works of art." The older homefront children over fifty are in Erikson's eighth stage: ego integrity versus despair. "It is the ego's accrued assurance of its proclivity for order and meaning. . . . It is the acceptance of one's one and only life cycle as something that had to be. . . ." It means, Erikson wrote, that the "possessor of integrity . . . knows that an individual life is the accidental coincidence of but one life cycle with but one segment of history. . . ."[25]

Thus, at this point in one's life, the psychologist Mihaly Csikszentmihalyi has written, the "task of making sense of the past" becomes very important, "bringing together what one has accomplished and what one has failed to accomplish in the course of one's life into a meaningful story that can be claimed as one's own." Not surprisingly, some of America's homefront children have taken up the job of reminding subsequent generations of their wartime history. ". . . I never, never imagined a future," lamented one homefront girl, "in which large numbers of people did not even remember the War!" Filmmakers, composers, and writers from this cohort have worked to fill that void. Some of the homefront children, notably Woody Allen (b. 1935), have made movies about childhood on the homefront. Steve Reich (b. 1936), the composer, has written music of great orig- inality and power, such as *Different Trains* (1988), which contrasts the trains car- rying Jews to their deaths in Nazi camps with those transporting Americans across the country during the Second World War. Reich has recalled that as a boy during the war he took long train trips across the United States. "You know that famous photo of the little kid in the Warsaw ghetto with his hands up in the air?" he asked. "He looks just like me! . . . had I been across the ocean, I would have been on another train. . . . and I would be dead." Other homefront children have writ- ten fiction, plays, and autobiographies depicting the war years; these writers include Charles Fuller, Ellen Gilchrist, Gail Godwin, Vivian Gornick, N. Scott Momaday, Willie Morris, Marge Piercy, and Philip Roth, all born between 1934 and 1939.[26]

When war broke out again, in 1991 in the Persian Gulf, the minds of many homefront children flashed back to their childhood years. And they pondered how the Second World War had affected not only them, but their children. "I wonder how many of us, now in our fifties, walk around not wounded or scarred but certainly marked by such experiences," asked a homefront girl, "—and I won- der how our being marked affected the way we treat our own children." Many expressed gratitude that their children would not be participating in the Persian Gulf conflict. A woman, whose son had lost his leg in a motorcycle accident while in the Army, wrote that for the first time she saw some "good" in this tragedy: "He is not driving a Bradley in Saudi Arabia." Another drew upon her homefront recollections in observing that "not everyone wearing a yellow ribbon" actually agreed with President George Bush's decision for war; instead, they were patri- otically supporting the wartime unity they had learned fifty years earlier. Gener- ativity infused the response of Ann David to the war. Born in 1937, she is a first- grade teacher. "When the Gulf War started this year," she wrote, "I found my memories from WWII helped me to know what to say to the children." Contin- ually reassuring them that "the war was too far away for them to get bombed or shot at," she showed them "over and over again on world maps how far away the war was." As Ann David knew, their fears were real.[27]

While the revived patriotism of the 1980s and 1990s has exalted the Second World War, this is not a unanimous opinion among the homefront children. They remember the era's shortcomings even as they laud the oneness of purpose that inspired them. In their letters, none of the homefront children ventured any counterfactual, or "What if?" history, but they evidently possess inner conflict about these years. What if they had been children during a peaceful era? Would they have preferred that? The homefront boys and girls believe they probably would have enjoyed greater emotional security had they grown up during peacetime. They are ambivalent, however, for they are still proud to be members of "the homefront generation" and to have been the bearers of the high ideals for which America's children stood between 1941 and 1945.

Some homefront children have a unique vantage point on the war years. One is Tatsuko Anne Tachibana, who was nine when her family was removed from its home in Oceanside, California, after the Pearl Harbor attack. The Federal Bureau of Investigation, which had identified her father as a "dangerous, enemy alien," imprisoned him at a detention center in New Mexico. Meanwhile, the Army took the rest of the family to the Poston internment camp in Arizona, where they stayed for most of the war. "As I look back over the years," Anne wrote, "I certainly harbor no bitterness or hatred toward this country, just a bit of sadness and many unanswered questions. . . ." She hopes "our country has learned a lesson," but she is doubtful. "Is it cynical," she asks, "to doubt the power of history and its lessons? We, as a people, seem doomed to repeat our mistakes."[28]

It is not cynical but realistic to doubt history's power. Children have never asked to be born into situations of war, and yet they repeatedly have been, and they have suffered whether in the war zone or on the homefront. And it is because of adults' repeated mistakes that children, who understand war least, have been so deeply marked by it. For their sake, as well as our own, we must strive for enlightenment in reversing history.

NOTES

Preface

1. Paul Buhle, *History and the New Left: Madison, Wisconsin, 1950–1970* (Philadelphia: Temple Univ. Press, 1990).

In addition, the early 1960s marked the beginning of a time of exciting changes in the historiography of the United States. Bracketing this era are two articles published ten years apart in the *American Historical Review*. In 1963, during my first year in graduate school, Carl Bridenbaugh, the colonial historian, warned that many students and younger professors were "products of lower middle class or foreign origins, and their emotions not infrequently get in the way of historical reconstructions. They find themselves in a very real sense outsiders on our past and feel themselves shut out"; the result was that they could not "communicate to and reconstruct the past." Budding scholars disregarded these concerns and began studying ethnic and racial culture, racial and class conflict, gender issues in American life, popular culture, and other topics. Ten years later, in 1973, the *American Historical Review* published an eloquent rebuttal to Bridenbaugh's ethnocentrism. Written by Herbert G. Gutman, an outsider whose parents were Jewish immigrants from Poland, the article examined ethnic culture and its interplay with, and resistance to, the industrial society that emerged in America during the nineteenth and early twentieth century. See Bridenbaugh, "The Great Mutation," *American Historical Review* 68 (Jan. 1963), 315–31; Gutman, "Work, Culture, and Society in Industrializing America, 1815–1919," *American Historical Review* 78 (June 1973), 531–88; Ira Berlin, "Introduction: Herbert G. Gutman and the American Working Class," in Berlin, ed., *Power & Culture: Essays on the American Working Class: Herbert G. Gutman* (New York: Pantheon, 1987), 3–69; Peter Novick, *That Noble Dream: The "Objectivity Question" and the American Historical Profession* (Cambridge: Cambridge Univ. Press, 1988), 469–521.

2. Tuttle, *Race Riot: Chicago in the Red Summer of 1919* (New York: Atheneum, 1970).

3. Christoph Conrad, "Conference on the Elderly in a Bureaucratic World: New Directions for History of Old Age," *Perspectives* (American Historical Association) 21 (Oct. 1983), 5.

4. Geertz, "The Impact of the Concept of Culture on the Concept of Man," in John R. Platt, ed., *New Views of the Nature of Man* (Chicago: Univ. of Chicago Press, 1965), 117; Geertz, *The Interpretation of Cultures: Selected Essays* (New York: Basic Books, 1973); Ronald G. Walters, "Signs of the Times: Clifford Geertz and Historians," *Social Research* 47 (Autumn 1980), 537–56.

5. Berlin, "The Concept of Scientific History," in Henry Hardy, ed., *Concepts and Categories: Philosophical Essays by Isaiah Berlin* (New York: Penguin, 1978), 132.

6. Arendt, *Men in Dark Times* (New York: Harcourt, Brace and World, 1968), 25.

Chapter 1. Pearl Harbor: Fears and Nightmares

1. Letter 423. See also letters 62, 158.

2. Letter 58; poem published in *Boston Herald*, April 12, 1942.

3. Letter 423; *Kansas City Star*, Jan. 14, 1989; *New York Times*, June 4, 1987.

4. Letter 16.

5. Letters 410, 42.

6. Letters 45, 131, 179.

7. Letters 16, 208.

8. Letter 256.

9. Letters 136, 191, 270, 342; Mrs. Arthur Gill to Children's Bureau, Jan. 12, 1942, in Records of the Children's Bureau, Record Group 102, National Archives, Washington, D.C.

10. Letters 246, 0.3.

11. Dr. Leslie B. Hohman, "How Should Children Be Trained for a War Situation," *Ladies' Home Journal* 59 (April 1942), 110; letters 12, 28, 342.

12. Joseph C. Solomon, "Reactions of Children to Black-Outs: A Preliminary Note," *American Journal of Orthopsychiatry* 12 (April 1942), 361–62; Eleanor Clifton, "Some Psychological Effects of the War As Seen by the Social Worker," *The Family: Journal of Case Work* 24 (June 1943), 127; Dorothy W. Baruch, *You, Your Children, and the War* (New York: D. Appleton-Century, 1942), 34–35; J. Louise Despert, "School Children in Wartime," *Journal of Educational Sociology* 16 (Dec. 1942), 222–24; Florene M. Young, "A Juvenile Case of War-Connected Trauma," *Journal of Psychology* 19 (Jan. 1945), 31–42. J. Louise Despert wrote of her clients: "In every case where anxiety in relation to the war was reported or observed, the child had previously presented an anxiety problem." Despert, *Preliminary Report on Children's Reactions to the War, Including a Critical Survey of the Literature* (New York: Josiah Macy, Jr., Foundation, 1942), 50.

13. Letters 11, 442B.

14. Letters 21, 32, 62, 66, 67, 83A, 122, 206C, 240C, 246, 316, 320, 440.

15. Letter 88; interview with Nick D. Vaccaro.

16. Letter 270; also letters 143, 206J, 219, 222, 267.

17. Letters 21, 28, 25, 63, 186, 240C, 244, 355; Lauretta Bender and John Frosch, "Children's Reactions to the War," *American Journal of Orthopsychiatry* 12 (Oct. 1942), 573, 579–80; Elisabeth R. Geleerd, "Psychiatric Care of Children in Wartime," *American Journal of Orthopsychiatry* 12 (Oct. 1942), 589–91; Mildred Burgum, "The Fear of Explosion," *American Journal of Orthopsychiatry* 14 (April 1944), 349–57.

18. "Air Raids and the Schools," *Journal of the National Education Association* 31 (Feb. 1942), 57–58; "Air Raid Protection and Evacuation Planning," *Education for Victory* 1 (Jan. 15, 1943), 23–24; Catherine Mackenzie, "Repertoire of Games," *New York Times*, May 31, 1942; letters 228, 267, 292, 332. A guide that stressed children's emotional as well as physical needs is Alice Brady, *Children Under Fire* (Los Angeles: Columbia, 1942), 171–82.

19. Letter 240.

20. Jean A. Thompson, "Pre-school and Kindergarten Children in Wartime," *Mental Hygiene* 26 (July 1942), 415; letters 25, 146, 312, 429.

21. Letters 21, 25, 27, 39, 83E, 340, 392; August A. Bronner, "Child Guidance in the Crisis," *American Journal of Orthopsychiatry* 12 (Oct. 1942), 594–97.

22. Letters 73, 95, 198, 206E, 292, 446; Dr. Martha M. Eliot to Dr. Lowell J. Reed, Oct. 17, 1942, in Record Group 102..

23. Letter 62.

24. Letters 36, 434.

25. Bert Webber, *Silent Siege: Japanese Attacks Against North America in World War II* (Fairfield, WA: Ye Galleon Press, 1983), 226–31; Robert C. Mikesh, *Japan's World War II Balloon Bomb Attacks on North America* (Washington, D.C.: Smithsonian Institution Press, 1990).

26. Webber, *Silent Siege*, 226–27; *New York Times*, May 6, 1985; Ventura County (California) *Star*, Aug. 12, 1985.

27. Webber, *Silent Siege*, 227–29.

28. *Ibid.*, 229–33; *New York Times*, May 6, 1985; letter 412. There is still a danger from unexploded balloon-bombs lying in remote areas of the United States and Canada. As of the mid-1980s, only about 300 of the 6000 bombs released had been found, but the only deaths to result so far have come from the Oregon tragedy.

29. Letter 82A; Sidonie Matsner Gruenberg, ed., *The Family in a World at War* (New York: Harper & Brothers, 1942), 266–67.

30. Letters 37, 48, 51, 174, 199, 316, 447.

31. Letters 129, 55D. For an extended discussion of the impact of the newsreels on children, see Chapter 9.

32. Letters 56, 61, 195, 286, 316, 354.

33. Letters 29, 271, 311, 385, 391, 442A, 442C, 442D, 442E.

34. Letters 29, 219, 253, 442F, 447.

35. Solomon, "Reactions of Children to Black-Outs," 361–62; letters 355, 420; Sherman Little, "Children in Wartime: A Mental Hygiene Approach to Their Care," *American Journal of Nursing* 43 (April 1943), 347–50; Charlotte Towle, "The Effect of the War upon Children," *Social Service Review* 17 (June 1943), 152; Eugene C. Ciccarelli, "Measures for the Prevention of Emotional Disorders," *Mental Hygiene* 26 (July 1942), 384.

36. Hohman, "How Should Children Be Trained for a War Situation," 110; *New York Times*, Feb. 8, 1942.

37. For a summary of this research, see Despert, *Preliminary Report on Children's Reactions to the War*, 3–36. Decades later, many of these children were still angry about being taken away from their families; see Ben Wicks, *No Time to Wave Goodbye* (London: Bloomsbury, 1988), a collection of interviews with and letters from the former evacuees.

38. *New York Times*, May 9, Aug. 6, 1942; Ciccarelli, "Measures for the Prevention of Emotional Disorders," 386–88; "Children after Air Raids," *The Spectator* 168 (Jan. 9, 1942), 26; Freud and Burlingham, *War and Children* (New York: Medical War Books, 1943); Freud and Burlingham, *Infants Without Families* (New York: International Univ. Press, 1944).

39. Martha M. Eliot, *Civil Defense Measures for the Protection of Children: Reports of Observations in Great Britain, Feb., 1941*, Children's Bureau Publication No. 279 (Washington: GPO, 1942); Despert, *Preliminary Report on Children's Reactions to the War*, 85.

40. "Care of Babies During Air Raids Is Explained," *Science News Letter* 41 (March 28, 1942), 200; Hazel Corbin, "The Baby in a Blackout," *Public Health Nursing* 35 (Jan. 1943), 26–27; "Answers Given to Child's Questions about War," *Science News Letter* 41 (June 20, 1942), 395.

41. *New York Times*, Dec. 14, 1941.

42. *New York Times*, June 18, 1944, quoted in David S. Bertolotti, Jr., "The Press and Technology III: World War II and Aviation Technology," *Journal of American Culture* 7 (Spring/Summer 1984), 108.

43. Lois Meek Stolz, *Our Changing Understanding of Young Children's Fears, 1920–1960* (New York: National Association for the Education of Young Children, 1964), 8; Gruenberg, *Family in a World at War*, 256; Despert, *Preliminary Report on Children's Reactions to the War*, 68; letters 256, 228.

Chapter 2. Depression Children and War Babies

1. Letters 32, 197, 345, 413; Michael Zuckerman, "Dr. Spock: The Confidence Man," in Charles E. Rosenberg, ed., *The Family in History* (Philadelphia: Univ. of Pennsylvania Press, 1975), 192–96.

2. Gesell and Ilg, *The Child from Five to Ten* (New York: Harper & Brothers, 1946), 447–49.

3. Lois Meek Stolz, *Our Changing Understanding of Young Children's Fears, 1920–1960* (New York: National Association for the Education of Young Children, 1964), 8; Sidonie Matsner Gruenberg, ed., *The Family in a World at War* (New York: Harper & Brothers, 1942), 256, 261.

4. Elder, "Social History and Life Experience," in Dorothy Eichorn et al., eds., *Present and Past in Middle Life* (New York: Academic, 1981), 3; Elder, "The Timing of Lives" (unpub. ms.), 1; Elder and Avshalom Caspi, "Human Development and Social Change: An Emerging Perspective on the Life Course," in Nial Bolger et al., eds., *Persons in Context: Developmental Processes* (New York: Cambridge Univ. Press, 1988), 77–113.

5. Elder, "Families and Lives: Some Developments in Life-Course Studies," in Tamara Hareven and Andrejs Plakans, eds., *Family History at the Crossroads: A Journal of Family History Reader* (Princeton: Princeton Univ. Press, 1987), 179–99; Elder, "Family History and the Life Course," in Tamara K. Hareven, ed., *Transitions: The Family and the Life Course in Historical Perspective* (New York: Academic, 1978), 21; Elder, "Perspectives on the Life Course," in Elder, ed., *Life Course Dynamics: Trajectories and Transitions, 1968–1980* (Ithaca: Cornell Univ. Press, 1985), 23–49; Elder and Caspi, "Human Development and Social Change," 77–79. For contemporary discussions of the relationship of development stages to the psychological impact of wartime change, see Lois Barclay Murphy, "The Young Child's Experience in War Time," *American Journal of Orthopsychiatry* 12 (July 1943), 497–501; Helen Ross, "Emotional Forces in Children As Influenced by Current Events," *American Journal of Orthopsychiatry* 12 (July 1943), 502–4; Bert I. Beverly, "Effect of War Upon the Minds of Children," *American Journal of Public Health* 33 (July 1943), 793–98; Francis E. Merrill, *Social Problems on the Home Front: A Study of War-time Influences* (New York: Harper & Brothers, 1948), 55–59; Lois Meek Stolz, "The Effect of Mobilization and War on Children," *Social Casework* 32 (April 1951), 143–49.

6. Ryder, "The Cohort As a Concept in the Study of Social Change," *American Sociological Review* 30 (Dec. 1965), 845; Glen H. Elder, Jr., *Children of the Great Depression: Social Change in Life Experience* (Chicago: Univ. of Chicago Press, 1974), 15–17; John Modell, *Into One's Own: From Youth to Adulthood in the United States, 1920–1975* (Berkeley: Univ. of California Press, 1989), 165–68; Elder and Caspi, "Human Development and Social Change," 80–82.

7. Ruth Shonle Cavan, *The American Family* (New York: Thomas Y. Crowell, 1953), 552–53; William Fielding Ogburn, "Marriages, Births, and Divorces," *Annals of the American Academy of Political and Social Science* 229 (Sept. 1943), 21–22; Andrew G. Truxall and Francis E. Merrill, *The Family in American Culture* (New York: Prentice-Hall, 1947), 292. On the subject of the returning war veterans and their war-born children, see Lois Meek Stolz et al., *Father Relations of War-Born Children* (Stanford: Stanford Univ. Press, 1954).

8. U.S. Bureau of the Census, *Historical Statistics of the United States, Colonial Times to 1970, Bicentennial Edition* (Washington: GPO, 1975), Part 1, 64; Modell, *Into One's Own*, 162–64; Virginia L. Galbraith and Dorothy S. Thomas, "Birth Rates and the Interwar Business Cycles," *Journal of the American Statistical Association* 36 (Dec. 1941), 465–76.

9. U.S. Census Bureau, *Historical Statistics of the United States, Bicentennial Edition*, Part 1, 64; Philip M. Hauser, "Population and Vital Phenomena," *American Journal of Sociology* 48 (Nov. 1942), 310–13; Modell, *Into One's Own*, 172–73, 179; Ogburn, "Marriages, Births, and Divorces," 20–22; Hugh Carter and Paul C. Glick, *Marriage and Divorce: A Social and Economic Study* (Cambridge: Harvard Univ. Press, 1976), 40–43; Truxall and Merrill, *The Family in American Culture*, 290; Karen Anderson, *Wartime Women: Sex Roles, Family Relations, and the Status of Women During World War II* (Westport, CT: Greenwood Press, 1981), 76–77.

10. U.S. Census Bureau, *Historical Statistics of the United States, Bicentennial Edition*, Part 1, 64. The marriage rates (per 1000 unmarried females over age 14) for 1943, 1944, and 1945 were 83.0, 76.5, and 83.6. See also Carter and Glick, *Marriage and Divorce*, 40–43.

11. U.S. Bureau of the Census, *The Wartime Marriage Surplus*, Population Series PM-1, No. 3 (Nov. 12, 1944); Bureau of the Census, *Marital Status of the Civilian Population: Feb., 1946*, Population Series P-S, No. 10 (Oct. 14, 1946); and Bureau of the Census, *Marriage Licenses Issued in Cities of 100,000 Inhabitants or More, 1939 to 1944, with Statistics by Months, 1941 to 1944*, Population Series PM-1, No. 4 (Dec. 14, 1945), all cited in Truxall and Merrill, *The Family in American Culture*, 282, 287–89; *New York Times*, Sept. 12, 1943, Oct. 31, 1944.

12. Truxall and Merrill, *The Family in American Culture*, 282.

13. Samuel Tenenbaum, "The Fate of Wartime Marriages," *American Mercury* 56 (Nov. 1945), 531; U.S. Children's Bureau, Statistical Series Number 1, *Maternal and Infant Mortality* (Washington: Federal Security Agency, 1947), 2–4; J. Yerushalmy, "Births, Infant Mortality, and Maternal Mortality in the United States—1942," *Public Health Reports* 59 (June 23, 1944), 798–801; Truxal and Merrill, *The Family in American Culture*, 289.

14. Wilson H. Grabill et al., *The Fertility of American Women* (New York: Wiley [for the Social Science Research Council in cooperation with the Bureau of the Census],

1958), 26; Ogburn, "Marriages, Births, and Divorces," 20–21; Hauser, "Population and Vital Phenomena," 313; U.S. Census Bureau, *Marriage Licenses Issued in Cities of 100,000 Inhabitants or More, 1939 to 1944*, quoted in Truxall and Merrill, *The Family in American Culture*, 291; Katharine Whiteside Taylor, "Shall They Marry in Wartime?," *Journal of Home Economics* 34 (April 1942), 215–16; Ernest W. Burgess and Harvey J. Locke, *The Family: From Institution to Companionship* (New York: American Book Company, 1945), 673–74; Tenenbaum, "The Fate of American Marriages," 531; J. Garry Clifford and Samuel R. Spencer, Jr., *The First Peacetime Draft* (Lawrence: Univ. Press of Kansas, 1986), 200–225.

15. John Modell and Duane Steffey, "Waging War and Marriage: Military Service and Family Formation, 1940–1950" (unpub. paper), 3; John Modell, "Normative Aspects of American Marriage Timing since World War II," *Journal of Family History* 5 (Summer 1980), 210–34; *New York Times*, Dec. 30, 1941, Jan. 10, 16, Feb. 6, 1942.

16. Tenenbaum, "The Fate of Wartime Marriages," 531; Wilson H. Grabill, "Effect of the War on the Birth Rate and Postwar Fertility Prospects," *American Journal of Sociology* 50 (Sept. 1944), 108; *New York Times*, Nov. 14, Dec. 23, 24, 30, 1941, Jan. 10, 16, July 18, 1942, April 4, 1943; U.S. Census Bureau, *Marriage Licenses Issued in Cities of 100,000 Inhabitants or More, 1939 to 1944*, quoted in Truxall and Merrill, *The Family in American Culture*, 292; Ogburn, "Marriages, Births, and Divorces," 21–23; U.S. Census Bureau, *Historical Statistics of the United States, Bicentennial Edition*, Part 1, 49. The figure for 1943 was 1,577,000 marriages; for 1944, 1,452,000; and for 1945, 1,613,000.

17. Modell, *Into One's Own*, 189–91; Ogburn, "Marriages, Births, and Divorces," 21; Truxall and Merrill, *The Family in American Culture*, 292.

18. During the war Ernest R. Burgess, the sociologist, directed his students to interview a number of women whose husbands were serving in the military (see Modell, *Into One's Own*, 182–84). These interviews, which are on deposit in the Burgess Papers, Regenstein Library, University of Chicago, deal with "adjustment problems of servicemen's wives," among other topics. See, for example, interviews 2530, 2550, 2570, 2603–5, 2607–8, 2700, 2705, 3001, 3308.   I wish to thank John Modell for bringing these interviews to my attention and for lending me his microfilm copies.

19. Truxall and Merrill, *The Family in American Culture*, 292; John F. Cuber, "Changing Courtship and Marriage Patterns," *Annals of the American Academy of Political and Social Science*, 229 (Sept. 1943), 35–36.

20. Sydney J. Harris, "Why Many Women Won't Marry," *Science Digest* 17 (May 1945), 79–82; *New York Times*, June 17, 1945; Anderson, *Wartime Women*, 77–78; Modell, *Into One's Own*, 186; Beth L. Bailey, *From Front Porch to Back Seat: Courtship in Twentieth-Century America* (Baltimore: Johns Hopkins Univ. Press, 1988), 35–36, 114–15.

21. Bossard, "The Hazards of War Marriages," *Psychiatry* (Feb. 1944), condensed in *Science Digest* 16 (July 1944), 1–2.

22. Florence Greenhoe Robbins, "Reasons for and against War Marriages," *Sociology and Social Research* 29 (Sept.-Oct. 1944), 25–34; Taylor, "Shall They Marry in Wartime?," 213–19; Evelyn Millis Duvall, "Marriage in War Time," *Marriage & Family Living* 4 (Autumn 1942), 73–74; Ruth Zurfluh, "The Impact of War on Family Life. III: Wartime Marriages and Love Affairs," *Family* 23 (Dec. 1942), 304–12; H. J. Locke, "Family Behavior in Wartime," *Sociology and Social Research* 27 (March-April 1943), 281–

84; S. Reimer, "War Marriages Are Different," *Marriage & Family Living* 5 (Nov. 1943), 84–85, 87; C. R. Rogers, "War Challenges Family Relationships," *Marriage & Family Living* 5 (Nov. 1943), 86–87; *New York Times*, Feb. 2, 1942; Cavan, *The American Family*, 550–51; Ruth Shonle Cavan, *The Family* (New York: Thomas Y. Crowell, 1948), 404–7; Tenenbaum, "The Fate of Wartime Marriages," 530; James H. S. Bossard, "What War Is Still Doing to the Family," in Howard Becker and Reuben Hill, eds., *Family, Marriage and Parenthood* (Boston: D. C. Heath, 1948), 712–15.

23. Truxall and Merrill, *The Family in American Culture*, 562; Modell and Steffey, "Waging War and Marriage," 8. "World War II," Ernest W. Burgess and Harvey J. Locke have written, "rather than reversing or introducing new trends, tended to speed up fundamental transformations in family structure, functions, and relationships which have been in the process of change for years" (*The Family*, 691). Other social scientists tend to agree; see Truxall and Merrill, *The Family in American Culture*, 296.

24. Judy Barrett Litoff, David C. Smith, Barbara Wooddall Taylor, and Charles E. Taylor, *Miss You: The World War II Letters of Barbara Wooddall Taylor and Charles E. Taylor* (Athens: Univ. of Georgia Press, 1990), 5, 291, 293, 303.

25. Robbins, "Reasons for and against War Marriage," 34; H. L. J. Carter and L. Foley, "What Are Young People Asking about Marriage?," *Journal of Applied Psychology* 27 (June 1943), 283–87.

26. Charles E. King, "Attitudes toward Marriage and Motherhood of 183 College Women," *Social Forces* 22 (Oct. 1943), 89–91; Ellen K. Rothman, *Hands and Hearts: A History of Courtship in America* (New York: Basic Books, 1984), 299–301; Truxall and Merrill, *The Family in American Culture*, 562–69; Anne Maxwell, "Should Marriage Wait?," *Woman's Home Companion* 69 (Nov. 1942), quoted in Modell and Steffey, "Waging War and Marriage," 8.

27. Popenoe, "Now Is the Time to Have Children," *Ladies' Home Journal* 59 (July 1942), 61; Bailey, *From Front Porch to Back Seat*, 102.

28. Taylor, "Shall They Marry in Wartime?," 218–19.

29. Taeuber and Taeuber (for the Social Science Research Council in cooperation with the Bureau of the Census), *People of the United States in the 20th Century* (Washington: GPO, 1971), 142–48, 171–73; U.S. Children's Bureau Publication 272, *White House Conference on Children in a Democracy: Final Report* (Washington: GPO, 1942), 12.

30. U.S. House of Representatives, 77th Cong., 2nd Sess., Subcommittee of the Committee on Appropriations, *Department of Labor—Federal Security Agency Appropriation Bill for 1943*, Part 1 (Washington: GPO, 1942), 374; *New York Times*, June 4, Nov. 12, 1942, Jan. 12, March 24, Oct. 1, 1944; U.S. Children's Bureau Publication 302, Edward E. Schwartz and Eloise R. Sherman, *Community Health and Welfare Expenditures in Wartime: 1942 and 1940—30 Urban Areas* (Washington: GPO, 1944), 2; U.S. Public Health Service, *Public Health Reports* 58 (April 16, 1943), 647; 59 (June 23, 1944), 800–801; Alfred Toombs, "War Babies," *Woman's Home Companion* 71 (April 1944), 32.

31. Taeuber and Taeuber, *People of the United States in the 20th Century*, 156; Susan Householder Van Horn, *Women, Work, and Fertility, 1900–1986* (New York: New York Univ. Press, 1988), 83–100, 116–23; Steven Mintz and Susan Kellogg, *Domestic Revolutions: A Social History of American Family Life* (New York: Free Press, 1988), 153–54; *New York Times*, Jan. 11, 1987.

32. Hauser, "Population and Vital Phenomena," 310–13; Hauser, "Population,"

American Journal of Sociology 47 (May 1942), 816–28; Grabill et al., *Fertility of American Women*, 27–28; *New York Times*, Oct. 16, Dec. 8, 1944, Jan. 24, 1945; George S. Masnick and Joseph A. McFalls, Jr., "A New Perspective on the Twentieth-Century American Fertility Swing," *Journal of Family History* 1 (Winter 1976), 227; Ogburn, "Marriages, Birth, and Divorces," 24–27; Truxall and Merrill, *The Family in American Culture*, 292–97; U.S. Census Bureau, *Historical Statistics of the United States, Bicentennial Edition*, Part 1, 49. Earlier, during the Great Depression, the birth rate also responded to events, but negatively. Demographer Wilson H. Grabill has written: "The trend in the birth rate between 1934 and 1940 bears a striking resemblance to trends in economic activities 9–10 months earlier" (Grabill, "Effect of the War on the Birth Rate," 108).

33. Nathan Sinai and Odin W. Anderson, *EMIC (Emergency Maternity and Infant Care): A Study of Administrative Experience* (Ann Arbor: School of Public Health, Univ. of Michigan, 1948), 22–53, 81–88; *The Child* 7 (Nov. 1942, April 1943), 64, 148; 8 (Oct. 1943), 55; 10 (July 1945), 11; Martha M. Eliot, "Maternity Care for Service Men's Wives," *Survey Midmonthly* 79 (April 1943), 113–14; "Uncle Sam Provides Obstetric Care," *American Journal of Nursing* 43 (May 1943), 470–71; Martha M. Eliot, "Experience with the Administration of a Medical Care Program for Wives and Infants of Enlisted Men," *American Journal of Public Health* 34 (Jan. 1944), 34–39; Amy Porter, "Babies for Free," *Collier's* 116 (Aug. 4, 1945), 18–19, 28. For a discussion of EMIC as well as of childbirth, abortion, and infant and maternal mortality during the war, see Chapter 11.

34. "Are Children Necessary?," *Better Homes & Gardens* 23 (Oct. 1944), 7; Mark Jonathan Harris et al., eds., *The Homefront: America During World War II* (New York: Putnam's, 1984), 228. During the war many Americans viewed the family as both the embodiment and savior of democracy; see Sonya Michel, "American Women and the Discourse of the Democratic Family in World War II," in Margaret Randolph Higonnet et al., eds., *Behind the Lines: Gender and the Two World Wars* (New Haven: Yale Univ. Press, 1987), 154–67. On the postwar reproductive ideology in the United States, see Elaine Tyler May, *Homeward Bound: American Families in the Cold War Era* (New York: Basic Books, 1988).

35. Amram Scheinfeld, "Motherhood's Back in Style," *Ladies' Home Journal* 61 (Sept. 1944), 136–37; "Is the Wartime Baby Out of Luck?" *Parents' Magazine* 19 (Dec. 1944), 20, 100, 102.

36. "How Many Do *You* Want?," *Woman's Home Companion* 72 (Dec. 1945), 7–8.

37. Hauser, "Population and Vital Phenomena," 318; *New York Times*, Sept. 21, Dec. 24, 1941, Jan. 6, Oct. 4, 1944, April 7, Oct. 25, 1945; Grabill et al., *Fertility of American Women*, 27; Michael S. Teitelbaum and Jay M. Winter, *The Fear of Population Decline* (Orlando: Academic, 1985); "Preview of the Postwar Generation: It Will Be Smaller; But It Can Be 'Demographically' Sounder," *Fortune* 27 (March 1943), 116–17, 130, 132; "Roughly 226,504,825 Noses," *Newsweek* (Jan. 12, 1981), 30; Ogburn, "Marriages, Births, and Divorces," 26–27; Meyer F. Nimkoff, "The Family," *American Journal of Sociology* 47 (May 1942), 865–67; Grabill, "Effect of the War on the Birth Rate," 111; "World Populations," *Life* 19 (Sept. 3, 1945), 45–48, 51; Frank W. Notestein, "Facts of Life," *Atlantic Monthly* 177 (June 1946), 75–83.

38. Jones, *Great Expectations: America and the Baby Boom Generation* (New York:

Ballantine Books ed., 1981), 19; D'Ann Campbell, *Women at War with America: Private Lives in a Patriotic Era* (Cambridge: Harvard Univ. Press, 1984), 91–92.

39. For a discussion of illegitimacy's effects on families and especially returning veterans, see Chapter 12.

40. Phillips Cutright, "Illegitimacy in the United States: 1920–1968," and Kingsley Davis, "The American Family in Relation to Demographic Change," both in Charles F. Westoff and Robert Parke, Jr., eds., Commission on Population Growth and the American Future, Research Reports, Volume 1: *Demographic and Social Aspects of Population Growth* (Washington: GPO, 1972), 383–91, 250–54; U.S. Census Bureau, *Historical Statistics of the United States, Bicentennial Edition,* Part 1, 52; *New York Times,* Dec. 6, 1945; U.S. House of Representatives, 79th Cong., 1st Sess., Subcommittee of the Committee on Appropriations, *Department of Labor–Federal Security Agency Appropriation Bill for 1946* (Washington: GPO, 1945), 270–71, 292–93; Maud Morlock, "Wanted: A Square Deal for the Baby Born Out of Wedlock," *The Child* 10 (May 1946), 167–69; Bossard, "What the War Is Still Doing to the Family," 716; Lottie S. Levi, "Unmarried Mothers in Wartime," in Child Welfare League of America, *The Impact of War on Children's Services* (New York: Child Welfare League of America, 1943), 4–8.

41. Truxall and Merrill, *The Family in American Culture,* 296–97. John Modell argues that while there was an increase in the percentage of first born during the war, the numbers were smaller than in 1946, when "there was a genuine baby boom" (*Into One's Own,* 200–201).

42. "Out of the Nowhere," *Newsweek* 25 (Jan. 22, 1945), 38, 40; *New York Times,* Jan. 2, 6, 21, 22, Feb. 18, March 8, 1945; U.S. House of Representatives, 78th Cong., 1st Sess., Subcommittee of the Committee on Naval Affairs, *Investigation of Congested Areas,* Part 3 (Washington: GPO, 1943), 653; Maud Morlock, "Babies on the Market," *Survey Midmonthly* 81 (March 1945), 67–69.

Chapter 3. "Daddy's Gone to War"

1. Letter 8.
2. Letters 72, 156, 220, 294.
3. Letters 284, 324C, 345, 446.
4. Letter 83; *Selective Service as the Tide of War Turns: The 3rd Report of the Director of Selective Service 1943–1944* (Washington: GPO, 1945), 53, 73–79, 131–35, 164–70, 586; John Modell and Duane Steffey, "Waging War and Marriage: Military Service and Family Formation, 1940–1950" (unpub. paper), 4; *Congressional Record,* Volume 88, pp. A3799–801; Volume 89, pp. 7625, 7628, 7802–3, 7821–22, 7846–57, 8118, 8750–51, 9801–2, 9811, A3334, A3811, A3875, A3881, A4012, A5104; Volume 90, pp. 155–59; *New York Times,* Aug. 19, 1942, March 11, Oct. 17, Dec. 11, 1943.
5. *Selective Service and Victory: The 4th Report of the Director of Selective Service* (Washington: GPO, 1948), 597; Selective Service System, Special Monograph No. 8, *Dependency Deferment* (Washington: Selective Service System, 1947), 79–80; John Modell, *Into One's Own: From Youth to Adulthood in the United States, 1920–1975* (Berkeley: Univ. of California Press, 1989), 188–89.
6. U.S. Bureau of the Census, *Historical Statistics of the United States, Colonial Times to 1970, Bicentennial Edition* (Washington: GPO, 1975), Part 2, p. 1141; Army

Service Forces, Office of Dependency Benefits, *Annual Report for the Fiscal Year 1944* (Newark, NJ: Office of Dependency Benefits, 1945), 72–73; Army Service Forces, Office of Dependency Benefits, *Annual Report for the Fiscal Year 1945* (Newark, NJ: Office of Dependency Benefits, 1946), 15; Creighton J. Hill, "Congress and the Fathers' Draft," *Senior Scholastic* 42 (Oct. 11–16, 1943), 15; William G. Truxall and Francis E. Merrill, *The Family in American Culture* (New York: Prentice-Hall, 1947), 299, 570; Steven Mintz and Susan Kellogg, *Domestic Revolutions: A Social History of American Family Life* (New York: Free Press, 1988), 167, 174.

7. Letters 40, 200A, 380, 413; and Whortle to Geneva D. Whortle, Aug. 6, 1943, copy in letter 413.

8. Letters 91, 206N, 297, 298, 307, 380.

9. Letters 23, 69, 101, 232, 239.

10. Letters 40, 202, 432.

11. Letters 139, 160.

12. Letters 15, 33, 257, 354.

13. Hareven, "The History of the Family as an Interdisciplinary Field," *Journal of Interdisciplinary History* 11 (Autumn 1971), 400; Bossard, "Family Backgrounds of Wartime Adolescents," *Annals of the American Academy of Political and Social Science* 236 (Nov. 1944), 33; Gerald D. Nash, *The American West Transformed: The Impact of the Second World War* (Bloomington: Indiana Univ. Press, 1985); Morton Sosna, "More Important Than the Civil War? The Social Impact of World War II on the South" (paper presented at the Southern Historical Association meeting, 1982).

14. Eleanor Clifton, "Some Psychological Effects of the War: As Seen by the Social Worker," *The Family* 24 (June 1943), 125; Hill et al., *Families under Stress: Adjustment to the Crises of War Separation and Reunion* (New York: Harper & Brothers, 1949), 57, 73; Truxall and Merrill, *Family in American Culture*, 571; "When Father's Away," *Child Study* 22 (Spring 1945), 113, 118; J. Louise Despert, "School Children in Wartime," *Journal of Educational Sociology* 16 (Dec. 1942), 227.

15. Glen H. Elder, Jr., *Children of the Great Depression: Social Change in Life Experience* (Chicago: Univ. of Chicago Press, 1974), 28, 44–46, 58–59, 61, 99–102; George E. Gardner and Harvey Spencer, "Reactions of Children with Fathers and Brothers in the Armed Forces," *American Journal of Orthopsychiatry* 14 (Jan. 1944), 40, 41–43.

16. Anne Kelton, "A Boy Needs a Man," *Parents' Magazine* 18 (April 1943), 31, 96; Peggy Robbins, "What Shall I Tell Him?," *Parents' Magazine* 19 (April 1944), 21, 94; Floyd B. Nichols, "Sons of Victory," *Hygeia* 22 (Oct. 1944), 748–49, 799. An exception is the article by Janet V. Landes, "Daddy Comes First," *Parents' Magazine* 19 (Nov. 1944), 153–54.

17. For a discussion of these studies of father absence, see Chapter 12.

18. Anne Kelton, "A Boy Needs a Man," *Parents' Magazine* 18 (April 1943), 31, 96; Peggy Robbins, "What Shall I Tell Him?," *Parents' Magazine* 19 (April 1944), 21, 94; Floyd B. Nichols, "Sons of Victory," *Hygeia* 22 (Oct. 1944), 748–49, 799. An exception is the article by Janet V. Landes, "Daddy Comes First," *Parents' Magazine* 19 (Nov. 1944), 153–54.

19. F. A. Pederson, "Does Research on Children Reared in Father-Absent Families Yield Information on Father Influence?," *The Family Coordinator* 25 (1976), 459–64; quoted in Ross D. Parke, *Fathers* (Cambridge: Harvard Univ. Press, 1981), 59–60.

20. See John A. Clausen, "American Research on the Family and Socialization," *Children Today* 46 (1978), 7–10.

21. Ruth Shonle Cavan, *The American Family* (New York: Crowell, 1953), 553; Hill et al., *Families under Stress*, 50–74; Florence Hollis, "The Impact of the War on Marriage Relationships," National Conference of Social Work, *Proceedings*, 70 (1943), 109–10. See also Chapter 12 for a discussion of the returning fathers and their problems as well as the impact of their homecoming on the children.

22. Letter 156.

23. Letter 178.

24. Letters 14, 97D, 433, 435.

25. Letter 298. These examples all deal with girls living in "families of women," but what about the boys? How did father absence fathers affect the homefront boys? Again, see Chapter 12, but the following article is suggestive: Lyn Carlsmith, "Effect of Early Father Absence on Scholastic Aptitude," *Harvard Educational Review* 34 (Winter 1964), 3–21.

26. Maryalyce Barrett, "Lest They Forget," *Parents' Magazine* 19 (Nov. 1944), 132; Juliet Danziger, "Life without Father," *New York Times Magazine* (May 7, 1944), 16; Landes, "Daddy Comes First," *Parents' Magazine* 19 (Nov. 1944), 153–54; Ethel Gorman, *So Your Husband's Gone to War!* (Garden City, NY: Doubleday, Doran, 1942); Barbara S. Bosanquet, "Far-away Father," *Woman's Home Companion* 70 (June 1943), 6; Charlotte Towle, "The Effect of the War upon Children," *Social Service Review* 17 (June 1943), 154; Gladys Denny Shultz, "If Daddy's Gone to War," *Better Homes & Gardens* 22 (Oct. 1943), 14; Leslie G. Hohman, "How to Live Without Your Husband," *Ladies' Home Journal* 61 (March 1944), 148–49; Gladys Denny Shultz, "Life without Father," *Better Homes & Gardens* 22 (June 1944), 22, 76–78; Dorothy W. Baruch, "Home without Father," *Woman's Home Companion* 71 (Aug. 1944), 69; Rosemary L. Donoghue, "Is the Wartime Baby Out of Luck?," *Parents' Magazine* 19 (Dec. 1944), 20, 100, 102; John F. Day, "While Dad's Away He Becomes a Paragon among Papas to His Child," *Life* (Aug. 13, 1945), 13–14; Marjorie Floyd, "While Daddy's Away," *Parents' Magazine* 20 (Sept. 1945), 143.

27. "Answers Given to Child's Questions about War," *Science News Letter* 41 (June 20, 1942), 395.

28. Dorothy Humphrey to the author April 13, 1989, July 23, Aug. 4, Sept. 3, 29, Oct. 8, 1990; "Freddie and Ellen, Childhood—World War II," which includes copies of scores of letters written by Dorothy Humphrey to Dr. Stanley G. Humphrey.

29. Letter 27; a similar letter is 373.

30. Letter 376.

31. Letter 380.

32. Letters 83, 232, 403, 436.

33. Letters 83F, 119, 377, 380. See also Evelyn Millis Duvall, "Loneliness and the Serviceman's Wife," *Marriage and Family Living* 7 (Autumn 1945), 77–81; Edward and Louise McDonagh, "War Anxieties of Soldiers and Their Wives," *Social Forces* 24 (Dec. 1945), 195–200.

34. Letters 15, 102.

35. Letters 157, 188.

36. Letters 237, 352.

37. Letter 7.

38. Letter 361.

39. Letters 83F, 312, 408.

40. Letters 55A, 97J, 157, 206, 213, 218, 237, 352, 357.

41. Letters 19, 21, 55A, 218, 442.

42. Letter 160; Hill et al., *Families under Stress*, 64–65; *New York Times*, Aug. 18, 24, 1942; May 16, 1943; Clifford A. Strauss, "Grandma Made Johnny Delinquent," *American Journal of Orthopsychiatry* 13 (April 1943), 343–46; Sherna Berger Gluck, *Rosie the Riveter Revisited: Women, the War, and Social Change* (Boston: Twayne, 1987), 64, 208, 231.

43. Letters 91, 135, 160, 216, 319.

44. Letters 256, 371.

45. Letter 427.

46. Ibid.; letters 89, 413; "Be a Wartime Foster Daddy," *Better Homes and Gardens* 22 (Oct. 1943), 7; Catherine Mackenzie, "Absent Fathers," *New York Times Magazine* (April 9, 1944), 29; Laura C. Reynolds, "Calling All Fathers," *Parents' Magazine* 20 (March 1945), 33, 102.

47. Letter 60.

48. Letter 64.

49. Letters 14, 178, 195.

50. Letters 5, 27, 123, 367.

51. Letters 206C, 270, 349, 416.

52. Letters 88, 208.

53. Truxall and Merrill, *Family in American Culture*, 571; letters 15, 187, 413, 434; Annette Tapert, ed., *Lines of Battle: Letters from American Servicemen, 1941–1945* (New York: Times Books, 1985), 86–87; Whortle to Susan W. Whortle, Aug. 9, 1943, in letter 413. On the importance of letters during the war, see the following by Judy Barrett Litoff and David C. Smith, "'Will He Get My Letter?' Popular Portrayals of Mail and Morale During World War II," *Journal of Popular Culture* 23 (Spring 1990), 21–43; with Barbara Wooddall Taylor and Charles E. Taylor, *Miss You: The World War II Letters of Barbara Wooddall Taylor and Charles E. Taylor* (Athens: Univ. of Georgia Press, 1990); *Since You Went Away: World War II Letters from American Women on the Home Front* (New York: Oxford Univ. Press, 1991); *Dear Boys: World War II Letters from a Woman Back Home* (University: Univ. Press of Mississippi, 1991).

54. Letters 7, 15, 55E, 73, 97A, 97C, 97D, 133, 160, 187; interview with Nick D. Vaccaro; Octavia Capuzzi Locke, "Mamma's Letter Writing," *Johns Hopkins Magazine* 39 (June 1987), 16–19.

55. Letters 101, 298.

56. Whortle to Geneva D. Whortle, Aug. 2, 6, 9, 11, 1943, and undated APO letter, all in letter 413.

57. Letters 199, 206K, 206O, 347.

58. Letter 187; Litoff and Smith, *Since You Went Away*, 134–35, 140–43.

59. The "lil Lulu" cartoons and Joe Wally's 1945 diary have been made available by Diane E. Tompkins, Kansas City, MO; also Diane Tompkins to the author, Feb. 19, Sept. 12, 1989, and 1990 interview with Diane Tompkins.

60. Randall Dwyer to "Putsie," Maureen Dwyer, undated; made available by Mau-

reen Dwyer Cytron, Dittmer, MO; letters from Charles R. Kolb to Cheryl Kolb, Sept. 2, Oct. 25, 1942, Oct. 10, Nov. 18, Dec. 12, 1943, Feb. 5, April 26, May 30, July 11, Aug. 22, Nov. 2, 30, Dec. 25, 1944, Jan. 1, 22, July 31, Sept. 9, 1945, and Cheryl Kolb to Charles R. Kolb, Oct. 22, Nov. 12, 1944, Jan. 1, 14, 29, March 30, 1945; copies of these letters made available by Cheryl Kolb Folley, Columbia, SC; also Cheryl Kolb Folley to the author, June 13, Aug. 21, Sept. 5, 1990; Tapert, *Lines of Battle*, 177–78; Litoff and Smith, *Since You Went Away*, 15–18, 139.

61. U.S. Bureau of the Census, *Historical Statistics of the United States, Bicentennial Edition*, Part 2, 1140; interview with Ann Mix, American W.W. II Orphans Network, Bellingham, WA; Thomas J. Woofter, Jr., "Paternal Orphans," *Social Security Bulletin* (Oct. 1945), 5–6; Edward E. Schwartz to Dr. Eliot, Dec. 26, 1944, in Records of the Children's Bureau, Record Group 102, National Archives, Washington, D.C.

62. Letters 52, 87, 220, 248, 259, 278, 322, 332, 363, 434, 442H; Eugene Lerner and Lois B. Murphy, "Further Report on Committee for Information on Children in Wartime," *Journal of Social Psychology* 18 (Nov. 1943), 416; Ruth Heller Freund, "When Father Stays Home from War," *Parents' Magazine* 20 (April 1945), 152–53.

63. Gesell, Ilg, and colleagues, *The Child from Five to Ten* (New York: Harper & Brothers, 1946), 449–50; Arnold Gesell, Frances L. Ilg, and Louise Bates Ames, *Youth: The Years from Ten to Sixteen* (New York: Harper & Brothers, 1956), 64, 102–3, 137, 172, 491–92; *Kansas City Star*, Oct. 1, 1990; Lawrence S. Wrightsman, "Parental Attitudes and Behaviors as Determinants of Children's Responses to the Threat of Nuclear War," *Vita Humana* 7 (1964), 178–85.

64. Ellen J. O'Leary, "In the Midst of Life," *Parents' Magazine* 20 (Sept. 1945), 142; Christine Lyons, "No Man in the House," *Parents' Magazine* 21 (July 1946), 28, 68.

65. Letters 62, 382.

66. Letters 52, 96, 177, 206E, 278, 284, 357, 389, 431.

67. Letters 322, 405.

68. Letters 26, 83F, 208, 312, 320, 381, 389.

69. Letter 55A, 177, 206O, 289; Gladys Denny Shultz, "'We Regret to Inform You,'" *Better Homes & Gardens* 22 (Aug. 1944), 12, 59–60.

70. Letters 2, 76, 206A, 298.

Chapter 4. Homefront Families on the Move

1. Lowell Juilliard Carr and James Edson Stermer, *Willow Run: A Study of Industrialization and Cultural Inadequacy* (New York: Harper & Brothers, 1952), 3–4, 41; Alan Clive, *State of War: Michigan in World War II* (Ann Arbor: Univ. of Michigan Press, 1979), 21, 24; "Work and Wage Experience of Willow Run Workers," *Monthly Labor Review* (Dec. 1945, reprinted with additional data as Serial No. R.1809), 1–2; Agnes E. Meyer, *Journey through Chaos* (New York: Harcourt, Brace, 1944), 31, 33–39, 102–3; *New York Times*, March 2, 1941.

2. U.S. Bureau of the Census, *Characteristics of the Population, Labor Force, Families, and Housing, Detroit-Willow Run Congested Production Area: June 1944*, Series CA-3, No. 9 (Washington: Bureau of the Census, 1944), 7.

3. Mrs. Castle's diary is quoted in Carr and Stermer, *Willow Run*, 96–100, 356–63. See also Karen Anderson, *Wartime Women: Sex Roles, Family Relations, and the Status*

of Women During World War II (Westport, CT: Greenwood, 1981), 85–86; Louise Olson and Ruth Schrader, "The Trailer Population in a Defense Area," *Sociology and Social Research* 27 (March-April 1943), 295–302. No first name was given for Mrs. Castle, so the author has supplied one.

4. Carr and Stermer, *Willow Run*, 96–110; Agnes E. Meyer, "Detroit's Willow Run Area Is a Housing Nightmare," *Washington Post*, March 3, 1943; and newspaper clippings in Records of the Committee for Congested Production Areas, Entry 16 (PI-128), Newspaper Clippings Folder: Detroit-Willow Run, in Record Group 212, National Archives, Washington, D.C.

5. Carr and Stermer, *Willow Run*, 96–110; Clive, *State of War*, 108; Judy Rosen, "Women in the Plant, the Community, and the Union at Willow Run" (M.A. thesis, Univ. of Michigan, 1978), 7–10, 12, 19–22; Anderson, *Wartime Women*, 85; Richard R. Lingeman, *Don't You Know There's a War On? The American Home Front, 1941–1945* (New York: Putnam's [Capricorn edition], 1976), 82–84; Frances E. Merrill, *Social Problems on the Home Front: A Study of War-time Influences* (New York: Harper & Brothers, 1948), 32.

6. Carr and Stermer, *Willow Run*, 102–3, 249; Clive, *State of War*, 174–75.

7. Carr and Stermer, *Willow Run*, 293–300; Malcolm Rogers, "For War Workers' Children," *Michigan Education Journal* 21 (1944), 444–48; Clive, *State of War*, 176; *New York Times*, Sept. 7, 1941, Oct. 31, 1943. An interesting comparison is the Plainview-Beechwood school district located near Wichita, Kansas, which increased from 50 people to 18,000, and from zero enrollment in the public schools to 4000: "Developing an Educational Philosophy in a 'Boomtown' Community," *Education for Victory* 3 (July 20, 1944), 19.

8. Carr and Stermer, *Willow Run*, 102, 106, 110.

9. Conrad Taeuber, "Wartime Population Changes in the United States," *Milbank Memorial Fund Quarterly* 24 (July 1946), 236–37; U.S. Bureau of the Census, *Historical Statistics of the United States, Colonial Times to 1970, Bicentennial Edition* (Washington: GPO, 1975), Part 2, p. 1141; Ewan Clague, "Problems of Migration," National Conference of Social Work, *Proceedings* 72 (1945), 66–67; H. J. Locke, "Family Behavior in Wartime," *Sociology and Social Research* 27 (March-April 1943), 278–79; Ernest W. Burgess and Harvey J. Locke, *The Family: From Institution to Companionship* (New York: American Book Company, 1945), 666–68; Philip M. Hauser, "Population and Vital Phenomena," *American Journal of Sociology* 48 (Nov. 1942), 314; U.S. Bureau of the Census, *Internal Migration in the United States: April 1940, to April 1947*, Series P-20, No. 14 (Washington: Bureau of the Census, 1948), 1–2; Stephen Mintz and Susan Kellogg, *Domestic Revolutions: A Social History of American Family Life* (New York: Free Press. 1988), 155–57; *New York Times*, April 11, July 13, 1944, March 10, 1945.

10. Letters 336, 429, 433.

11. Bureau of the Census, *Internal Migration in the United States: April 1940, to April 1947*, 1, 6; Henry S. Shryock, Jr., "Wartime Shifts of the Civilian Population," *Milbank Memorial Fund Quarterly* 25 (July 1947), 270; Henry S. Shryock, Jr., "Redistribution of Population: 1940 to 1950," *Journal of the American Statistical Association* 46 (Dec. 1951), 417–37.

12. Bureau of the Census, *Internal Migration in the United States: April 1940, to April 1947*, 4; Conrad Taeuber, "Recent Trends in Rural-Urban Migration in the United

States," *Milbank Memorial Fund Quarterly* 25 (April 1947), 203–13; Taeuber, "Wartime Population Changes in the United States," 236–39; Henry S. Shryock, Jr., and Hope Tisdale Eldridge, "Internal Migration in Peace and War," *American Sociological Review* 12 (Jan. 1947), 27–36; Hauser, "Population and Vital Phenomena," 314–15; *New York Times*, Nov. 7, 1944.

13. Bureau of the Census, *Internal Migration in the United States: April 1940, to April 1947*, 2, 4.

14. Taeuber, "Wartime Population Changes in the United States," 236–39; Locke, "Family Behavior in Wartime," 278; *New York Times*, July 13, Nov. 7, 1944, March 10, July 5, 1945.

15. U.S. House of Representatives, 78th Cong. 1st Sess., Subcommittee of the Committee on Naval Affairs, *Investigation of Congested Areas* (Washington: GPO, 1943), 836, 1780; "Educational Planning in Areas of Intensive War Activities," *Education for Victory* 3 (Sept. 4, 1944), 4.

16. Zelma Parker, "Strangers in Town," *Survey Midmonthly* 79 (June 1943), 170–71; Hubert Owen Brown, "The Impact of War Worker Migration on the Public School System of Richmond, California, from 1940 to 1945" (unpub. Ph.D. diss., Stanford Univ., 1973), 169–78, 227–31, 289–306; letter 206E.

17. Letter 365.

18. Letters 69, 135.

19. Letters 97H, 403.

20. Florence Ptacek, "We're in the Army Now," *Parents' Magazine* 18 (April 1943), 21, 70–71.

21. Letters 69, 123, 365, 433.

22. Arthur L. Rautman, St. Petersburg, FL, to the author, March 18, 1990; Rautman, "Children of War Marriages," *Survey Midmonthly* 80 (July 1944), 198–99; Rautman with Edna Brower, "War Themes in Children's Stories," *Journal of Psychology* 19 (Jan. 1945), 191–202; Rautman with Edna Brower, "War Themes in Children's Stories: II. Six Years Later," *Journal of Psychology* 31 (Jan. 1951), 263–70.

23. Edward and Louise McDonagh, "War Anxieties of Soldiers and Their Wives," *Social Forces* 24 (Dec. 1945), 196; Juliet Danziger, "Daddy Comes Home on Leave," *Parents' Magazine* 19 (Oct. 1944), 72, 74.

24. Letter 298.

25. U.S. Census Bureau, *Historical Statistics of the United States, Bicentennial Edition*, part 2, pp. 1114–16.

26. Ibid., part 2, p. 1141.

27. Ibid., part 1, p. 126.

28. Hauser, "Population and Vital Phenomena," 310; Sparkman, "Two Years of Work by the Tolan Committee," *Congressional Record* 91 (July 20, 1942), A2849; U.S. House of Representatives, 76th and 77th Congs., Select Committee to Investigate the Interstate Migration of Destitute Citizens (later renamed Select Committee Investigating National Defense Migration; hereinafter both committees are referred to as Tolan Committee), *Hearings*, Parts 1–34 (Washington: GPO, 1940–42).

29. Tolan Committee, *Hearings*, part 11, pp. 4321–23; part 25, pp. 9774–75; part 27, pp. 10317–24.

30. Ibid., part 18, p. 7609; part 27, pp. 10286–99; part 28, pp. 10783–89. See the valuable WPA Surveys on Defense Migration to various cities and towns; 52 of these surveys were reprinted in ibid., part 27, p. 10449 ff.

31. Ibid., part 20, pp. 8109–19, 8123–27; part 21, pp. 8350–51; part 25, p. 9767; part 27, pp. 10256–75, 10294–95, 10298–99; Sparkman, "Two Years of Work by the Tolan Committee," A2850; Sparkman, "National Defense Migration and Unemployment," A326–27; Clive, *State of War,* 18–30.

32. Tolan Committee, *Hearings,* part 11, p. 4326; part 20, p. 8110. See also Loula Friend Dunn, "The Powder-Mill Town," *Journal of Educational Sociology* 15 (April 1942), 460–72.

33. Tolan Committee, *Hearings,* part 11, p. 4256; part 17, pp. 6833–39, 6844–53; part 32, pp. 11983, 12204 ff.; Sparkman, "National Defense Migration and Unemployment," A326–27; Lorraine Garkovich, *Population and Community in Rural America* (New York: Praeger, 1989), 109; Dunn, "Powder-Mill Town," 462–63; Paul H. Landis, "Internal Migration by Subsidy," *Social Forces* 22 (Dec. 1943), 183–87; letter 426.

34. Tolan Committee, *Hearings,* part 22, pp. 8534, 8563.

35. See especially "Case Histories of Agricultural Displacement," in ibid., part 22, pp. 8435–49; for other case histories as well as personal depositions, see ibid., part 21, pp. 8222–23, 8259–65, 8311–15, 8342–44, 8349–51; part 22, pp. 8435–49, 8467–82, 8532–41; *New York Times,* Sept. 30, 1941.

36. Sparkman, "Two Years of Work by the Tolan Committee," A2849; E. D. Tetreau, "Wartime Changes in Arizona Farm Labor," *Sociology and Social Research* 28 (May–June 1944), 385; Shryock and Eldridge, "Internal Migration in Peace and War," 29; Tolan Committee, *Hearings,* part 20, pp. 8208–10.

37. Taeuber, "Wartime Population Changes in the United States," 240–41.

38. Nash, *The American West Transformed: The Impact of the Second World War* (Bloomington: Indiana Univ. Press, 1985), 107–27; Harry Schwartz, *Seasonal Farm Labor in the United States* (New York: Columbia Univ. Press, 1945), 23–25; Carey McWilliams, *North from Mexico: The Spanish-Speaking People of the United States* (Philadelphia: Lippincott, 1949), 265–69; Ernesto Galarza, *Merchants of Labor: The Mexican Bracero Story* (Charlotte, CA: McNally & Loftin, 1964), 46–57; Richard B. Craig, *The Bracero Program: Interest Groups and Foreign Policy* (Austin: Univ. of Texas Press, 1971), 47–48; Vernon M. Briggs, Jr., et al., *The Chicano Worker* (Austin: Univ. of Texas Press, 1977), 81–82; Erasmo Gamboa, *Mexican Labor and World War II: Braceros in the Pacific Northwest, 1942–1947* (Austin: Univ. of Texas Press, 1990), 22–73.

39. Tolan Committee, *Hearings,* Part 33, especially pp. 12603–12, 12625–32, 12664–77, 12717–23, 12745–49, 12758–67, 12859–63, 12943–51, 13032–33. See also Philip M. Hauser, "Population and Vital Phenomena," *American Journal of Sociology* 48 (Nov. 1942), 314; Carey McWilliams, *Factories in the Field: The Story of Migratory Farm Labor in California* (Boston: Little, Brown, 1938), 171–73.

40. Carr and Stermer, *Willow Run,* 39, 49, 356–57; Bureau of the Census, *Characteristics of the Population, Labor Force, Families, and Housing, Detroit-Willow Run Congested Production Area: June 1944,* 19. The African-American migration to Detroit and elsewhere is detailed in Chapter 10.

41. Carr and Stermer, *Willow Run,* 39, 49, 356–57; Clive, *State of War,* 170–74,

179; Lewis M. Killian, *White Southerners* (Amherst: Univ. of Massachusetts Press, rev. ed., 1985), 98.

42. Mrs. Cuff is quoted in Carr and Stermer, *Willow Run*, 239–40; see also Clive, *State of War*, 179–80. For other studies of the wartime southern white migration, see Lewis M. Killian, "Southern White Laborers in Chicago's West Side" (unpub. Ph.D. diss., Univ. of Chicago, 1949), especially 113, 117–49, and Chapter 6: "Stereotypes of Southern Whites," 118–49; Robert J. Havighurst with Hugh Gerthon Morgan, *The Social History of a War-Boom Community* (New York: Longmans, Green, 1951), 102–8; Hugh Gerthon Morgan, "Social Relationships of Children in a War-Boom Community," *Journal of Educational Research* 40 (Dec. 1946), 276–86; Bernice L. Neugarten, "Social Class and Friendship among School Children," *American Journal of Sociology* 51 (Jan. 1946), 305–13; James S. Brown and George A. Hillery, Jr., "The Great Migration, 1940–1960," in Thomas R. Ford, ed., *The Southern Appalachian Region: A Survey* (Lexington: Univ. of Kentucky Press, 1962), 55, 58–59, 72.

43. Carr and Stermer, *Willow Run*, 241; Clive, *State of War*, 101. For the Michigan Department of Health's assessment of conditions at Willow Run, see Tolan Committee *Hearings*, part 18, pp. 7552–53.

44. Carr and Stermer, *Willow Run*, 238–70.

45. Ibid., 242–43; Ford, "The Passing of Provincialism," in Ford, ed., *Southern Appalachian Region*, 34.

46. Rogers, "For War Workers' Children," 446; also letter 55D.

47. David M. Katzman and William M. Tuttle, Jr., eds., *Plain Folk: The Life Stories of Undistinguished Americans* (Urbana: Univ. of Illinois Press, 1982), ix–xx; John B. Stephenson, *Shiloh: A Mountain Community* (Lexington: Univ. of Kentucky Press, 1968), 23.

48. Clive, *State of War*, 172–73; Harry K. Schwarzweller et al., *Mountain Families in Transition: A Case Study of Appalachian Migration* (University Park: Pennsylvania State Univ. Press, 1971), 70, 75–78, 102–11; Killian, *White Southerners*, 91–119.

49. Elmora Messer Matthews, *Neighbor and Kin: Life in a Tennessee Ridge Community* (Nashville: Vanderbilt Univ. Press, 1965), x; Clive, *State of War*, 175–78; Roscoe Griffin, "Appalachian Newcomers in Cincinnati," and William E. Cole, "Social Problems and Welfare Services," both in Ford, ed., *Southern Appalachian Region*, 79–84, 246–47; Schwarzweller, *Mountain Families in Transition*, 9; Harriet Arnow, *The Dollmaker* (New York: Avon Books edition, 1972).

50. Brown and Hillery, "The Great Migration, 1940–1960," 55, 58–61, 71–72, 74, 77; Killian, *White Southerners*, 92–94, 100–103, 106–13; John C. Belcher, "Population Growth and Characteristics," and Rupert B. Vance, "The Region's Future: A National Challenge," both in Ford, ed., *Southern Appalachian Region*, 37–38, 45, 289–90; U.S. Bureau of the Census, *Internal Migration in the United States: April, 1940, to April, 1947*, 2–4, 7.

51. Carr and Stermer, *Willow Run*, 128–29; Clive, *State of War*, 182.

52. U.S. Bureau of the Census, *Total Population of Ten Congested Production Areas, 1944* (Washington: Bureau of the Census [Population Series CA-1, No. 11], 1944), summarized in Carr and Stermer, *Willow Run*, 363; Bureau of the Census, *Characteristics of the Population, Labor Force, Families, and Housing, Detroit-Willow Run Congested Production Area: June 1944*, 7.

53. "War Creates New Manufacturing Areas," *Business Week* (Aug. 11, 1945), 44; Dunn, "The Powder-Mill Town," 460–72; "Over 25 Per Cent Increase in Population of Six Areas," *Public Management* 25 (Jan. 1943), 21; *New York Times*, Dec. 12, 1942, Jan. 28, 1945; letter 206H.

54. U.S. House of Representatives, 78th Cong., 1st Sess., Subcommittee of the Committee on Naval Affairs, *Investigation of Congested Areas* (Washington: GPO, 1943), esp. 340–41, 538–41, 652–55, 658–63, 666–68, 817–19, 833–36, 1773–80, 2036–42.

55. Letters 22, 32, 46; House Subcommittee on Naval Affairs, *Investigation of Congested Areas*, 661–62, 789–93, 798–99, 802–3, 833–34. Regarding housing problems in war-boom communities, see the wealth of testimony in Tolan Committee, *Hearings*, parts 11, 13–15, 17–18, 25, 32; pp. 4498–4505, 4551–60, 4878–83, 4878–4904, 5131–44, 5561–65, 5738–43, 5889–90, 5937, 6248–53, 6429–43, 6626–33, 6851–52, 6904–13, 7109–17, 7239–57, 7568–69, 9888–91, 11983–88 *passim*, 11993–95, 12251–52. The Tolan Committee also published migration surveys done by the Work Projects Administration of 52 cities; see *Hearings*, part 27, pp. 10449–627.

56. Tolan Committee, *Hearings*, parts 13–14, 17; pp. 5131–41, 5561–65, 6909–10; letters 27, 206H; D'Ann Campbell, *Women at War with America: Private Lives in a Patriotic Era* (Cambridge: Harvard Univ. Press, 1984), 169–72; *New York Times*, March 4, 1943; Clive, *State of War*, 103.

57. U.S. Senate, 78th Cong., 1st Sess., Subcommittee of the Committee on Education and Labor (Pepper Committee), *Wartime Health and Education* (Washington: GPO, 1943–44), 617–18, 621, 733–35, 742, 737, 739, 744–46, 768–69, 772–73, 784–93, 805–8, 810–25, 854–57, 874–77, 914–21, 924–33, 1022–24. On children's health in war-boom communities, see also the testimony of Katherine F. Lenroot, chief of the Children's Bureau, in U.S. House of Representatives, 77th Cong., 2nd Sess., Subcommittee of the Committee on Appropriations, *Department of Labor-Federal Security Agency Appropriation Bill for 1943* (Washington: GPO, 1942), 374–75; Tolan Committee, *Hearings*, parts 11–12, 17–19; pp. 4340–78, 4571–79, 4896–4905, 6982–87, 7552–53, 7791–94; "Infant Deaths," *Newsweek* 21 (April 19, 1943), 70; Katharine F. Lenroot, "Children in a Time of Crisis," *Parents' Magazine* 17 (Jan. 1942), 28–29, 60; John L. Springer, "The Battle for Child Health," *Parents' Magazine* 18 (Jan. 1943), 30–31, 60–61; U.S. Children's Bureau Bulletin 302, *Community Health and Welfare Expenditures in Wartime 1942 and 1940—30 Urban Areas* (Washington: GPO, 1944), 18–22.

58. Pepper committee, *Wartime Health and Education*, 734–35.

59. Ibid., 767–68, 773–74.

60. Letters 62, 97B, 165, 190, 199, 211, 247, 261, 263, 266, 310, 318, 324, 336, 362; Clive, *State of War*, 104; Mintz and Kellogg, *Domestic Revolutions*, 156; Mark Jonathan Harris et al., *The Homefront: America during World War II* (New York: Putnam's 1984), 42–43.

61. Letter 72.

62. Letters 55D, 120, 379, 429; Clive, *State of War*, 105–6; Faith M. Williams, "The Standard of Living in Wartime," *Annals of the American Academy of Political and Social Science* 229 (Sept. 1943), 120–21; John B. Blandford, Jr., "Wanted: 12 Million New Houses," *National Municipal Review* 34 (Sept. 1945), 376–77; Stanley Elkins and Eric

McKitrick, "A Meaning for Turner's Frontier, Part 1: Democracy in the Old Northwest," *Political Science Quarterly* 69 (Sept. 1954), 321–53; Harris et al., *The Homefront*, 41–44.

63. Blandford, "Wanted: 12 Million New Houses," 376–77; Rilla Schroeder, "Housing in the Battle for Production," *Independent Woman* 21 (Jan. 1942), 4–6, 27; Kathryn R. Murphy, "Housing for War Workers," *Monthly Labor Review* 54 (June 1942), 1257–77; "Housing Provided in 138 Defense Areas," *Monthly Labor Review* 55 (Dec. 1942), 1203–12; Edward Weinfeld, "New York State Faces War Housing Issues," *National Municipal Review* 31 (Dec. 1942), 584–90; "New Housing in Nonfarm Areas," *Monthly Labor Review* 56 (April 1943), 652–60; Edmund N. Bacon, "Wartime Housing," *Annals of the American Academy of Political and Social Science* 229 (Sept. 1943), 128–37; Helen Weigel Brown, "Uncle Sam Houses His Children," *Parents' Magazine* 19 (Jan. 1944), 39, 79–81.

64. Pepper Committee, *Wartime Health and Education*, 748, 782.

65. Rachel Dunaway Cox, "Can Families Take It?," *Parents' Magazine* 18 (Aug. 1943), 19, 91.

66. Pepper Committee, *Wartime Health and Education*, 617–19; Tolan Committee, *Hearings*, part 32, pp. 12223–24; Cox, "Can Families Take It?," 94–95; letter 408; Rosen, "Women in the Plant, the Community, and the Union at Willow Run," 16.

67. Cox, "Can Families Take It?," 95.

Chapter 5. Working Mothers and Latchkey Children

1. Letter 164.

2. G. G. Wetherill, "Health Problems in Child Care," *Hygeia* 21 (Sept. 1943), 634. For information about "Rosie the Riveter," see Paddy Quick, "Rosie the Riveter: Myths and Realities," *Radical America* 9 (July-Oct. 1975), 115–31; Leila J. Rupp, *Mobilizing Women for War: German and American Propaganda, 1939–1945* (Princeton: Princeton Univ. Press, 1978); Miriam Frank et al., *The Life and Times of Rosie the Riveter: The Story of Three Million Working Women during World War II* (Emeryville, CA: Clarity Educational Productions, 1982); Maureen Honey, *Creating Rosie the Riveter: Class, Gender, and Propaganda during World War II* (Amherst: Univ. of Massachusetts Press, 1984); Sherna Berger Gluck, *Rosie the Riveter Revisited: Women, the War, and Social Change* (Boston: Twayne, 1987).

3. U.S. Senate, 78th Cong., 1st Sess., Subcommittee of the Committee on Education and Labor (Pepper Committee), *Wartime Health and Education* (Washington: GPO, 1943–44), 68, 262–63; *New York Times*, July 2, 1943. See also Bishop John A. Duffy's comments in *New York Times*, June 25, 1942; and Joseph B. Schuyler, S.J., "Women at Work," *Catholic World* 157 (April 1943), 26–30.

4. Hoover, "Wild Children," *American Magazine* 136 (July 1943), 40, 105; *New York Times*, June 14, 1943; Hoover, "There Will Be a Postwar Crime Wave Unless —," *The Rotarian* 66 (April 1945), 12–14; James Gilbert, *A Cycle of Outrage: America's Reaction to the Juvenile Delinquent in the 1950s* (New York: Oxford Univ. Press, 1986), 24–41; Pepper Committee, *Wartime Health and Education*, 100–120. For a helpful discussion of the varying interpretations of women's role in child care, see Judith D. Auerbach, *In the Business of Child Care: Employer Initiatives and Working Women* (New York: Praeger, 1988), 3–12.

5. "Allowances for Servicemen's Dependents, 1942," *Monthly Labor Review* 55 (Aug. 1942), 226–28; Marietta Stevenson, "New Governmental Services for People in Wartime," *Social Service Review* 16 (Dec. 1942), 595–602; Denzel C. Cline, "Allowances to Dependents of Servicemen in the United States," *Annals of the American Academy of Political and Social Science* 227 (May 1943), 1–8; Helen Tarasov, "Family Allowances: An Anglo-American Contrast," *Annals of the American Academy of Political and Social Science* 227 (May 1943), 9–21; Robert E. Bondy, "Special Welfare Services to Families of Men in Service," National Conference of Social Work *Proceedings* 70 (1943), 76–79. See also Chapter 11.

6. U.S. Senate, 78th Cong., 1st Sess., Committee on Education and Labor (Thomas Committee), *Wartime Care and Protection of Children of Employed Mothers* (Washington: GPO, 1943), 91; Phyllis Aronson, "The Adequacy of the Family Allowance System As It Affects the Wives and Children of Men Drafted into the Armed Forces" (unpub. M.S.W. thesis, Wayne Univ., 1944), 87; D'Ann Campbell, *Women at War with America: Private Lives in a Patriotic Era* (Cambridge: Harvard Univ. Press, 1984), 198–99; "Sources of Wartime Labor Supply in the United States," *Monthly Labor Review* 59 (Aug. 1944), 264–78; Army Service Forces, *Annual Report of the Office of Dependency Benefits for the Fiscal Year 1944* (Newark, NJ: Office of Dependency Benefits, 1945), 34; Harold N. Gilbert, "Green Checks for the Folks Back Home," *American Magazine* 137 (Jan. 1944), 94; *New York Times*, Oct. 16, 1942, Nov. 27, 1943.

7. U.S. Women's Bureau Bulletin No. 211, Mary Elizabeth Pidgeon, *Employment of Women in the Early Postwar Period: With Background of Prewar and War Data* (Washington: GPO, 1946), 7; Campbell, *Women at War with America*, 72–84; Modell, *Into One's Own: From Youth to Adulthood in The United States, 1920–1975* (Berkeley: Univ. of California Press, 1989), 170.

8. Pidgeon, *Employment of Women in the Early Postwar Period*, 7; Campbell, *Women at War with America*, 72–84, 224; U.S. Women's Bureau Bulletin No. 192–8, *Employment of Women in Army Supply Depots in 1943* (Washington: GPO, 1945), 22–23; *New York Times*, March 16, 1986; Rupp, *Mobilizing Women for War*, 177, 240–54; Lynn Y. Weiner, *From Working Girl to Working Mother: The Female Labor Force in the United States, 1820–1980* (Chapel Hill: Univ. of North Carolina Press, 1985), 95, 97, 110–12; Suzanne M. Bianchi and Daphne Spain, *American Women in Transition* (New York: Russell Sage Foundation [*The Population of the United States in the 1980s: A Census Monograph Series*], 1986), 148, 166; Jean Bethke Elshtain, *Women and War* (New York: Basic Books, 1987), 189–91.

9. U.S. Women's Bureau Bulletin No. 225, *Handbook of Facts on Women Workers* (Washington: GPO, 1948), 12–13; William H. Chafe, *The American Woman: Her Changing Social, Economic, and Political Roles, 1920–1970* (New York: Oxford Univ. Press, 1972), 158; Ruth Milkman, *Gender at Work: The Dynamics of Job Segregation by Sex during World War II* (Urbana: Univ. of Illinois Press, 1987), 49–83; Hazel A. Fredericksen, "The Program for Day Care of Children of Employed Mothers," *Social Service Review* 17 (June 1943), 162; Ruth A. Kasman, "Employed Mothers of Children in the A.D.C. Program Cook County Bureau of Public Welfare," *Social Service Review* 19 (March 1945), 96–110.

10. Irene E. Murphy, "Detroit's Experience with the Wartime Care of Children,"

National Conference of Social Work *Proceedings* 70 (1943), 133. See also Beatrice Gray Cook, "Mother—1943 Model," *American Home* 29 (March 1943), 29.

11. *New York Times*, Sept. 17, Oct. 24, Dec. 18, 1943; and Office of Labor Production, War Production Board, "Problems of Women War Workers in Detroit," Aug. 20, 1943, and Anne L. Gould, Office of Labor Production, War Production Board, "The Child Care Program and Its Relation to War Production," Oct. 30, 1943, both in UAW War Policy Division—Women's Bureau, Box 1, United Auto Workers Papers, Archives of Labor History and Urban Affairs, Reuther Library, Wayne State University.

12. Pepper Committee, *Wartime Health and Education*, 753–54.

13. Ibid., 754–55. See also U.S. Women's Bureau, *Bibliography on Night Work for Women* (Washington: U.S. Dept. of Labor, 1946); James L. Hymes, Jr., "Child Care Problems of the Night Shift Worker," *Journal of Consulting Psychology* 8 (July-Aug. 1944), 225–28.

14. Pepper Committee, *Wartime Health and Education*, 621, 736–42, 754–56, 926–32.

15. Ibid., 781.

16. Meyer, *Journey through Chaos* (New York: Harcourt, Brace, 1944), 60; Meyer, "War Orphans, U.S.A.," *Reader's Digest* 43 (Aug. 1943), 98–102; Meyer's testimony before the Thomas Committee, *Wartime Care and Protection of Children of Employed Mothers*, 71–74.

17. Gladys Denny Shultz, "Who's Going to Take Care of Me, Mother[,] If You Take a War-Plant Job?," *Better Homes & Gardens* 21 (May 1943), 9. See also Jessie Binford, "The War Must Be Won, but—Don't Forget the Children," *The Rotarian* 62 (Feb. 1943), 38–40; Jane Lynott Carroll, "Raising a Baby on Shifts," *Parents' Magazine* 18 (Oct. 1943), 20, 77–78, 80.

18. *New York Times*, Nov. 27, 1942; Chafe, *The American Woman*, 164; Carl N. Degler, *At Odds: Women and the Family in America from the Revolution to the Present* (New York; Oxford Univ. Press, 1980), 420; Alan Clive, *State of War: Michigan in World War II* (Ann Arbor: Univ. of Michigan Press, 1979), 197; National Manpower Council, *Womanpower* (New York: Columbia Univ. Press, 1957), 327.

19. Karen Anderson, *Wartime Women: Sex Roles, Family Relations, and the Status of Women During World War II* (Westport, CT: Greenwood, 1981), 140–42, 144–45; *New York Times*, Sept. 17, Oct. 24, Dec. 18, 1943; Thomas Committee, *Wartime Care and Protection of Children of Employed Mothers*, 15; Alice Kessler-Harris, *Out to Work: A History of Wage-Earning Women in the United States* (New York: Oxford Univ. Press, 1982), 294; letter 97D; Clive, *State of War*, 197; "Wartime Care of Working Mothers' Children in Minneapolis," *Monthly Labor Review* 57 (July 1943), 107–8.

20. *New York Times*, Sept. 24, Oct. 1, 18, 24, Dec. 6, 18, 1943; March 9, 1944.

21. Pepper Committee, *Wartime Health and Education*, 737, 928.

22. Richard L. Pifer, "A Social History of the Home Front: Milwaukee Labor during World War II" (Ph.D. diss., Univ. of Wisconsin, 1983), 324–29.

23. U.S. Woman's Bureau Bulletin No. 209, *Women Workers in Ten War Production Areas and Their Postwar Employment Plans* (Washington: GPO, 1946), 21–22, 56; U.S. Women's Bureau Bulletin No. 208, *Women's Wartime Hours of Work: The Effect on Their Factory Performance and Home Life* (Washington: GPO, 1947), 5. For working mothers'

child-care arrangements in Los Angeles, see Thomas Committee, *Wartime Care and Protection of Children of Employed Mothers*, 13; and for Detroit, see Committee on Day Care of Children, Wayne County Council of Defense, *Day Care News* (April 1943), 1, in Records of the U.S. Office of Education, "Day Care" files, Record Group 12, National Archives.

24. Letters 55D, 167, 240C, 311, 369.

25. Quoted in Marc Scott Miller, *The Irony of Victory: World War II and Lowell, Massachusetts* (Urbana: Univ. of Illinois Press, 1988), 87.

26. Letters 24, 127, 160, 336, 377; Carroll, "Raising a Baby on Shifts," 20, 77–78, 80; *New York Times*, Aug. 18, 24, 1942, May 16, 1943; Clifford A. Strauss, "Grandma Made Johnny Delinquent," *American Journal of Orthopsychiatry* 13 (April 1943), 343–46; Gluck, *Rosie the Riveter Revisited*, 64, 208, 231. On changes in grandparenthood since 1940, see Andrew J. Cherlin and Frank F. Furstenberg, Jr., *The New American Grandparent: A Place in the Family, a Life Apart* (New York: Basic Books, 1986), 24–51.

27. Letters 83, 243, 342.

28. Letters 38, 97E, 249, 250.

29. Jacqueline Jones, *Labor of Love, Labor of Sorrow: Black Women, Work, and the Family from Slavery to the Present* (New York: Basic Books, 1985), 105, 129, 184–85, 232–40, 251–56; Karen Tucker Anderson, "Last Hired, First Fired: Black Women Workers during World War II," *Journal of American History* 69 (June 1982), 82–97; Karen Anderson, *Wartime Women*, 36–42; U.S. Women's Bureau Bulletin No. 205, Kathryn Blood, *Negro Women War Workers* (Washington: GPO, 1945), 17–19; Chafe, *The American Woman*, 142–43; Kessler-Harris, *Out to Work*, 279.

30. Jones, *Labor of Love, Labor of Sorrow*, 234; *New York Times*, Oct. 23, 1943.

31. Letters 60, 64, 178, 336, 377.

32. Letters 77A, 83C, 223, 243.

33. Mark Jonathan Harris et al., eds., *The Homefront: America during World War II* (New York: Putnam's, 1984), 133, 191–94; Lyn Childs quoted in *The Life and Times of Rosie the Riveter*, transcript from the WGBH Educational Foundation for "The American Experience," 105 (1988), 12.

34. "Maternity Leave for Working Women," *Public Health Nursing* 37 (Nov. 1945), 587; Palmer, "Your Baby or Your Job," *Woman's Home Companion* 70 (Oct. 1943), 4, 137–38; *New York Times*, March 31, 1943, Nov. 15, 1944.

35. Palmer, "Your Baby or Your Job," 4, 137–38; Laura Vitray, "'Era of the Stork,'" *New York Times*, Nov. 21, 1943.

36. *New York Times*, Dec. 1, 1944; R. J. Thomas, UAW, to All Local Unions, Nov. 18, 1943, in UAW War Policy Division—Women's Bureau, Box 1, United Auto Workers Papers.

37. Gesell quoted in Miss Trout to Miss Lenroot, "Meeting of United Federal Workers of America, CIO, Aug. 6, 1944," Aug. 10, 1944, in Records of the Children's Bureau, Record Group 102, National Archives; Hartmann, *The Home Front and Beyond: American Women in the 1940s* (Boston: Twayne, 1982), 146–47; Hartmann, "Women's Organizations during World War II: The Interaction of Class, Race, and Feminism," in Mary Kelley, ed., *Woman's Being, Woman's Place: Female Indentity and Vocation in American History* (Boston: G. K. Hall, 1980), 321; Sonya Michel, "American Women and the

Discourse of the Democratic Family in World War II," in Margaret Randolph Higonnet et al., eds., *Behind the Lines: Gender and the Two World Wars* (New Haven: Yale Univ. Press, 1987), 165–67; Anderson, *Wartime Women*, 139; Child Welfare League of America, *The Impact of the War on Children's Services* (New York: Welfare League of America, 1943), 3–4, 15–19; Howard Dratch, "The Politics of Child Care in the 1940's," *Science and Society* 38 (Summer 1974), 177–80; Thomas Committee, *Wartime Care and Protection of Children of Employed Mothers*, 9–23; L. R. Goldmintz, "The Growth of Day Care: 1890–1946" (D.S.W. diss., Yeshiva Univ., 1987).

38. In this regard, the FWA program, "since it bypassed other federal- and state-level agencies . . . may have served as a model in the mid-1960s for how to . . . develop and expand local programs on a national scale" (*Bernard Greenblatt, Responsibility for Child Care* [San Francisco: Jossey-Bass, 1977], 62).

39. Thomas Committee, *Wartime Care and Protection of Children of Employed Mothers*, 1–5, 9–33; Michel, "American Women and the Discourse of the Democratic Family," 160–63; Dratch, "Politics of Child Care in the 1940's," 180–84; Susan B. Anthony, II, *Out of the Kitchen—Into the War: Women's Winning Role in the Nation's Drama* (New York: Stephen Daye, 1943), 144–45; "Whose Baby?," *Business Week* (March 20, 1943), 32, 34, 37; Kathryn Close, "Day Care Up to Now," *Survey Midmonthly* 79 (July 1943), 197; *New York Times*, June 9, 22, July 8, 10, Nov. 27, 1943, Jan. 20, 1944.

40. Pepper Committee, *Wartime Health and Education*, 287–300; *New York Times*, June 30, Sept. 17, Dec. 6, 1943, May 18, 1944; "Concerning Children," *Survey Midmonthly* 80 (Feb. 1944), 58; *Congressional Record* 90 (Feb. 25, 1944), A983.

41. Dratch, "The Politics of Child Care in the 1940's," 184–86; *Congressional Record* 90 (Feb. 25, March 9, 1944), A983, 2449–57; *New York Times*, March 8, 10, 29, May 1, 1944.

42. Women's Bureau, *Employed Mothers and Child Care*, 19; Greenblatt, *Responsibility for Child Care*, 59–61; Anderson, *Wartime Women*, 146; *New York Times*, May 1, 18, 1944; Kathryn Close, "After Lanham Funds—What?," *Survey Midmonthly* 81 (May 1945), 131.

43. According to Bernard Greenblatt, the WPA and Lanham Act programs "represented a major shift of national policy on preschool progams. The federal government had essentially declared that . . . if the national interest is served, public funds should help wage-earning mothers carry the burden of the cost of day care. . . ." (Greenblatt, *Responsibility for Child Care*, 63). See also Virginia Kerr, "One Step Forward—Two Steps Back: Child Care's Long American History," in Pamela Roby, ed., *Child Care—Who Cares? Foreign and Domestic Infant and Early Childhood Development Policies* (New York: Basic Books, 1973), 162–65; Auerbach, *In the Business of Child Care*, 40–44.

44. *New York Times*, March 14, 1942, March 18, 1943; Ruth Schooler, "Child Care Program in a City of Steel Mills," *Journal of Home Economics* 36 (June 1944), 331; Ruth Carson, "Minding the Children," *Collier's* 3 (Jan. 30, 1943), 46, 48; Ann Ross, "What Seven Mothers Did," *Parents' Magazine* 18 (May 1943), 32, 97; Anderson, *Wartime Women*, 137; Elspeth Bragdon, "A Day Care Project," in *The Impact of War on Children's Services* (New York: Child Welfare League of America, 1943), 10–12; Mary Elizabeth Evans, "Nursery School Lessons Learned in Wartime," *Journal of Home Economics* 38 (May 1946), 257–60. There was a growing interest during the war in the training of

child-care teachers and aides; see the articles by Dorothy W. Baruch, "Are Teaching Techniques Meant for Children?" and "More Teacher Training Wanted," *Journal of Consulting Psychology* 8 (March-April, Sept.-Oct., Nov.-Dec. 1944), 107–17, 323–26, 360–66.

45. *New York Times*, Aug. 24, 1942; "Marvelous for Terry?," *Time* (March 22, 1943), 40.

46. Carol Slobodin, "When the U.S. Paid for Day Care," *Day Care and Early Education* (Sept.-Oct. 1975), 23; Karen Beck Skold, "The Job He Left Behind: American Women in the Shipyards during World War II," in Carol R. Berkin and Clara M. Lovett, eds., *Women, War, and Revolution* (New York: Holmes & Meier, 1980), 55–72; Amy Kesselman, *Fleeting Opportunities: Women Shipyard Workers in Portland and Vancouver During World War II and Reconversion* (Albany: State Univ. of New York Press, 1990), 65–89.

47. "The Kaiser Child Service Centers: An Interview with Lois Meek Stolz," in James L. Hymes, Jr., ed., *Early Childhood Education: Living History Interviews*, Book 2 (Carmel: Hacienda Press, 1978), 27–32; Ruby Takanishi, "An American Child Development Pioneer, Lois Meek Stolz" (1977 oral history on deposit in University Archives, Stanford University Libraries) 185–225, 241–58; Lois Meek Stolz, "War Invades the Children's World," *National Parent-Teacher* 36 (May 1942), 4–6; *Congressional Record* 89, pp. A4729–30; Hymes, "Child Care Problems of the Night Shift Worker," 225–28; Opal Rae Weimar, "Vanport City Extends Its School Service," *Recreation* 37 (Dec. 1943), 510–12.

48. Wolff and Phillips, "Designed for 24-Hour Child Care," *Architectural Record* 95 (March 1944), 84–88; "The Kaiser Child Service Centers: An Interview with Lois Meek Stolz," 32–33, 53; Dratch, "Politics of Child Care in the 1940's," 195–201; Auerbach, *In the Business of Child Care*, 44–47.

49. Kaiser Company, Inc.—Portland Yard and the Oregon Shipbuilding Corporation, Portland, Oregon, "Child Service Center, 1943–1945: Final Report" (mimeo., 1945), 1; "The Kaiser Child Service Centers: An Interview with Lois Meek Stolz," 48–56; Slobodin, "When the U.S. Paid for Day Care," 23; *New York Times*, Nov. 12, 17, 1944.

50. Idella Purnell Stone to Eleanor Roosevelt, March 9, 1942, in Records of the Children's Bureau, Record Group 102, National Archives.

51. "Policy of the War Manpower Commission on Employment in Industry of Women with Young Children," *The Child* 7 (Oct. 1942), 49–50; "Extended School Services for Children of Working Mothers," *Education for Victory* 1 (Nov. 16, 1942), 13, 14, 22.

52. U.S. House of Representatives, 78th Cong., 1st Sess., Committee on Appropriations, *Hearings on the First Deficiency Appropriation Bill for 1943* (Washington: GPO, 1943), 733; "Employment in War Work of Women with Young Children," *Monthly Labor Review* 55 (Dec. 1942), 1184–85.

53. *Education for Victory* 1 (Oct. 15, 1942), 1–2; ibid. (Nov. 16, 1942), 13–14, 22; ibid. (Jan. 15, 1943), 1, 14; Thomas Committee, *Wartime Care and Protection of Children of Employed Mothers*, 11; U.S. House of Representatives, 78th Cong., 2nd Sess., Subcommittee of the Committee on Appropriations, *Department of Labor-Federal Security Agency Appropriation Bill for 1945* (Washington: GPO, 1944), 301–2; I. L. Kandel, *The Impact of the War upon American Education* (Westport, CT: Greenwood reprint, 1974), 48–51.

54. "Extended School Services for Young Children," *Elementary School Journal* 43 (April 1943), 439–41; *Education for Victory* 1 (Feb. 15, 1943), 28–29, 30; ibid. (April 15, 1943), 27, 32; ibid. (June 1, 1943), 27–28; ibid. (June 15, 1943), 15; ibid. (July 15, 1943), 23–24.

55. U.S. Women's Bureau Bulletin No. 246, Mary Elizabeth Pidgeon, *Employed Mothers and Child Care* (Washington: GPO, 1953), 17; *New York Times*, June 30, 1943; Thomas Committee, *Wartime Care and Protection of Children of Employed Mothers*, 28.

56. *Education for Victory* 2 (Aug. 16, 1943), 22; ibid. (Dec. 15, 1943), 3–4. I wish to thank Howard Shorr for sending me a copy of the Bell Aircraft flyer.

57. Ibid. 2 (April 3, 1944), 14; ibid. (May 20, 1944), 5–6; ibid. 3 (March 3, 1945), 12–13.

58. Ibid. 3 (Feb. 3, 1945), 9; Thomas Committee, *Wartime Care and Protection of Children of Employed Mothers*, 24–25; Dorothy W. Baruch, "Extending Extended School Services to Parents," *Journal of Consulting Psychology* 8 (July-Aug. 1944), 241–52.

59. Ernest R. Groves and Gladys Hoagland Groves, *The Contemporary American Family* (Chicago: Lippincott, 1947), 669–70.

60. The School District of Kansas City, MO, Board Minutes, July 13, Aug. 19, Sept. 16, 1943, Feb. 15, 1945; Herold C. Hunt to Board of Education, July 15, 1943; Board of Education, Resolution of July 23, 1943; Children's War Service Program, Report to the Advisory Committee, Aug. 2, 1943; Superintendent's Report, Jan. 20, March 16, 1944, March 15, April 19, June 21, Aug. 2, Sept. 20, 1945: all from the files of the Board of Education of the School District of Kansas City, MO.

For detailed information about the Kansas City, Missouri, programs, I thank Laurelle O'Leary, J. Glenn Travis, and Mildren U. LaBouff, secretary of the Kansas City, MO, Board of Education.

61. Will S. Denham to Herold C. Hunt, Feb. 23, 1945, in files of the Board of Education of the School District of Kansas City, MO.

62. *New York Times*, June 30, 1943; Women's Bureau, *Employed Mothers and Child Care*, 13–14, 17, 19. Before- and after-school care is at the heart of a proposal made in 1991 by Edward F. Zigler of Yale University: "We can solve the child-care crisis by implementing a second system within already existing elementary school buildings, where formal education takes place, and create the school of the 21st century." See David A. Hamburg, *Fundamental Building Blocks of Early Life* (New York: Carnegie Corporation of New York, 1987), 16; Zigler and Mary E. Lang, *Child Care Choices: Balancing the Needs of Children, Families, and Society* (New York: Free Press, 1991).

63. Eleanor Clifton, "Some Psychological Effects of the War: As Seen by the Social Worker," *The Family* 24 (June 1943), 126; Margaret W. Gerard, "Psychological Effects of War on the Small Child and Mother," *American Journal of Orthopsychiatry* 13 (July 1943), 493–96; Dorothy W. Baruch, "Child Care Centers and the Mental Health of Children in This War," *Journal of Consulting Psychology* 7 (Nov.-Dec. 1943), 252–66; Anna Hartoch Schachtel and Marjorie B. Levi, "Character Structure of Day Nursery Children in Wartime As Seen Through the Rorschach," *American Journal of Orthopsychiatry* 15 (April 1945), 213–22.

64. Zucker, "Working Parents and Latchkey Children," *Annals of the American Academy of Political and Social Science* 236 (Nov. 1944), 43–47; "Eight-Hour Orphans,"

Saturday Evening Post 215 (Oct. 10, 1942), 105–6; "Door-Key Children," *Science News Letter* 42 (Oct. 10, 1942), 230.

65. See Chapter 12.

66. Stolz and collaborators, *Father Relations of War-Born Children* (Stanford: Stanford Univ. Press, 1954), 316–26; Stolz, "Effects of Maternal Employment on Children: Evidence from Research," *Child Development* 31 (Dec. 1960), 749–82; Stolz, *Our Changing Understanding of Young Children's Fears, 1920–1960* (New York: National Association for the Education of Young Children, 1964), 19.

67. Denise Riley, *War in the Nursery: Theories of Child and Mother* (London: Virago, 1983), especially 42–79.

68. Stolz, "Effects of Maternal Employment on Children," 779.

69. Ibid.; *Lawrence Journal-World*, Dec. 2, 1985.

70. *New York Times*, March 15, May 22, Aug. 21, 28, Sept. 20, Oct. 18, 1945; Dorothy W. Baruch, "When the Need for War-time Services for Children Is Past—What of the Future?," *Journal of Consulting Psychology* 9 (Jan.-Feb. 1945), 45–57; *Congressional Record*, vol. 91, pp. 8657, 9337, A700, A943, A3868–69, A3928–29, A3998–4002, A4010, A4015, A4025–26, A4076–77, A4155–57, A4194–95, A4290–91; Close, "After Lanham Funds—What?," 131–35; Monica B. Owen, "Save Our Child Care Centers," *Parents' Magazine* 21 (March 1946), 20; "Concerning Children," *Survey Midmonthly* 82 (Nov. 1946), 300–301; "Child-Care Programs," *School Life* 29 (Feb. 1947), 26; Dratch, "Politics of Child Care in the 1940s," 190–95; Chafe, *The American Woman*, 186–87; Greenblatt, *Responsibility for Child Care*, 64–78, 229.

71. Elta S. Pfister, "The Case for Child Care Centers," *Journal of Consulting Psychology* 8 (July-Aug. 1944), 199–205.

72. *New York Times*, Oct. 15, 1944.

73. The nation was a presidential signature away from such a policy in 1971, but Richard M. Nixon vetoed the Child Development Act of that year, claiming that funding child-care centers would "commit the vast moral authority of the National Government to the side of communal approaches to child rearing over against the family-centered approach" (Sheila M. Rothman, *Woman's Proper Place: A History of Changing Ideals and Practices, 1870 to the Present* [New York: Basic Books, 1978], 275–76; Carole E. Jaffe, *Friendly Intruders: Childcare Professionals and Family Life* [Berkeley: Univ. of California Press, 1977], ix). An understanding of the history of child care on the American homefront would have provided President Nixon with examples of innovative, government-financed but family-supportive programs. In 1990, Congress passed a child-care program worth $2.5 billion over three years, most of it in the form of income tax credits for the working poor; this was the first major child-care legislation since the Second World War. Still, even in the face of a much more urgent problem in the 1990s, there continues to be no national policy.

Chapter 6. Rearing Preschool Children

1. Lewis, *Children of the Cumberland* (New York: Columbia Univ. Press, 1946), xvii.

2. See Lewis Jacobs, "World War II and the American Film," *Cinema Journal* 7 (Winter 1967–68), 18–19.

Despite cries for national unity, the American children's homefront was badly fractured by racial and ethnic hostility; see Chapter 10.

3. William M. Tuttle, Jr., and Catharine Tuttle, "Heartland Memories: A Less Lonely Crowd Speaks for Itself," *Explore* (Fall 1988), 16–17; Robert Hendrickson, *American Talk: The Words and Ways of American Dialects* (New York: Viking, 1986); Frederic G. Cassidy, chief ed., *Dictionary of American Regional English*, vol. 1: *Introduction and A-C* (Cambridge: Harvard Univ. Press, 1985); Craig M. Carver, *American Regional Dialects: A Word Geography* (Ann Arbor: Univ. of Michigan Press, 1987); *New York Times*, Dec. 3, 1986.

4. "Text Books in Mississippi," *Opportunity* 18 (April 1940), 99; C. Vann Woodward, *American Counterpoint: Slavery and Racism in the North-South Dialogue* (Boston: Little, Brown, 1971), 6–11.

5. Odum, *Southern Regions of the United States* (Chapel Hill: Univ. of North Carolina Press, 1936), 5–11, 245–90.

6. Cavan, *The Family* (New York: Crowell, 1942), 436–44; "The Relative Educational Standing of the 48 States," *School and Society* 60 (Sept. 16, 1944), 190–92.

Long after the Second World War was over, the work of cultural geographers continues to demonstrate the regional nature of American life; see Wilbur Zelinsky, *The Cultural Geography of the United States* (Englewood Cliffs, NJ: Prentice-Hall, 1973), 117–34; Raymond D. Gastil, *Cultural Regions of the United States* (Seattle: Univ. of Washington Press, 1975); Joel Garreau, *The Nine Nations of North America* (Boston: Houghton Mifflin, 1981); Ann Markusen, *Regions: The Economics and Politics of Territory* (Totowa, NJ: Rowman and Littlefield, 1987).

7. Chapters 7, 8, and 9 focus on the homefront lives of the school-age children.

8. The exception to this generalization is father absence, which had significant developmental consequences for the preschoolers; see Chapters 3 and 12.

9. Lewis, *Children of the Cumberland*, xv–xvi. Contrary to some historical interpretations, there was in medieval Europe the conception of childhood as a distinct stage in human development, one requiring special care; see Shulamith Shahar, *Childhood in the Middle Ages* (London: Routledge, 1990).

10. See Roger G. Barker and Herbert F. Wright, *One Boy's Day: A Specimen Record of Behavior* (New York: Harper & Brothers, 1951); Barker and Wright, *Midwest and Its Children* (New York: Harper & Row, 1955); Barker and Paul V. Gump, *Big School, Small School* (Stanford: Stanford Univ. Press, 1964); Barker, *Ecological Psychology: Concepts and Methods for Studying the Environment of Human Behavior* (Stanford: Stanford Univ. Press, 1968); Urie Bronfenbrenner, *The Ecology of Human Development: Experiments by Nature and Design* (Cambridge: Harvard Univ. Press, 1979).

11. Alex Inkeles, "Social Structure and Socialization," in David A. Goslin, ed., *Handbook of Socialization Theory and Research* (Chicago: Rand McNally, 1969), 621; Anne Buttimer, "Social Space in Interdisciplinary Perspective," *Geographical Review* 59 (1969), 417–26.

12. Farm children's lives in the 1940s also varied from one family to another, depending upon additional factors such as locale, race, and ethnicity: see Harvey J. Locke, "Contemporary American Farm Families," *Rural Sociology* 10 (June 1945), 142–51.

13. U.S. Children's Bureau Publication 272, *White House Conference on Children in a Democracy: Final Report* (Washington: GPO, 1942), 14, 20.

14. *Congressional Record* 91 (July 26, 1945), 8056. See also U.S. Children's Bureau Publication 294, *Facts about Child Health 1946* (Washington: GPO, 1946), 1–2.

15. Children's Bureau, *White House Conference on Children in a Democracy*, 17, 18.

16. Clausen, "Family Structure, Socialization, and Personality," in Lois W. Hoffman and Martin L. Hoffman, eds., *Review of Child Development Research*, vol. 2 (New York: Russell Sage Foundation, 1966), 2–3, 9–15.

17. Lewis, *Children of the Cumberland*, xii.

18. Ibid., xii–xiii.

19. I am indebted to Victor Barnouw's discussion in *Culture and Personality* (Homewood, IL: Dorsey, third ed., 1979), 5; Tylor, *The Origins of Culture* (New York: Harper Torchbooks, 1958), 1; Geertz, "The Impact of the Concept of Culture on the Concept of Man," in John R. Platt, ed., *New Views on the Nature of Man* (Chicago: Univ. of Chicago Press, 1965), 106–7.

20. Barnouw, *Culture and Personality*, 5; Tamara Hareven, "The History of the Family as an Interdisciplinary Field," *Journal of Interdisciplinary History* 11 (Autumn 1971), 400.

21. See Manford Hinshaw Kuhn, "American Families Today: Development and Differentiation of Types," in Howard Becker and Reuben Hill, eds., *Family, Marriage, and Parenthood* (Boston: D. C. Heath, 1948), 131–68.

22. Valentine, *Culture and Poverty: Critique and Counter-Proposals* (Chicago: Univ. of Chicago Press, 1968), 106.

23. Quoted is Robert LeVine, "Culture, Personality, and Socialization: An Evolutionary View," in Goslin, *Handbook of Socialization Theory and Research*, 518; Barnouw, *Culture and Personality*, 33.

24. John W. M. Whiting et al., "The Learning of Values," in Evon Z. Vogt and Ethel M. Albert, eds., *People of Rimrock: A Study of Values in Five Cultures* (Cambridge: Harvard Univ. Press, 1966), 83–125. See also Evon Z. Vogt, *Modern Homesteaders: The Life of a Twentieth-Century Frontier Community* (Cambridge: Harvard Univ. Press, 1955); Florence Rockwood Kluckhohn and Fred L. Strodtbeck, *Variations in Value Orientations* (Evanston, IL: Row, Peterson, 1961).

25. Elkin and Handel, *The Child and Society: The Process of Socialization* (New York: Random House, second ed., 1972), 67–70.

Culture thus has an important effect on the child's developing personality. Definitions of personality vary widely, but I have found the following to be helpful: personality "is a more or less enduring organization of forces within the individual. . . . The forces of personality are not responses but *readiness for response*. . . ." (Theodore Adorno et al., *The Authoritarian Personality* [New York: Harper, 1950], 5; Alex Inkeles and Daniel J. Levinson, "National Character: The Study of Modal Personality and Sociocultural Systems," in Gardner Lindzey, ed., *Handbook of Social Psychology*, vol. 2: *Special Fields and Applications* [Cambridge, MA.: Addison-Wesley, 1954], 977–1020).

26. E. P. Hutchinson, *Immigrants and Their Children, 1850–1950* (New York: Wiley [for the Social Science Research Council in cooperation with the Bureau of the Census], 1956), 1–7; Roger Daniels, *Coming to America: A History of Immigration and Ethnicity in American Life* (New York: HarperCollins, 1990), 265–306.

27. Hutchinson, *Immigrants and Their Children*, 5–6.

28. Bruno Bettelheim and Morris Janowitz, *Social Change and Prejudice* (includ-

ing *Dynamics of Prejudice*) (London: Free Press of Glencoe, 1964), 3–24, 78–83; Daniel Katz and Kenneth W. Braly, "Verbal Stereotypes and Racial Prejudice," in Eleanor E. Maccoby et al., eds., *Readings in Social Psychology* (New York: Holt, third ed., 1958), 40–46.

29. Otto Klineberg, ed., *Characteristics of the American Negro* (New York: Harper, 1944), 277–78.

30. Ruby J. R. Kennedy, "Single or Triple Melting Pot? Intermarriage Trends in New Haven, 1870–1940," *American Journal of Sociology* 49 (Jan. 1944), 331–39.

31. Ruth Shonle Cavan, "Subcultural Variations and Mobility," in Harold T. Christensen, ed., *Handbook of Marriage and the Family* (Chicago: Rand McNally, 1964), 553–57; John L. Thomas, "The Factor of Religion in the Selection of Marriage Mates," *American Sociological Review* 16 (Aug. 1951), 487–91.

32. Kennedy, "Single or Triple Melting Pot?," 331–39; Cavan, "Subcultural Variations and Mobility," 557.

33. The wartime figures on enemy aliens were: 599,111 Italians, 263,930 Germans, and 47,963 Japanese: Joseph S. Roucek, "Italo-Americans and World War II," *Sociology and Social Research* 29 (July–Aug. 1945), 470.

34. Jill S. Quadagno, "The Italian American Family," in Charles H. Mindel and Robert W. Habenstein, eds., *Ethnic Families in America* (New York: Elsevier, second ed., 1981), 79.

35. Earl Lomon Koos, *Families in Trouble* (New York: King's Crown Press, 1946), 77.

36. Roucek, "Italo-Americans and World War II," 465–66, 467. See also Jerome S. Bruner and Jeannette Sayre, "Shortwave Listening in an Italian Community," *Public Opinion Quarterly* 5 (Dec. 1941), 640–56; Ronald H. Bayor, *Neighbors in Conflict: The Irish, Germans, Jews, and Italians of New York City, 1929–1941* (Baltimore: Johns Hopkins Univ. Press, 1978), 118–20, 123–25; Gary Ross Mormino, *Immigrants on the Hill: Italian-Americans in St. Louis, 1882–1982* (Urbana: Univ. of Illinois Press, 1986), 217–19.

37. Roucek, "Italo-Americans and World War II," 466. In 1940 an estimated 80 percent of Italian-language publications were pro-Mussolini, 10 percent anti-Mussolini, and the rest neutral; Joseph C. Roucek, "Foreign-Language Press in World War II," *Sociology and Social Research* 27 (July–Aug. 1943), 464–65.

38. Roucek, "Italo-Americans and World War II," 466; William Foot Whyte, *Street Corner Society: The Social Structure of an Italian Slum* (Chicago: Univ. of Chicago Press, second ed., 1955), xv–xviii.

39. Whyte, *Street Corner Society*, xvii, xix–xx. On the importance of the *paesani*, see Grazia Dore, "Some Social and Historical Aspects of Italian Emigration to America," *Journal of Social History* 2 (Winter 1968), 95–122; Herbert J. Gans, *The Urban Villagers: Group and Class in the Life of Italian-Americans* (Glencoe: Free Press, 1962).

40. Quadagno, "The Italian American Family," 62; Leonard Covello, *The Social Background of the Italo-American School Child* (Leiden, the Netherlands: Brill, 1967), 149–54.

41. Quadagno, "The Italian American Family," 64–65, 73–74; Richard Gambino, *Blood of My Blood* (New York: Doubleday, 1974), 119–55; Covello, *The Social Background of the Italo-American School Child*, 284–96.

42. Hutchinson, *Immigrants and Their Children*, 6; Greeley, *Why Can't They Be Like*

Us? (New York: Wiley, 1971), 77; Quadagno, "The Italian American Family," 67–72; Glazer and Moynihan, *Beyond the Melting Pot: The Negroes, Puerto Ricans, Jews, Italians and Irish of New York City* (Cambridge: MIT Press, 1963), 187.

43. Gans, *The Urban Villagers,* 56.

44. Colleen Leahy Johnson, "Family Support Systems of Elderly Italian Americans," *Journal of Minority Aging* 4 (1979), 38.

45. Gans, *The Urban Villagers,* 59–60; Eleanor Pavenstadt, "A Comparison of the Child-Rearing Environment of Upper Lower and Very Lower Class Families," *American Journal of Orthopsychiatry* 35 (Jan. 1965), 89–98.

46. W.E.B. Du Bois, the black scholar and civil rights leader, wrote of African Americans' identity crisis: "One ever feels his two-ness,—An American, a Negro; two souls, two thoughts, two unreconciled strivings; two warring ideals in one dark body, whose dogged strength alone keeps it from being torn asunder." One wishes "to merge his double self into a better and truer self." Granting the immense historical differences between the experiences of Italian Americans and African Americans, Italian Americans along with other ethnic Americans still shared a sense of "two-ness" (Du Bois, "Strivings of the Negro People," *Atlantic Monthly* 80 [Aug. 1897], 194–98).

47. Irvin L. Child, *Italian or American? The Second Generation in Conflict* (New Haven: Yale Univ. Press, 1943), 152.

48. Yarrow et al., "Social Perceptions and the Attitudes of Children," *Genetic Psychology Monographs* 40 (Nov. 1949), 444. See also Chapter 10.

49. Dollard, *Criteria for the Life History, with Analyses of Six Notable Documents* (New Haven: Yale Univ. Press, 1935), 15.

50. Cavan, "Subcultural Variations and Mobility," 536–37.

51. W. Lloyd Warner and Paul S. Lunt, *The Social Life of a Modern Community* (New Haven: Yale Univ. Press, 1941), 211–26; Cavan, "Subcultural Variations and Mobility," 540; Bingham Dai, "Some Problems of Personality Development among Negro Children," in Clyde Kluckhohn and Henry A. Murray, eds., *Personality in Nature, Society, and Culture* (New York: Knopf, 1949), 437–59. See also Chapter 10.

52. James R. Shortridge, "Patterns of Religion in the United States," *Geographical Review* 66 (Oct. 1976), 421–28; Wilbur Zelinsky, "An Approach to the Religious Geography of the United States: Patterns of Church Membership in 1952," *Annals of the Association of American Geographers* 51 (1961), 139–93; N. J. Demerath III, *Social Class in American Protestantism* (Chicago: Rand McNally, 1965); Gary Schwartz, *Sect Ideologies and Social Status* (Chicago: Univ. of Chicago Press, 1970).

53. Demos, *Past, Present, and Personal: The Family and the Life Course in American History* (New York: Oxford Univ. Press, 1986), 197.

54. Clausen, "Family Structure, Socialization, and Personality," 21–22.

55. Stone, "The Play of Little Children," in R. E. Herron and Brian Sutton-Smith, eds., *Child's Play* (New York: Wiley, 1971), 11.

56. Kinsey et al., *Sexual Behavior in the Human Male* (Philadelphia: W. B. Saunders, 1948), 440–47.

For the relationship between social class and mental health, see August B. Hollingshead and Frederick C. Redlich, "Schizophrenia and Social Structure," *American Journal of Psychiatry* 110 (March 1954), 695–701; John A. Clausen and Melvin L. Kohn, "The Ecological Approach in Social Psychiatry," *American Journal of Sociology* 60 (Sept.

1954), 140–51; Hollingshead and Redlich, *Social Class and Mental Illness* (New York: Wiley, 1958); Jerome K. Myers and Bertram H. Roberts, *Family and Class Dynamics in Mental Health* (New York: Wiley, 1959); Jerome K. Myers and Lee L. Bean, *A Decade Later: A Follow-Up of Social Class and Mental Illness* (New York: Wiley, 1968).

57. No book makes this point better than Justin Wise Polier, *Everyone's Child, Nobody's Child: A Judge Looks at Underprivileged Children in the United States* (New York: Scribner's, 1941), especially 269–77.

58. Davis and Havighurst, "Social Class and Color Differences in Child-Rearing," *American Sociological Review* 11 (Dec. 1946), 698–710; Davis and Havighurst, *Father of the Man: How Your Child Gets His Personality* (Boston: Houghton Mifflin, 1947), 215–19; W. Allison Davis, "American Status Systems and the Socialization of the Child," *American Sociological Review* 6 (June 1941), 345–54.

59. Davis and Havighurst, "Social Class and Color Differences in Child-Rearing," 707–10. See also Allison Davis and John Dollard, *Children of Bondage: The Personality Development of Negro Youth in the Urban South* (Washington, D.C.: American Council on Education, 1940), 256–78; Davis and Havighurst, *Father of the Man*, Appendix I: "Social-Class and Color Differences in Practices of Child-Rearing," 215–19; Allison Davis, *Social-Class Influences upon Learning* (Cambridge: Harvard Univ. Press, 1965), 12–13; John H. Rohrer and Munro S. Edmonson, eds., *The Eighth Generation: Cultures and Personalities of New Orleans Negroes* (New York: Harper, 1960).

60. Davis and Havighurst, "Social Class and Color Differences in Child-Rearing," 699 (emphasis in original).

61. Ten years later, developmental psychologists argued that Davis and Havighurst's findings were wrong and that "the middle-class mothers were generally more permissive . . . toward their young children than were working-class mothers." A decade can make a big difference, however; and this was a decade of major changes in child rearing in the United States. For the postwar research done on this topic, see, for example, Robert R. Sears et al., *Patterns of Child Rearing* (Evanston: Row, Peterson, 1957), esp. 442–47; Eleanor E. Maccoby and Patricia K. Gibbs and the Staff of the Laboratory of Human Development, Harvard University, "Methods of Child Rearing in Two Social Classes," in William E. Martin and Celia Burns Stendler, eds., *Readings in Child Development* (New York: Harcourt Brace, 1954), 380–96; Havighurst and Davis, "A Comparison of the Chicago and Harvard Studies of Social Class Differences in Child Rearing," *American Sociological Review* 20 (Aug. 1955), 438–42; Urie Bronfenbrenner, "Socialization and Social Class through Time and Space," in Eleanor E. Maccoby et al., eds., *Readings in Social Psychology* (New York: Holt, third ed., 1958), 400–425.

62. Duvall, "Conceptions of Parenthood," *American Journal of Sociology* 52 (Nov. 1946), 193–203; Martha C. Ericson, "Child-Rearing and Social Status," *American Journal of Sociology* 52 (Nov. 1946), 190–92; Gertrude Gilmore Lafore, *Practices of Parents in Dealing with Preschool Children*, Child Development Monograph No. 31 (New York: Bureau of Publications, Teachers College, Columbia Univ., 1945).

Three comprehensive discussions of the research done on social class and childhood socialization from the 1940s to the 1960s are Ruth Shonle Cavan, "Subcultural Variations and Mobility," 540–51, 559–64; Edward Zigler and Irvin L. Child, "Socialization," in Gardner Lindzey and Elliot Aronson, eds., *The Handbook of Social Psychology*, vol. 3: *The Individual in a Social Context* (Reading, MA.: Addison-Wesley, second ed.,

1968–69), 483–501; Robert D. Hess, "Social Class and Ethnic Influences upon Social-ization," in Paul H. Mussen, ed., *Carmichael's Manual of Child Psychology*, vol. 2 (New York: Wiley, third ed., 1970), 457–557.

63. Littledale, "Then and Now," *Parents' Magazine* 21 (Oct. 1946), 127; Catherine Mackenzie, "Bringing Up Baby—Then and Now," *New York Times Magazine* (Feb. 8, 1943), 14, 23; Terry Strathman, "From the Quotidian to the Utopian: Child Rearing Lit-erature in America, 1926–1946," *Berkeley Journal of Sociology* 29 (1984), 1–34.

64. Littledale, "Then and Now," 127, 128.

65. Watson, *Psychological Care of Infant and Child* (London: Allen & Unwin, 1928), 73.

66. According to Urie Bronfenbrenner's survey of social class and child-rearing from 1933 and 1958, "There is ample evidence that . . . middle-class mothers were much more likely than working-class mothers to be exposed to current informaton on child care" (Bronfenbrenner, "Socialization and Social Class through Time and Space," 411). See also Elise P. Fishbein, "New Ideas in Child Development," *Hygeia* 25 (March 1947), 201–2; Martha Wolfenstein, "Trends in Infant Care," *American Journal of Orthopsychiatry* 23 (Jan. 1953), 120–30; Wolfenstein, "Fun Morality: An Analysis of Recent American Child-Training Literature," in Margaret Mead and Martha Wolfenstein, eds., *Childhood in Contemporary Cultures* (Chicago: Univ. of Chicago Press, 1955), 168–78; Bronfenbren-ner, "The Changing American Child—A Speculative Essay," *Journal of Social Issues* 17 (1961), 6–18; L. J. Borstelmann, Changing Concepts of Childrearing, 1920s–1950s: Parental Research, Parental Guidance, Social Issues" (paper delivered at Southeastern Conference on Human Development, 1976); Nancy Pottishman Weiss, "Mother, the Invention of Necessity: Dr. Benjamin Spock's *Baby and Child Care*," *American Quarterly* 29 (Winter 1977), 519–29; Weiss, "The Mother-Child Dyad Revisited: Perceptions of Mothers and Children in Twentieth Century Child-Rearing Manuals," *Journal of Social Issues* 34 (Spring 1978), 31–36; Duane F. Alwin, "From Obedience to Autonomy: Changes in Traits Desired in Children," *Public Opinion Quarterly* 52 (Spring 1988), 33–52. On the need for historians to be cautious in using child-rearing manuals, see Jay Mechling, "Advice to Historians on Advice to Mothers," *Journal of Social History* 9 (Fall 1975), 44–63.

67. Murphy, "Cultural Factors in the Development of Children," *Childhood Edu-cation* 23 (Oct. 1946), 53; Clausen, "Family Structure, Socialization, and Personality," 16–19; Celia B. Stendler, "Sixty Years of Child Training Practices: Revolution in the Nurs-ery," *Journal of Pediatrics* 36 (Jan. 1950), 129–31; Clark E. Vincent, "Trends in Infant Care Ideas," *Child Development* 22 (Sept. 1951), 199–207.

68. Hamilton Cravens, "Child Saving in the Age of Professionalism, 1915–1930"; and Leroy Ashby, "Partial Promises and Semi-Visible Youths: The Depression and World War II," both in Joseph M. Hawes and N. Ray Hiner, eds., *American Childhood: A Research Guide and Historical Handbook* (Westport, CT: Greenwood, 1985), 448–49, 502; Robert R. Sears, *Your Ancients Revisited: A History of Child Development* (Chicago: Univ. of Chi-cago Press, 1975), 19–21.

69. Sears, *Your Ancients Revisited*, 43–45; Borstelmann, "Changing Concepts of Childrearing," 10–12; Littledale, "Then and Now," 130, 132; Ashby, "Partial Promises and Semi-Visible Youths," 502–3; Michael Gordon, "*Infant Care* Revisited," *Journal of*

Marriage and the Family 30 (Nov. 1968), 578–79; Mary Cable, *The Little Darlings: A History of Child Rearing in America* (New York: Scribner's, 1975), 182–85; Neil M. Cowan and Ruth Schwartz Cowan, *Our Parents' Lives: The Americanization of Eastern European Jews* (New York: Basic Books, 1989), 177–209.

70. Fishbein, "New Ideas in Child Development," 224; Terry Strathman, "From the Quotidian to the Utopian," 6–8; Littledale, "Then and Now," 130, 132, 136; Borstelmann, "Changing Concepts of Childrearing," 8–10; A. Michael Sulman, "The Humanization of the American Child: Benjamin Spock As a Popularizer of Psychoanalytic Thought," *Journal of the History of the Behavioral Sciences* 9 (July 1973), 258–65; William G. Bach, "The Influence of Psychoanalytic Thought on Benjamin Spock's *Baby and Child Care*," *Journal of the History of the Behavioral Sciences* X (Jan. 1974), 91–94; Gesell et al., *The First Five Years of Life: A Guide to the Study of the Preschool Child* (New York: Harper & Brothers, 1940), 6–9, 319–43; Spock, "A Babies' Doctor Advises on Toilet Training," *Parents' Magazine* 20 (March 1945), 20.

71. Ashby, "Partial Promises and Semi-Visible Youth," 503; Borstelmann, "Changing Concepts of Childrearing," 12–14; Murphy, "Cultural Factors in the Development of Children," 53; Steven Mintz and Susan Kellogg, *Domestic Revolutions: A Social History of American Family Life* (New York: Free Press, 1988), 164–65.

72. Clark E. Vincent, "Trends in Infant Care Ideas," 205; Littledale, "Then and Now," 136; Stendler, "Sixty Years of Child Training Practices," 126, 131; Kenyon, "Less Rigid Schedules for Baby," *Good Housekeeping* 110 (1940), 92; Herman N. Bundesen, "Bad Habits in Children," *Ladies' Home Journal* 58 (Aug. 1941), 83–84; A. H. Maslow and I. Szilagyi-Kessler, "Security and Breast-feeding," *Journal of Abnormal and Social Psychology* 41 (Jan. 1946), 83–85; Ralph H. Ojetmann and associates, "A Functional Analysis of Child Development Materials in Current Newspapers and Magazines," *Child Development* 19 (March-June 1946), 76–92.

73. *Infant Care* (Washington: GPO, 1942), 41, 59–60; *Infant Care* (Washington: GPO, 1945), 52; Wolfenstein, "Fun Morality," 169–74.

74. Shultz, "Watch Out for War Nerves!," *Better Homes & Gardens* 22 (Nov. 1943), 12.

75. Travers, "Children Are Tough," *Woman's Home Companion* 69 (June 1942), 48.

76. Clara Savage Littledale, "Explaining the War to Our Children," *Parents' Magazine* 17 (Feb. 1942), 17; Mary Shattuck Fisher, "What Shall We Tell Children about War?," *Journal of Home Economics* 34 (May 1942), 277–79; Evelyn Ardis Whitman, "Keeping Young Chins Up," *Parents' Magazine* 17 (Sept. 1942), 29, 94–95; Anna W. M. Wolf, "War and Discipline," *Parents' Magazine* 17 (Oct. 1942), 20–21 ff.; Frederick H. Allen, "Can the Youngest Take It?," *Parents' Magazine* 17 (Nov. 1942), 26–27 ff.; Barbara Biber, "Childhood As Usual?," *Parents' Magazine* 18 (Feb. 1943), 26–27 ff.; James Benjamin, "Let Your Child Know What to Expect," *Parents' Magazine* 19 (Nov. 1944), 20–21 ff.

77. Ruth E. Beckey, "Your Child As a Postwar Citizen," *Hygeia* 21 (June 1943), 460.

78. Helen Steers Burgess, "Facing War with Our Children," *Parents' Magazine* 17 (March 1942), 19; Whitman, "Keeping Young Chins Up," 95.

79. Packard, "Give the War Babies a Break," *American Magazine* 139 (May 1945),

24–25 ff. See also two articles by Gladys Denny Shultz: "If Daddy's Gone to War," *Better Homes & Gardens* 22 (Oct. 1943), 14 ff.; "Life without Father," *Better Homes & Gardens* 22 (June 1944), 22 ff.

80. Fishbein, "New Ideas in Child Development," 224–25; Horowitz to the author, Aug. 5, 1983.

81. "Lullabies and Play Needed by Infants," *Science News Letter* 45 (April 1, 1944), 216; Burgess, "Facing War with Our Children," 36; Mary Jo Bane, "A Review of Child Care Books," *Harvard Educational Review* 43 (Nov. 1973), 669; Ashby, "Partial Promises and Semi-visible Youths," 503; Child Study Association of America, *The Child, the Family, the Community: A Classified Booklist* (New York: Child Study Association of America, undated), 14–20, 51–52; book reviews in *American Journal of Nursing* 42 (Oct. 1942), 1222; 44 (Feb., June, Oct. 1944), 195, 524, 1015; Michael Zuckerman, "Dr. Spock: The Confidence Man," in Charles E. Rosenberg, ed., *The Family in History* (Philadelphia: Univ. of Pennsylvania Press, 1975), 182–84; William Graebner, "The Unstable World of Benjamin Spock: Social Engineering in a Democratic Culture, 1917–1950," *Journal of American History* 67 (Dec. 1980), 612–29.

Chapter 7. School-age Children Fight the War

1. Letters 146, 148, 210, 228, 332, 443.

2. Letters 10, 81. Americans of all ages talked of their national pride and of the need for personal sacrifice for the war effort; see Robert B. Westbrook, "'I Want a Girl, Just Like the Girl That Married Harry James': American Women and the Problem of Political Obligation in World War II," *American Quarterly* 42 (Dec. 1990), 587–614; Mark H. Leff, "The Politics of Sacrifice on the American Home Front in World War II," *Journal of American History* 77 (March 1991), 1296–1318; Leff, "Merchandizing Images of Homefront Sacrifice: Advertising and Propaganda in the United States and Great Britain in World War II" (paper delivered at Indiana University's conference on the American Home Front during World War II, Oct. 1991).

3. Letter 408.

4. Greenstein, *Children and Politics* (New Haven: Yale Univ. Press, rev. ed., 1969), 1.

5. Ibid., 4–5. See also J. Cayce Morrison, "The Teaching of Patriotism," *School and Society* 56 (Oct. 3, 1942), 281–86.

6. Greenstein, *Children and Politics*, 9–15; Easton and Hess, "The Child's Political World," *Midwest Journal of Political Science* 6 (Aug. 1962), 237–38; Richard E. Dawson and Kenneth Prewitt, *Political Socialization: An Analytic Study* (Boston: Little, Brown, 1969), 44–48; Robert Hess and Judith V. Torney, *The Development of Political Attitudes in Children* (Chicago: Aldine, 1967), 220–21; Howard Tolley, Jr., *Children and War: Political Socialization to International Conflict* (New York: Teachers College Press, 1973), 5–6; Easton and Hess, "Youth and the Political System," in Seymour Lipset and Leo Lowenthal, eds., *Culture and Social Character: The Work of David Riesman Reviewed* (New York: Free Press, 1961), 226–51; Hess and Easton, "The Child's Changing Image of the President," *Public Opinion Quarterly* 24 (Winter 1960), 632–44; Hess and Easton, "The Role of the Elementary School in Political Socialization," *School Review* 70 (Autumn 1962), 253–65.

7. David Easton and Jack Dennis, *Children in the Political System: Origins of Political Legitimacy* (New York: McGraw-Hill, 1969), 107; Tolley, *Children and War*, 6; Easton and Dennis, "The Child's Acquisition of Regime Norms: Political Efficacy," *American Political Science Review* 61 (March 1967), 25–38; Dawson and Prewitt, *Political Socialization*, 48–50.

8. Piaget focused on mental schemata that are unobservable, but this is not to say that the functional results of these structures are untestable. Piagetian psychologists have shown the changes in mental structures in a series of fascinating tests in which children, by mentally reversing sequences and reconstructing events, eventually master such concepts as quantity, length, shape, and color, and then "justify" or explain their conclusions and manipulations.

9. There is an immense body of literature by and about Piaget; helpful summaries and bibliographies can be found in John H. Flavell, *The Developmental Psychology of Jean Piaget* (Princeton: Van Nostrand, 1963); Jonas Langer, *Theories of Development* (New York: Holt, Rinehart and Winston, 1969), 87–156; Charles J. Brainerd, *Piaget's Theory of Intelligence* (Englewood Cliffs, NJ: Prentice-Hall, 1978); Henry W. Maier, *Three Theories of Child Development* (New York: Harper & Row, 1978), 12–70; Rochel Gelman and Renee Baillargeon, "A Review of Some Piagetian Concepts," in Paul H. Mussen, ed., *Handbook of Child Psychology* (fourth ed.), vol. 3: *Cognitive Development*, John H. Flavell and Ellen M. Markman, eds. (New York: Wiley, 1983), 167–230.

10. Neil J. Salkind, *Theories of Human Development* (New York: Van Nostrand, 1981), 208; Patricia H. Miller, *Theories of Developmental Psychology* (San Francisco: Freeman, 1983), 62.

11. Representative writings of Kohlberg include *Stages in the Development of Moral Thought and Action* (New York: Holt, Rinehart and Winston, 1969); "Stage and Sequence: The Cognitive-Developmental Approach to Socialization," in David A. Goslin, ed., *Handbook of Socialization Theory and Research* (Chicago: Rand McNally, 1969), 347–480; "From Is to Ought," in T. Mischel, ed., *Cognitive Development and Epistemology* (New York: Academic Press, 1971), 151–235; *The Philosophy of Moral Development*, vol. 1: *Moral Stages and the Idea of Justice* (New York: Harper & Row, 1981). In moral development, Gilligan's most significant publication is *In a Different Voice: Psychological Theory and Women's Development* (Cambridge: Harvard Univ. Press, 1982).

12. Lee, *Personality Development in Childhood* (Monterey, CA: Brooks/Cole, 1976), 132–33; James R. Rest, "Morality," in Mussen, *Handbook of Child Psychology*, vol. 3, pp. 566–629.

13. See Chapter 14.

14. Tolley, *Children and War*, 11–12; Hess and Torney, *The Development of Political Attitudes in Children*, 92–115.

15. See Patricia P. Minuchin and Edna K. Shapiro, "The School As a Context for Social Development," in Paul H. Mussen, ed., *Handbook of Child Psychology* (fourth ed.), vol. 4: *Socialization, Personality, and Social Development*, E. Mavis Hetherington, ed. (New York: Wiley, 1983), esp. 230–38.

16. "Educational Policy Concerning Young Children and the War," *Education for Victory* 1 (April 15, 1942), 8–9. For federal discussions in this area, see Timothy De Witt Connelly, "Education for Victory: Federal Efforts to Promote War-Related Instructional

Activities by Public School Systems, 1940–1945" (Ph.D. diss., Univ. of Maryland, 1982), 74–117.

17. *Journal of the National Educational Association*, 32 (Sept. 1943), 173–74; Carter V. Good, "Educational Issues of 1942 and the Task Ahead," *School and Society* 57 (March 27, 1943), 344; Dorothy W. Baruch, *You, Your Children, and War* (New York: D. Appleton-Century, 1942), 219–22; Anna W. M. Wolf, "The Home Front in Wartime," *Journal of Educational Sociology* 16 (Dec. 1942), 202–18; Angelo Patri, *Your Children in Wartime* (Garden City, NY: Doubleday, Doran, 1943), 89–97.

18. "The Elementary School and the War," *Elementary School Journal* 43 (June 1943), 568; Agnes Benedict, "Towards Democracy," *Parents' Magazine* 19 (May 1944), 30–31, 128–29; James Marshall, "Wars Are Made in Classrooms," *Saturday Review of Literature* 27 (Nov. 11, 1944), 24; *New York Times*, Jan. 17, 1943.

19. For the wartime experiences of German-, Italian-, and Japanese-American children, see Chapter 10.

20. Letters 302, 370, 373, 408; Allport, *The Nature of Prejudice* (Cambridge: Addison-Wesley, 1954), 307–10.

21. Howard Lane, "The Good School for the Young Child in Wartime," *Education* 63 (Feb. 1943), 352; Jeanie G. Lee, "Every Classroom a Citadel," *Journal of the National Education Association* 32 (Sept. 1943), 173–74.

22. Cremin, *The Transformation of the School: Progressivism in American Education, 1876–1957* (New York: Knopf, 1961), 328–32; L. Thomas Hopkins, "The War and the Curriculum," *Education* 63 (Feb. 1943), 351; Charles I. Glicksberg, "Discipline *Versus* Freedom in Wartime Education," *School and Society* 57 (Feb. 27, 1943), 243–46; "On Coordinating Experiences," *School and Society* 59 (April 1, 1944), 237; James M. Wallace, *The Promise of Progressivism in Journalism and Education, 1914–1941* (New Brunswick, NJ: Rutgers Univ. Press, 1991).

23. Cremin, *The Transformation of the School*, 328–30; and the following, published by the NEA and the American Association of School Administrators, for the Educational Policies Commission: *Education for All American Youth* (1944), *Educational Services for Young Children* (1945), and *Education for All American Children* (1948).

24. "A Wartime Program for Elementary Schools," *School and Society* 57 (June 19, 1943), 683–84; Harold S. Tuttle, "The Schools Can Help Win," *School and Society* 58 (Aug. 28, 1943), 129–31; George H. Deer, "Democracy's Children," *School and Society* 58 (Sept. 11, 1943), 188–90; "Education for Internal and International Peace," *The School Review* 53 (Feb. 1945), 67–74; Weber, *My Country School Diary: An Adventure in Creative Teaching* (New York: Harper & Brothers, 1946), 211; Violet Garrison Shankin, "War Effort Related Opinions and Practices of Detroit Elementary School Teachers" (M.Ed. thesis, Wayne Univ., 1943).

25. James, *Exile Within: The Schooling of Japanese Americans 1942–1945* (Cambridge: Harvard Univ. Press, 1987), 8, 38–41, 63–67, 71, 78–80, 113–14, 135, 170–71.

26. *The Code of the Good American* (Washington: NEA [Personal Growth Leaflet No. 62], n.d.), 3–15.

27. See Chapter 10.

28. "School Health and Physical Education: Developing an Educational Philosophy in a 'Boom-Town' Community," *Education for Victory* 3 (July 20, 1944), 19.

29. Letters 72, 146, 251, 282; Jeannett Terrill Murray, "Home Front Hard for Montana Girl," Clarksville, AK, *The Graphic*, Feb. 7, 1990.

30. Letters 379, 440; Ernest G. Hesser, "Music Education in Wartime," *School and Society* 58 (July 3, 1943), 9–11.

31. Letters 24, 32, 140, 257, 332, 359.

32. Letters 97A, 185, 210, 284, 426.

33. Letters 27, 83B, 83I, 182, 187, 365.

34. Letters 53, 96, 208; *New York Times*, Dec. 13, 1942, Jan. 17, Aug. 26, 1943, Oct. 4, 5, 8, 1944, July 23, 1945; Ralph W. Tyler, "The Importance in Wartime of Co-operation between Schools and Parents," *Elementary School Journal* 43 (Feb. 1943), 332–33; Willard Waller, "The Family and Other Institutions," *Annals of the American Academy of Political and Social Science* 229 (Sept. 1943), 109–12; Carter V. Good, "Education after Two Years of War," *School and Society* 59 (May 13, 1944), 337–39; Joy Elmer Morgan, "The Fight for the Rights of Children," *Journal of the National Education Association* 34 (Feb. 1945), 25; Agnes E. Meyer, "Equal Rights for Children," *Parents' Magazine* 20 (Feb. 1945), 91.

35. Letters 226, 350.

36. Federal report quoted in Alan Clive, *State of War: Michigan in World War II* (Ann Arbor: Univ. of Michigan Press, 1979), 117; Benjamin Fine, *Our Children Are Cheated: The Crisis in American Education* (New York: Henry Holt, 1947), 186; Frazier, "Wartime Changes in Teacher Certification," *Education for Victory* 3 (Oct. 3, 1944), 9–12; Connelly, "Education for Victory," 251–96.

37. Frazier, "Wartime Changes in Teacher Certification," 12; letter 104.

38. "Schools at War Program," *Education for Victory* 1 (Oct. 15, 1942), 7; "Activities of School Children Related to the War Effort," *Education for Victory* 1 (April 15, 1942), 5–8; "War Duties of Schools," *Elementary School Journal* 43 (Oct. 1942), 67–69; *New York Times*, Nov. 12, 1942; "Elementary Schools in War Program," *Education for Victory* 2 (Aug. 2, 1943), 5–6.

39. "Schools-at-War Scrapbooks Go to England," *Education for Victory* 2 (Nov. 1, 1943), 12; "Schools at War," *Ladies' Home Journal* 119 (Oct. 1942), 24–25.

40. Cohen, *Children of the Mill: Schooling and Society in Gary, Indiana, 1906–1960* (Bloomington: Indiana Univ. Press, 1990), 168–69; letter 146; Clive, *State of War*, 205.

41. Johnson, "Chicago's Public Schools Contribute to the War Effort," *Chicago Schools Journal* (Jan.-June 1943), 49–51.

42. "Education's Part in the War Effort," *Journal of the National Education Association* 35 (May 1946), 250.

43. Patri, *Your Children in Wartime* (Garden City, NY: Doubleday, Doran, 1943), 89.

44. Letter 53.

45. Lawrence *Daily Journal World*, Feb. 28, 1942; National Recreation Association, *Summer Playground Notebook*, no. 9 (July 30, 1943), 3; letters 42, 316; *New York Times*, Jan. 17, 1944.

46. Letter 150; *New York Times*, Dec. 5, 14, 21, 26, 28, 29, 1942, March 14, Sept. 8, 1944; Ronald Lippitt and Alvin Zander, "A Study of Boys' Attitudes Toward Participation in the War Effort," *Journal of Social Psychology* 17 (May 1943), 309–25; "New Appeals to Schools in Wastepaper Campaign," *Elementary School Journal* 45 (Sept. 1944),

4; Anna W. M. Wolf and Irma Simonton Black, "What Happened to the Younger People," in Jack Goodman, ed., *While You Were Gone: A Report on Wartime Life in the United States* (New York: Simon and Schuster, 1946), 75. One of the most helpful publications for understanding the classroom during wartime is *Education for Victory*. See the following articles in *Education for Victory*: "Salvaging for Victory," 1 (May 1, 1942), 2; "Schools and the Salvage for Victory Campaign Are Victorious," 1 (Aug. 1, 1942), 22; "An Appeal to Schools to Intensify Salvage Efforts," 2 (Oct. 1, 1943), 16–17; "Schools Urged to Collect Waste Paper," 2 (Nov. 1, 1943), 21; "Waste Paper Collection," 2 (Jan. 20, 1944), 3–4; "Schools and Scrap," 2 (March 3, 1944), 13–14; "Schools Urged to Help Salvage Tin Cans," 3 (March 20, 1945), 2.

47. Letters 56, 150, 317, 322, 378, 438; *New York Times*, Aug. 26, Oct. 7, 11, 21, 1942; John W. Studebaker, "30,000,000 Soldiers for Our New Third Front," *Saturday Evening Post* (Sept. 26, 1942), 15, 75; "'They Also Serve,'" *Recreation* 38 (April 1944), 21–22.

48. Letters 63, 72, 97D, 136, 238, 262, 284, 302, 324B, 353, 386, 408, 440; Flora Wyatt, ed., *Tales from the One-Room Schoolhouse* (Lawrence: Univ. of Kansas Kappa Chapter of Phi Delta Kappa, 1989), 35.

49. National Recreation Association, *Summer Playground Notebook*, no. 9 (July 30, 1943), 3; letters 30, 38, 136, 137, 142, 176, 194, 254, 322, 336, 348, 383, 440; *New York Times*, Dec. 13, 1942; May 25, 1943; March 12, Dec. 27, 1945.

50. Julietta K. Arthur, "Never Too Young to Help," *American Home* 28 (Sept. 1942), 28; J. Louise Despert, "School Children in Wartime," *Journal of Educational Sociology* 16 (Dec. 1942), 226; Leaf, *A War-time Handbook for Young Americans* (Philadelphia: Frederick A. Stokes, 1942), 3–30; United States House of Representatives, 76th and 77th Congs., Select Committee to Investigate the Interstate Migration of Destitute Citizens (later renamed Select Committee Investigating National Defense Migration; hereinafter both committees are referred to as Tolan Committee), *Hearings*, part 25 (Washington: GPO, 1942), p. 9770; Stella Chess, "War Ideologies of Children," *American Journal of Orthopsychiatry* 13 (July 1943), 505–9.

51. *New York Times*, March 5, 1943; letter 260.

52. Letters 46, 87, 137, 150, 152, 270, 276, 317, 336; Grace McLean Abbate, "Group Procedures Found Effective in the Prevention and Handling of Emotional Disorders," *Mental Hygiene* 26 (July 1942), 401–2; *New York Times*, Jan. 1, 1942, Jan. 1, Feb. 17, 18, June 30, July 12, Oct. 21, 1944.

53. Mercedes Roseberry, *This Day's Madness: A Story of the American People Against the Background of the War Effort* (New York: Macmillan, 1944), 149; letters 74, 248, 270, 310.

54. John W. Studebaker, "Contribution of Education to the War Effort," *Education for Victory* 2 (July 15, 1943), 4; "Key Assignments in Third War Loan Drive," *Education for Victory* 2 (Oct. 1, 1943), 6; "School-children Expert Salesmen," *Education for Victory* 2 (Dec. 1, 1943), 28.

55. "The Schools and the 5th War Loan," *Education for Victory* 2 (May 20, 1944), 19; "News Notes from the Schools," *Education for Victory* 3 (July 20, 1944), 20; "Summer War Savings Activities," *Education for Victory* 3 (Sept. 4, 1944), 27; Wolf and Black, "What Happened to the Younger People," 74–75.

For a discussion of the Girl and Boy Scouts and the Camp Fire Girls, see Chapter 8.

56. Letters 31, 317, 363; *New York Times*, July 16, 23, 1943.

57. "A Victory Garden for Children," *House & Garden* 81 (June 1942), 21–23 in the supplement; J. W. Feldman, "Children's Victory Gardens," *Recreation* 36 (March 1943), 662–63; "Young Gardeners Strike Pay Dirt," *Better Homes & Gardens* 21 (April 1943), 58, 80–83. One of the most helpful sources on schools during the war is *Education for Victory*. In *Education for Victory*, see "Victory Gardening Through the Schools," 1 (March 3, 1942), 16–18; "Los Angeles Schools Launch Food for Freedom Program," 1 (April 1, 1942), 3–4; "State Education Departments Publish School Garden Aids," 1 (June 1, 1942), 24–25; (Feb. 15, 1943), 23–25; "Facts about School-Directed Gardens," 1 (July 1, 1942), 15; "School-Directed Victory Gardens," 1 (Feb. 15, 1943), 23–25; "Schools Alert for Victory Gardening," 1 (March 15, 1943), 1–2, 24; "School Garden Organization and Summer Care," 1 (May 15, 1943), 20–21; "Victory Gardens and the Schools in 1945," 3 (Jan. 20, 1945), 13.

58. Letters 172, 302, 427, 438.

59. Letters 0.1, 4, 16, 20, 63, 74, 129, 137, 146, 176, 182, 186, 206, 218, 228, 236, 270, 276, 298, 302, 314, 333, 373, 406, 412, 429, 436, 440, 443, 449. For a good popular history of wartime rationing, see Richard R. Lingeman, *Don't You Know There's a War On? The American Homefront, 1941–1945* (New York: Capricorn, 1976), 234–70.

60. Letters 60, 66, 113, 136, 206K, 222, 282, 299, 316.

61. Letters 341, 358, 406.

62. Letters 25, 118, 143, 156A, 156B, 206K, 228, 250, 297, 336, 385, 431, 436, 449.

63. Letters 97F, 129, 186, 204, 442D.

64. Letters 350, 363.

65. Letters 118, 131.

66. Letters 0, 292, 355.

67. Letters 42, 429, 443.

68. Letters 45, 85, 186, 222, 307, 324F, 372.

69. Letters 180, 206B, 406.

70. Morris, *North Toward Home* (Boston: Houghton Mifflin, 1967), 37.

71. Letters 50, 114, 171, 246, 394, 442.

72. Covert quoted in Mark Jonathan Harris et al., eds., *The Homefront: America During World War II* (New York: Putnam's, 1984), 74; letters 62, 376.

73. Letters 4, 30, 50, 154, 387, 456.

74. Letters 66, 74, 97C, 127, 129, 206, 220, 236, 248, 254, 276, 284, 290, 312, 328, 406, 408.

75. Letters 46, 50, 74, 129, 178, 209, 210, 220, 321, 324F, 374; Kleinberg, "When Roosevelt Died and a Boy Grew Up," in *Lawrence Journal-World*, April 12, 1989.

76. Letters 0.2, 50, 64; Kleinberg, "When Roosevelt Died and a Boy Grew Up."

77. See Chapter 10, which discusses the children of pacifists and Jehovah's Witnesses.

78. Letters 10, 32, 50, 53, 64, 87, 114, 146, 148, 166, 248, 257, 440.

79. Letters 228, 236, 240C, 302, 332, 370, 373, 398, 408, 443. See also Chapter 9.

Chapter 8. Children Play War Games

1. Letters 32, 137, 174, 413, 442D.

2. Letters 0.3, 26, 83; "Folk Rhymes for Children," *New York Times Magazine*, Nov. 5, 1944; Roger D. Abrahams, ed., *Jump-Rope Rhymes: A Dictionary* (Austin: Univ. of Texas Press, 1969), 47.

3. Frank Rowsome, Jr., *The Verse by the Side of the Road: The Story of the Burma-Shave Signs and Jingles* (Brattleboro, VT: Stephen Greene Press, 1965), 96, 98.

4. Letters 240D, 270, 278, 322, 438; Iona and Peter Opie, *The Lore and Language of Schoolchildren* (Oxford: Clarendon, 1960), 101–2.

5. Wolfenstein, *Children's Humor: A Psychological Analysis* (Glencoe: Free Press, 1954), 12, 16; Singer et al., *The Child's World of Make-Believe: Experimental Studies of Imaginative Play* (New York; Academic, 1973), 14. Other helpful studies are Kenneth H. Rubin et al., "Play," in Paul H. Mussen, ed., *Handbook of Child Psychology* (fourth edition); vol. 4: *Socialization, Personality, and Social Development*, E. Mavis Hetherington, ed. (New York: Wiley, 1983), 693–774; Singer, *Daydreaming: An Introduction to the Experimental Study of Inner Experience* (New York: Random House, 1966), 130–70; Elliott M. Avedon and Brian Sutton-Smith, eds., *The Study of Games* (New York: Wiley, 1971); R. E. Herron and Sutton-Smith, eds., *Child's Play* (New York: Wiley, 1971); Sutton-Smith, *The Folkgames of Children* (Austin: Univ. of Texas Press, 1972); Jerome Bruner et al., *Play—Its Role in Development and Education* (New York: Basic, 1976); Helen B. Schwartzman, *Transformations: The Anthropology of Children's Play* (New York: Plenum, 1978); Paul E. McGhee and Anthony J. Chapman, eds., *Children's Humor* (New York: Wiley, 1980); Sutton-Smith, *The Folkstories of Children* (Philadelphia: Univ. of Pennsylvania Press, 1981). The best introduction to Piaget's cognitive psychology is Piaget and Barbel Inhelder, *The Psychology of the Child* (New York: Basic Books, 1969). For an introduction to the historical study of play, see Bernard Mergen, "The Discovery of Children's Play," *American Quarterly* 28 (Oct. 1975), 399–420; and Bernard Mergen, ed., *Play and Playthings: A Reference Guide* (Westport, CT: Greenwood, 1982).

6. Baruch, *You, Your Children, and War* (New York: D. Appleton-Century, 1942), 31–34, 40–45, 67.

7. Letters 43, 81, 220, 225, 229, 238, 259, 259A, 316, 407, 436, 442; audiotape recollections of James F. Tompkins.

8. Letters 64, 74, 120, 130, 137, 171, 220, 234, 238, 272, 442; *New York Times*, April 2, Aug. 24, Dec. 9, 1942, Nov. 12, 1943, July 15, 1945; Elizabeth Hooper, "Dolls in War-time," *Hobbies* 46 (Sept. 1941), 12, 14–15; Caroline B. Piercy, "Dolls in the Service of Uncle Sam," *Hobbies* 47 (Aug. 1942), 28–29; "Toys Are Scarce," *Business Week* (Oct. 7, 1944), 34, 36; "Toy Walkie-Talkie Thrills Young Commandos," *Popular Science Monthly* 146 (March 1945), 170; "Toy Trouble," *Business Week* (March 23, 1946), 26, 28; Virginia W. Marx, "Toys—To Delight," *Parents' Magazine* 21 (Nov. 1946), 164–65.

9. Letters 24, 33, 42, 97E, 110, 154, 163, 220, 286, 298, 397; Charlotte Towle, "The Effect of the War upon Children," *Social Service Review* 17 (June 1943), 155; Iona

and Peter Opie, *Children's Games in Street and Playground* (Oxford: Clarendon, 1969), 338–40.

10. Momaday, *The Names: A Memoir* (New York: Harper & Row, 1976), 99.

11. Ralph C. Preston, "What Children Think of War Play," *Parents' Magazine* 18 (March 1943), 21; C. Madeleine Dixon, "When Play Goes Warlike," *Parents' Magazine* 17 (July 1942), 26–27, 48, 77; Rautman, "Children's Play in War Time," *Mental Hygiene* 27 (Oct. 1943), 551; "Children Should Not Be Encouraged in War Games," *Science News Letter* 45 (Feb. 12, 1944), 104.

12. Gorham, *So Your Husband's Gone to War!* (Garden City, NY: Doubleday, Doran, 1942), 89; *New York Times*, Aug. 9, 1943; "Concerning Children," *Survey Mid-monthly* 79 (Sept. 1943), 247; Child Study Association, "Children in Wartime: Parents' Questions," in Sidonie Matsner Gruenberg, ed., *The Family in a World at War* (New York: Harper & Brothers, 1942), 262–63; Anna W. M. Wolf, "War Begins in the Nursery," *Child Study* 20 (Fall 1942), 35–38; Agnes E. Benedict, "The Children's War," *New York Times Magazine* (Nov. 18, 1942), 25; "The Roots of War in Human Nature: Aggression and Hate in Childhood and Family Life" (a symposium), *Child Study* 20 (Spring 1943), 72–77, 91. For the role of hatred among children on the homefront, see Chapter 10.

13. Petronella R. Tacionis, "A Day in Pinneys Woods," *Recreation* 39 (May 1945), 83, 110.

14. Letter 110; Ickis, "What About Games in Wartime?," *Recreation* 36 (Jan. and Feb. 1943), 556–61, 594, 609–13.

15. Letters 24, 42, 43, 316, 322; John Morton Blum, *V Was for Victory: Politics and American Culture During World War II* (New York: Harcourt Brace Jovanovich, 1976), 20.

16. Gould, *Child Studies Through Fantasy: Cognitive-Affective Patterns in Development* (New York; Quadrangle, 1972), 21.

17. Letters 32, 66, 110, 236, 316, 407, 441.

18. Letters 24, 26, 33, 81, 127, 130, 286, 321, 324, 441, 454.

19. Maccoby, "Gender and Relationships: A Developmental Account," *American Psychologist* 45 (April 1990), 513–20.

20. Murphy, "Psychology of Preadolescent Children in Wartime: II. The Young Child's Experience in Wartime," *American Journal of Orthopsychiatry* 13 (July 1943), 497–501. While recent scholarship in developmental psychology does not confirm identification as the reason for sex-typed roles, I believe the concept provides helpful insights into children's learning; see the discussion of father absence in Chapter 12.

21. Maccoby and Jacklin, *The Psychology of Sex Differences* (Stanford: Stanford Univ. Press, 1974), 227–76, 349–74; Maccoby and Jacklin, "Sex Differences in Aggression: A Rejoinder and Reprise," *Child Development* 51 (Dec. 1980), 964–80; Ross D. Parke and Ronald G. Slaby, "The Development of Aggression," in Paul H. Mussen, ed., *Handbook of Child Psychology* (fourth ed.), vol. 4: *Socialization, Personality, and Social Development*, E. Mavis Hetherington, ed. (New York: Wiley, 1983), 547–641, esp. 561–62; Lawrence, KS, *Journal-World*, July 31, 1983. A helpful wartime study that focuses on gender roles is Florene M. Young, "Psychological Effects of War on Young Children," *American Journal of Orthopsychiatry* 17 (July 1947), 500–510.

22. Konner, "The Aggressors," *New York Times Magazine* (Aug. 14, 1988), 33–34.

23. Eleanor Palmer Bonte and Mary Musgrove, "Influences of War as Evidenced in Children's Play," *Child Development* 14 (Dec. 1943), 196–98; Amram Scheinfeld, "Play Reveals the Boy or Girl," *Parents' Magazine* 19 (April 1944), 137–38; Sutton-Smith, *The Folkgames of Children*, 412.

24. Gilligan, *In a Different Voice: Psychological Theory and Women's Development* (Cambridge: Harvard Univ. Press, 1982), 100; letter 328. There is valuable literature in psychology on children's sex-typing and development of gender roles. Good places to start are Aletha C. Huston, "Sex-Typing," in Mussen, ed., *Handbook of Child Psychology*, vol. 4: *Socialization, Personality, and Social Development*, 387–467; Sandra Lipsitz Bem, "Gender Schema Theory and Its Implications for Child Development: Raising Gender-Aschematic Children in a Gender-Schematic Society," *Signs* 8 (Summer 1983), 598–616. For a historian's insights into sex-typing, see Sheila M. Rothman, *Women's Proper Place: A History of Changing Ideals and Practices, 1870 to the Present* (New York: Basic, 1978).

25. Komarovsky, *Women in the Modern World: Their Education and Their Dilemmas* (Boston: Little, Brown, 1953), 53–67; Meyer Rabban, "Sex Role Identification in Two Diverse Social Groups," *Genetic Psychology Monographs* 42 (1950), 81–158.

26. Letters 20, 442, 450.

27. Letters 97E, 163, 286, 359.

28. Letters 148, 184, 185, 220, 234, 265, 276, 356, 384; "Our Sky Enemies," *Boys' Life* (April 1942), 40.

29. Letters 81, 165, 220, 264, 265, 302, 384, 397, 431.

30. Letters 154, 397; Catherine Mackenzie, "What Are Boys Made Of?," *New York Times Magazine* (Sept. 17, 1944).

31. Mark Jonathan Harris et al., eds., *The Homefront: America During World War II* (New York: Putnam's, 1984), 73; letters 81, 133, 144, 322, 328.

32. Letters 10, 33, 55D, 83I, 97L, 133, 206C, 254, 316, 332, 407.

33. Letter 442H; "Some Wartime Programs for Girls," *Recreation* 38 (April 1944), 23, 51.

34. Letters 87, 220; Filene, *Him/Her/Self: Sex Roles in Modern America* (Baltimore: Johns Hopkins Press, second ed., 1986), 162.

35. Letters 7, 70, 251, 328, 367, 454.

36. Letters 24, 25, 108, 121; *New York Times*, Feb. 13, 1942.

37. Smith-Rosenberg, *Disorderly Conduct: Visions of Gender in Victorian America* (New York: Alfred A. Knopf, 1985), esp. 12–19.

38. Strout, "Ego Psychology and the Historian," *History and Theory* 7, (No. 3, 1968), 296–97. For the historian, helpful discussions are: Robert R. Sears, "Identification as a Form of Behavioral Development," in Dale B. Harris, ed., *The Concept of Development* (Minneapolis: Univ. of Minnesota Press, 1957), 149–61; Jerome Kagan, "The Concept of Identification," *Psychological Review* 65 (Sept. 1958), 296–305; Sears et al., *Identification and Child Rearing* (Stanford: Stanford Univ. Press, 1965); David B. Lynn, *Parental and Sex-Role Identification: A Theoretical Formulation* (Berkeley, CA: McCutchan, 1969), 3–13.

39. Stone, "The Play of Little Children," in Herron and Sutton-Smith, *Child's Play*, 9–11; Ruth E. Hartley et al., *Understanding Children's Play* (New York: Columbia Univ. Press, 1952), 47–51; Sutton-Smith, "The Play of Girls," in Claire B. Kopp, ed., *Becoming Female: Perspectives on Development* (New York: Plenum, 1979), 238–44.

40. Letters 26, 55C, 66, 133, 209, 413; Helen Ross, "Psychology of Preadolescent Children in War Time: III. Emotional Forces in Children as Influenced by Current Events," *American Journal of Orthopsychiatry* 13 (July 1943), 502–4; Leanna Geddie and Gertrude Hildreth, "Children's Ideas about the War," *Journal of Experimental Education* 13 (Dec. 1944), 92–97; Ralph C. Preston, *Children's Reactions to a Contemporary War Situation* (New York: Teachers College, Bureau of Publications, 1942), 77–79.

41. Letters 200A, 239.

42. Letters 302, 322; *New York Times*, July 20, 1941.

43. *New York Times*, Dec. 8, 1942, July 19, 1943, Nov. 6, 1988; Virginia De Tar, "Before the Juvenile Court Steps In," *Journal of Home Economics* 35 (June 1943), 334.

44. Letters 97B, 251, 270, 328, 332.

45. Letters 129, 275, 407.

46. Rothschild, "A Million or More in '44: Girl Scouts, Girls' Roles, and the American War Effort" (paper presented at the Berkshire Conference on the History of Women, 1990), 2; Girl Scouts of the United States, *Leadership of Girl Scout Troops* (New York: Girl Scouts, 1943), 8–9; letters 12B, 55B, 72, 87, 156A, 339, 383, 446; *New York Times*, Feb. 1, Oct. 23, 1941, July 3, 11, Oct. 1, 1942, June 15, 20, 30, Aug. 16, Oct. 6, 1943, March 14, Aug. 12, Oct. 4, Dec. 14, 1944, Sept. 22, 1945; Catherine Mackenzie, "Line for the Girl Scouts," *New York Times Magazine*, Oct. 29, 1944.

47. Harold P. Levy, *Building a Popular Movement: A Case Study of the Public Relations of the Boy Scouts of America* (New York: Russell Sage Foundation, 1944), 130–31; James E. West, "Strong—For America," *Boys' Life* (Feb. 1942), 10–11, 48–49; "Boy Scouts at War," *Time* 39 (June 15, 1942), 14; "Can You Say . . . 'Today I Have Served My Country,'" *Boys' Life* (July 1942), 26; "Training Scouts for War Work," *Popular Mechanics* 78 (Aug. 1942), 78–79; Ronald Lippitt and Alvin Zander, "A Study of Boy Attitudes toward Participation in the War Effort," *Journal of Social Psychology* 17 (May 1943), 309–25; "Progress of the Boy-Scout Movement . . . ," *School and Society* 60 (Oct. 7, 1944), 230–31; Will Oursler, *The Boy Scout Story* (Garden City, NY: Doubleday, 1955), 246–47.

48. Rothschild, "A Million or More in '44," 2–12; letters 12B, 72, 206L, 218; Robert C. Ferguson, "Americanism in Late Afternoon Radio Adventure Serials, 1940–1945" (unpub. paper), 3.

49. *New York Times*, Nov. 5, 1943.

50. Letters 27, 182, 431, 435.

51. Patsy Fisher, Racine, WI, to "Mr. Roosevelt," Nov. 25, 1942, copy made available to me by Margaret Baker, Baldwin, KS.

Chapter 9. Children's Entertainment

1. Letter 236.

2. Letters 114, 408; "Jack the Nazi Killer," *Newsweek* 22 (Aug. 23, 1943), 80; Robert C. Ferguson, "Americanism in Late Afternoon Radio Adventure Serials, 1940–1945" (unpub. paper), 4–8; J. Fred MacDonald, *Don't Touch That Dial: Radio Programming in American Life, 1920–1960* (Chicago: Nelson-Hall, 1980), 68–69, 203–4; Thomas Whiteside, "Up, Up and Awa-a-y," *New Republic* 116 (March 3, 1947), 15–17; Jim Harmon, *The Great Radio Heroes* (Garden City, NY: Doubleday, 1967), 49–51.

3. MacDonald, *Don't Touch That Dial*, 68–69, 203–4; "Jack the Nazi Killer," 80.

4. Roy De Verl Willey and Helen Ann Young, *Radio in Elementary Education* (Boston: D.C. Heath, 1948), 10–13.

5. Letters 114, 278, 408, 413; Willey and Young, *Radio in Elementary Education*, 3–4; Weston R. Clark, "Radio Listening Habits of Children," *Journal of Social Psychology* 12 (Aug. 1940), 131–49; "Mack and the Beanstalk," *Newsweek* 21 (June 28, 1943), 108; *New York Times*, April 26, Oct. 18, Nov. 1, 8, 15, 1942, May 27, 1945; John K. Hutchens, "Tracy, Superman, et AL. Go to War," *New York Times Magazine*, Nov. 21, 1943; "Radio Programs for Children," *Child Study* 21 (Fall 1943), 28–29; "Radio Programs for Children," *Child Study* 22 (Winter 1944–45), 53; "It's Superfight," *Newsweek* 27 (April 29, 1946), 61; Albert N. Williams, "And a Little Child Shall Lead Them," *Saturday Review of Literature* 30 (Feb. 8, 1947), 26–27; Whiteside, "Up, Up and Awa-a-y," 15–17; Ferguson, "Americanism in Late Afternoon Radio Adventure Serials," 9–11.

6. Letters 166, 192, 228, 229, 300; "Something for the Boys & Girls," *Time* 46 (Sept. 24, 1945), 78; Arthur Frank Wertheim, *Radio Comedy* (New York: Oxford Univ. Press, 1979), 263–82; MacDonald, *Don't Touch That Dial*, 43–44, 140–41, 257–71; Harmon, *The Great Radio Heroes*, 85, 107, 231–33; George A. Willey, "The Soap Operas and the War," *Journal of Broadcasting* 7 (Fall 1963), 339–52; Paul A. Lazarsfeld and Frank N. Stanton, eds., *Radio Research 1942–1943* (New York: Duell, Sloan and Pearce, 1944), 34–69; Erik Barnouw, *The Golden Web: A History of Broadcasting in the United States*, vol. 2: *1933 to 1953* (New York: Oxford Univ. Press, 1968), 94–97.

7. Letters 43, 44, 53, 59, 97K, 131, 146, 171, 195, 206K, 208, 241, 246, 276, 323, 394, 408, 440; Clara Savage Littledale, "Radio Interprets the War to Children," *Parents' Magazine* 18 (March 1943), 17; MacDonald, *Don't Touch That Dial*, 288, 291–310.

8. Newman F. Baker, "Current Notes: Radio Crime Programs," *Journal of Criminal Law and Criminology* 31 (1940), 222–23; Azriel L. Eisenberg, *Children and Radio Programs* (New York: Columbia Univ. Press, 1936), 185–90; *New York Times*, Oct. 5, 1941, June 27, 1945; Willey and Young, *Radio in Elementary Education*, 8–9; Erle Kenney and Harriet E. Neall, "His Ear to the Radio," *Parents' Magazine* 19 (March 1944), 88; Howard Rowland, "Radio Crime Dramas," *Educational Research Bulletin* 23 (Nov. 15, 1944), 210–11; Sara Ann Fay, "Are Children's Radio Programs a Good Influence?," *Library Journal* 70 (Feb. 15, 1945), 175–76; Dorothy Gordon, *All Children Listen* (New York: George W. Stewart, 1942), 44–49, 52–61; Child Study Association, "Children in Wartime: Parents' Questions," in Sidonie Matsner Gruenberg, ed., *The Family in a World at War* (New York: Harper & Brothers, 1942), 261–62; MacDonald, *Don't Touch That Dial*, 43–46, 68–70.

9. Dorothea English Murphy, "A Survey of the Radio Interests and Listening Habits of 358 Elementary School Students" (unpub. M.A. thesis, Fordham Univ., 1946), 60–68; Jane Ferguson Porter, "Radio Interests of the Children in Mason County, West Virginia" (unpub. M.A. thesis, Marshall College, 1945), 10–17; Colleen Kelly Gery, "An Historical Study of American Radio's Role in Informing Children in the United States about World War II, 1941–1945" (unpub. M.A. thesis, Indiana Univ., 1968), 18–82.

10. "Jack the Nazi Killer," 80; MacDonald, *Don't Touch That Dial*, 257–71; Raymond William Stedman, *The Serials: Suspense and Drama by Installment* (Norman: Univ. of Oklahoma Press, 1977), 329–39.

11. Mark H. Leff, "The Politics of Sacrifice on the American Home Front in World

War II," *Journal of American History* 77 (March 1991), 1296–1318; Rubicam, "Advertising," in Jack Goodman, ed., *While You Were Gone: A Report on Wartime Life in the United States* (New York: Simon and Schuster, 1946), 421–46.

12. Harold T. Christensen and Doyle L. Green, "The Commercialization of Patriotism in World War II," *Sociology and Social Research* 27 (July-Aug. 1943), 447–52; Theodore S. Repplier, "Advertising Dons Long Pants," *Public Opinion Quarterly* 9 (Fall 1945), 269–78.

13. Letters 15, 25, 97D, 166, 203, 335, 385, 390, 426; War Finance Division, Department of the Treasury, *New Songs for Schools at War* (Washington: GPO, 1943); Joe Bookman, "The Truth about Silly War Songs," *Collier's* 112 (July 24, 1943), 20, 42–43; Robert Fyne, "'You're a Sap, Mr. Jap': Tin Pan Alley Fights the Axis" (unpub. paper), 1–6.

14. Mead is quoted in *Newsweek* 92 (Nov. 27, 1978), 75–76. See also see Kenneth Keniston and the Carnegie Council on Children, *All Our Children: The American Family Under Pressure* (New York: Harcourt Brace Jovanovich, 1977); David Elkind, *The Hurried Child: Growing Up Too Fast Too Soon* (Reading, MA: Addison-Wesley, 1981); Alice Miller, *Prisoners of Childhood: How Narcissistic Parents Form and Deform the Emotional Lives of Their Gifted Children* (New York: Basic Books, 1981); Valerie Polakow Suransky, *The Erosion of Childhood* (Chicago: Univ. of Chicago Press, 1982); Neil Postman, *The Disappearance of Childhood* (New York: Delacorte, 1982); Marie Winn, *Children without Childhood* (New York: Pantheon, 1983); Vance Packard, *Our Endangered Children: Growing Up in a Changing World* (Boston: Little, Brown, 1983); Richard Louv, *Childhood's Future: New Hope for the American Family* (Boston: Houghton Mifflin, 1990).

Television, which saw its first transmission in 1927, was not commercially broadcast during the war, but some people were already worrying about the effects of "blind broadcasting" on children. Would children stop using their imaginations? As Rudolph Arnheim, the psychologist, warned before the war, "The words of the story-teller . . . , the voices of dialogue, the complex sounds of music conjure up worlds of experience and thought that are easily disturbed by the undue addition of visual things." Arnheim's apprehension was not groundless; postwar research showed that radio stimulated the child's imagination "significantly more" than television did. (Arnheim is quoted in Marie Winn, "Why Has Radio Tuned Out Children?," *New York Times*, Sept. 25, 1983.)

What effects has television had on the generations born after the Second World War? Joshua Meyrowitz, the psychologist, has contended that the television years "have seen a remarkable change in the image and roles of children. Childhood as a protected and sheltered period of life has all but disappeared. . . . In the shared environment of television, children and adults know a great deal about one another's behavior and social knowledge—too much, in fact, for them to play out the traditional complementary roles of innocence versus omniscience" (Meyrowitz, "The Adultlike Child and the Childlike Adult: Socialization in an Electronic Age," *Daedalus* 113 [Summer 1984], 19–48; Meyrowitz, *No Sense of Place: The Impact of Electronic Media on Social Behavior* [New York: Oxford Univ. Press, 1985], 226–67).

15. Letters 12B, 187, 276, 316, 405.

16. Letters 97E, 442B.

17. Letters 150, 312, 318, 350, 435; Raymond Fielding, *The American Newsreel, 1911–1967* (Norman: Univ. of Oklahoma Press, 1972), 288–95; Kansas City *Star*, April

21, 1986; Roger Manvell, *Films and the Second World War* (South Brunswick, NJ: A. S. Barnes, 1974), 122–23, 183–85.

18. Letters 42, 150, 222; Fielding, *The American Newsreel*, 295; interview with Frances Degen Horowitz, Feb. 9, 1984.

19. See Allen L. Woll, *The Hollywood Musical Goes to War* (Chicago: Nelson-Hall, 1983); John H. Lenihan, *Showdown: Confronting Modern America in the Western Film* (Urbana: Univ. of Illinois Press, 1980); *New York Times*, May 19, 1943.

20. David Culbert, "'Why We Fight': Social Engineering for a Democratic Society at War," in K.R.M. Short, ed., *Film & Radio Propaganda in World War II* (London: Croon Helm, 1983), 173–91; Erik Barnouw, *Documentary: A History of the Non-Fiction Film* (New York: Oxford Univ. Press, 1983), 139–64; Manvell, *Films and the Second World War*, 167–83; "War Films Available to School and Adult Audiences," *The School Review* 51 (April 1943), 205; Harold Putnam, "The War Against War Movies," *The Educational Screen* 22 (May 1943), 162–63, 175; Thornton Delehanty, "The Disney Studio at War," *Theater Arts* 27 (Jan. 1943), 31–33; Richard Schickel, *The Disney Version: The Life, Times, Art, and Commerce of Walt Disney* (New York: Simon and Schuster, 1968), 270–71; Leonard Maltin, *The Disney Films* (New York: Bonanza, 1973), 16, 60–64; Maltin, *Of Mice and Magic: A History of American Animated Cartoons* (New York: McGraw-Hill, 1980), 70–71.

21. Garth Jowett, *Film: The Democratic Art* (Boston: Little, Brown, 1976), 316, 473–75, 483; "Big Movie Year," *Business Week* (Feb. 13, 1943), 37–38; Peter A. Soderbergh, "The Grand Illusion: Hollywood and World War II, 1930–1945," *University of Dayton Review* 5 (Winter 1968–69), 18.

22. Letters 228, 229, 243, 254, 300, 312, 350, 408; Morris, *North Toward Home* (Boston: Houghton Mifflin, 1967), 34–35; "How to Run a Theater," *Time* 42 (Nov. 22, 1943), 94.

23. Gladys Denny Shultz, "Comics—Radio—Movies," *Better Homes & Gardens* 24 (Nov. 1945), 22–23, 73–75, 108; Jim Harmon and Donald F. Glut, *The Great Movie Serials: Their Sound and Fury* (Garden City, NY: Doubleday, 1972), 217–19, 235–41, 244–61, 273–81; Stedman, *The Serials*, 330.

24. "The Production Code, 1930–1968," in Robert H. Bremner et al., eds., *Children and Youth in America: A Documentary History*, vol. 3: 1933–1973 (Cambridge: Harvard Univ. Press, 1974), 891–93; Catherine MacKenzie, "Movies—And Superman," *New York Times Magazine*, Oct. 12, 1941; Catherine C. Edwards, "Let's Talk about the Movies," *Parents' Magazine* 20 (Oct. 1945), 31 ff. The best studies of the wartime feature films are Jowett, *The Democratic Art*, 293–332; Manvell, *Films and the Second World War*, 176–203; Dorothy B. Jones, "The Hollywood War Film: 1942–1944," *Hollywood Quarterly* 1 (1945), 1–19; Bosley Crowther, "The Movies," in Goodman, *While You Were Gone*, 511–32; Lewis Jacobs, "World War II and the American Film," *Cinema Journal* 7 (Winter 1967–68), 1–21; Soderbergh, "The Grand Illusion: Hollywood and World War II, 1930–1945," 13–22; Colin Shindler, *Hollywood Goes to War: Films and American Society, 1939–1952* (London: Routledge & Kegan Paul, 1979); Bernard F. Dick, *The Star-Spangled Screen: The American World War II Film* (Lexington: Univ. of Kentucky Press, 1985); Otto Friedrich, *City of Nets: A Portrait of Hollywood in the 1940s* (New York: Harper & Row, 1986); Jeanine Basinger, *The World War II Combat Film* (New York: Columbia Univ. Press, 1986); Clayton R. Koppes and Gregory D. Black, *Hollywood Goes*

to *War: How Politics, Profits, and Propaganda Shaped World War II Movies* (New York: Free Press, 1987); Terry Christensen, *Reel Politics: American Political Movies from "Birth of a Nation" to "Platoon"* (London: Basil Blackwell, 1987), 63–101.

25. *New York Times*, Oct. 24, 1943; Florene M. Young, "Psychological Effects of War on Young Children," *American Journal of Orthopsychiatry* 17 (July 1947), 507; Josette Frank, "Chills and Thrills in Radio, Movies and Comics," *Child Study* 25 (Spring 1948), 42–46, 48; Manvell, *Films and the Second World War*, 199. For a summary of prewar studies of the psychological impact of films, see Franklin Fearing, "Influence of the Movies on Attitude and Behavior," *Annals of the American Academy of Political and Social Science* 254 (Nov. 1947), 70–79. A study of postwar films released between 1945 and 1949 is Martha Wolfenstien and Nathan Leites, *Movies: A Psychological Study* (Glencoe, IL: Free Press, 1950).

26. Momaday, *The Names: A Memoir* (New York; Harper & Row, 1976), 89; letters 32, 228, 443; Ken D. Jones and Arthur F. McClure, *Hollywood at War: The American Motion Picture and World War II* (New York: Castle, 1973), 198–99.

27. Letter 219; Ruth Weeden Stewart, "The Year's New Books for Boys and Girls Over Ten," *Library Journal* 68 (Oct. 15, 1943), 824–25; Lena Barksdale, "A Selected List of Children's Books," *The Nation* 157 (Nov. 20, 1943), 591–93; Margaret C. Scoggin, "Young People in a World at War," *Horn Book* 19 (Nov. 1943), 394–400; "Books: Hints to Santa," *Newsweek* 22 (Dec. 13, 1943), 105; "Children's Books of 1943–1944" and "Children's Books of 1944–45," both in *Journal of the National Educational Association* 33 (Nov. 1944), 195–96, and 34 (Nov. 1945), 167–68; "Most Widely Used Children's Books of 1939–1943," *Chicago Schools Journal* (Sept.-Dec. 1945), 33; Jane Cobb and Helen Dore Boylston, "What's Your Child Reading?," *The Atlantic*, 178 (Nov. 1946), 160 ff.; U.S. Children's Bureau Publication No. 304, *For the Children's Bookshelf: A Booklist for Parents* (Washington: GPO, 1946).

28. John J. DeBoer, "Children's Books in Wartime," *Chicago Schools Journal* (Sept.-Dec. 1942), 40–42; Marie Nelson Taylor, "Facing the War with Our Young People," *Wilson Library Bulletin* 17 (April 1943), 656–58; Vernon Ives, "Children's Books and the War," *Publishers Weekly* 144 (Oct. 23, 1943), 1592–93; letter 220.

29. Mason, *The Girl Sleuth: A Feminist Guide* (Old Westbury, NY: Feminist Press, 1975), 6, 8–12; Carol Billman, *The Secret of the Stratemeyer Syndicate: Nancy Drew, the Hardy Boys, and the Million Dollar Fiction Factory* (New York: Ungar, 1986).

30. Tunis, *Keystone Kids* (New York: Harcourt, Brace, 1943), 149–61, 188, 208; Stewart, "The Year's New Books for Boys and Girls Over Ten," 823; Alice Dalgliesh, "'To Light a Candle,'" *Publishers Weekly* 147 (April 28, 1945), 1736; Smith, "The Writer Who Taught Generations How to Play the Game," *Wall Street Journal*, Jan. 2, 1990.

31. Maltin, *Of Mice and Magic*, 116–17, 137–39, 163–67, 172, 246–47, 250–52, 284–85. For the portrayal of Japanese Americans in popular culture, see Chapter 10.

32. Carl A. Posey, "The Strange Case of Carter Hall," *Air & Space* (Nov. 1989), 69; Catherine MacKenzie, "Children and the Comics," *New York Times Magazine*, July 11, 1943; *New York Times*, Feb. 2, March 25, Dec. 15, 1944, Jan. 13, Dec. 16, 25, 1945; William W. Savage, Jr., *Comic Books and America, 1945–1954* (Norman: Univ. of Oklahoma Press, 1990), 3–13.

33. Laurance F. Shaffer, *Children's Interpretations of Cartoons* (New York: Bureau of Publications, Teachers College, Columbia Univ., 1930), 1–5, 21–41; Margaret K.

Thomas, "Superman Teaches School in Lynn, Mass.," *Magazine Digest* (April 1944), 5–7; Dorothy Canfield Fisher, "What's Good for Children," *Christian Herald* (May 1944), 26–28.

34. Harvey Zorbaugh, "Comics—Food for Half-Wits?," *Science Digest* 17 (April 1945), 79–82; Elliot M. Rosenberg, "Winning the War with Captain Marvel," *New York Times Magazine*, Sept. 3, 1985; "Up, Up and Awaaay!!!," *Time* 131 (March 14, 1988), 68; "The Comic Book (Gulp!) Grows Up," *Newsweek* 111 (Jan. 18, 1988), 70–71; Gweneira Williams and Jane Wilson, "They Like It Rough: In Defense of Comics," *Library Journal* 67 (March 1, 1942), 204–6; "The Comics and Their Audience," *Publishers Weekly* 141 (April 18, 1942), 1476–79; "Issues Relating to the Comics," *Elementary School Journal* 42 (May 1942), 641–44; "Profitable Reading Pleasure," *Parents' Magazine* 17 (Nov. 1942), 44; Sister M. Katharine McCarthy and Marion W. Smith, "The Much Discussed Comics," *Elementary School Journal* 44 (Oct. 1943), 97–101; Milton Caniff, "The Comics," in Goodman, ed., *While You Were Gone*, 488–510; Steve M. Barkin, "Fighting the Cartoon War: Information Strategies in World War II," *Journal of American Culture* 7 (Summer 1984), 113–17; Michael Uslan, ed., *America at War: The Best of DC War Comics* (New York: Simon and Schuster, 1979), 5–12.

35. Steinem, "Introduction," in *Wonder Woman* (New York: Holt, Rinehart and Winston, 1972), unpaginated; Gerhart Saenger, "Male and Female Relations in the American Comic Strip," *Public Opinion Quarterly* 19 (Summer 1955), 195–205.

36. Steinem, "Introduction."

37. Letters 74, 270.

38. Kathryn Weibel, *Mirror Mirror: Images of Women Reflected in Popular Culture* (Garden City, NY: Anchor, 1977), 116–19; Baker, *Images of Women in Film: The War Years, 1941–1945* (Ann Arbor, MI: UMI Research Press, 1981); Dick, *The Star-Spangled Screen*, 174–87; Leo A. Handel, *Hollywood Looks at Its Audience: A Report of Film Audience Research* (Urbana: Univ. of Illinois Press, 1950), 120–24; Carol Trayner Williams, *The Dream Beside Me: The Movies and the Children of the Forties* (Rutherford, NJ: Fairleigh Dickinson Univ. Press, 1980), 41–80; Barbara Deming, *Running Away from Myself: A Dream Portrait of America Drawn from the Films of the Forties* (New York: Grossman, 1969), 8–10.

39. Lawrence W. Levine, *Highbrow/Lowbrow: The Emergence of Cultural Hierarchy in America* (Cambridge: Harvard Univ. Press, 1988), esp. 243–56; Godfrey Hodgson, *America in Our Time: From World War II to Nixon* (Garden City, NY: Doubleday, 1976), 12–16.

40. Letter 120; Peter A. Soderbergh, "The War Films," *Discourse* 11 (Winter 1968), 87–91; Soderbergh, "On War and the Movies: A Reappraisal," *Centennial Review* 11 (Summer 1967), 405–18; Peter Roffman and Jim Purdy, *The Hollywood Social Problem Film: Madness, Despair, and Politics from the Depression to the Fifties* (Bloomington: Indiana Univ. Press, 1981), 227–35; Dick, *The Star-Spangled Screen*, 324–27.

41. William H. Chafe, *The Paradox of Change: American Women in the 20th Century* (New York: Oxford Univ. Press, 1991), 154–93; Betty Friedan, *The Feminine Mystique* (New York: Norton, 1963); Peter G. Filene, *Him/Her/Self: Sex Roles in Modern America*, second ed. (Baltimore: Johns Hopkins Univ. Press, 1986), 148–76; Elaine Tyler May, *Homeward Bound: American Families in the Cold War Era* (New York: Basic Books, 1988), 58–161.

42. Elshtain, *Women and War* (New York: Basic Books, 1987), xii, 14–17; Walter Goodman, "Romance Narrows the Generation Gap," *New York Times*, Aug. 10, 1986.

Chapter 10. The Fractured Homefront

1. Interview with Wanda Davis Newbrough.

2. *1975 Yearbook of Jehovah's Witnesses* ... (Brooklyn: Watch Tower Bible and Tract Societies of Pennsylvania and New York and International Bible Students Association, 1974), 188–90.

3. *West Virginia State Board of Education* v. *Barnette*, 319 U.S. 624 (1943), quoted in Paul L. Murphy, *The Constitution in Crisis Times* (New York: Harper Torchbook ed., 1972), 200; David R. Manwaring, *Render unto Ceasar: The Flag-Salute Controversy* (Chicago: Univ. of Chicago Press, 1962).

4. Interview with Wanda Davis Newbrough; letter 447.

5. Letter 83F.

6. Ibid.

7. Ibid.

8. Letter 18.

9. Donovan Senter, memorandum to Waldemar A. Nielsen, "Potential Revolution in the Deep South," Sept. 7, 1942, in Division of Program Surveys, Project Files, 1940–45, Special Reports re Negroes, Records of the Bureau of Agricultural Economics, Record Group 83, National Archives, Washington, D.C.

10. "Attitudes of White Interviewees toward Negro Occupancy of Sojourner Truth Project," in Records of the Bureau of Agricultural Economics, Division of Program Surveys, Project Files, 1940–45, Special Reports re Negroes, Record Group 83; Dominic J. Capeci, Jr., *Race Relations in Wartime Detroit: The Sojourner Truth Housing Controversy of 1942* (Philadelphia: Temple Univ. Press, 1984), 75–99; "Strangers That Sojourn," *Commonweal* 35 (March 20, 1942), 524–25.

11. Surveys Division, Bureau of Special Services, Office of War Information, Special Memorandum No. 64, "Opinions in Detroit Thirty-Six Hours after the Race Riots" (June 30, 1943), in Entry 13 (PI-128), Records Relating to the Operations of the Area Offices, Folder: Willow Run-Race Riots' Study, Records of Committee for Congested Production Areas, Record Group 212, National Archives; newspaper clippings in Tab V—Detroit Race Riot, Records of the Office of the Provost Marshall General, Record Group 389, National Archives; *New York Times*, June 23, 1943; Alfred McClung Lee and Norman Daymond Humphrey, *Race Riot* (New York: Dryden Press, 1943); Robert Shogan and Tom Craig, *The Detroit Race Riot: A Study in Violence* (New York: Chilton, 1964); Harvard Sitkoff, "The Detroit Race Riot of 1943," *Michigan History* 53 (Fall 1969), 183–206; Capeci, *Race Relations in Wartime Detroit*, 161, 167–70.

12. Harvard Sitkoff, "Racial Militancy and Interracial Violence in the Second World War," *Journal of American History* 58 (Dec. 1971), 661–81; Lee and Humphrey, *Race Riot*, 49, 98–104; Dominic J. Capeci, Jr., *The Harlem Riot of 1943* (Philadelphia: Temple Univ. Press, 1977); *PM*, Aug. 3, 1943; Kenneth B. Clark, "Group Violence: A Preliminary Study of the Attitudinal Pattern of Its Acceptance and Rejection: A Study of the 1943 Harlem Riot," *Journal of Social Psychology* 19 (May 1944), 319–37; Neil A. Wynn, *The Afro-American and the Second World War* (London: Paul Elek, 1976), 68–73;

James S. Olson and Sharon Phair, "The Anatomy of a Race Riot; Beaumont, Texas, 1943," *Texana* (Winter 1974), 64–72; James A. Burran, "Racial Violence in the South during World War II" (Ph.D. diss., Univ. of Tennessee, 1977); Carey McWilliams, *North from Mexico: The Spanish-Speaking People of the United States* (New York: Greenwood rep., 1968), 227–58; Chester B. Himes, "Zoot Riots Are Race Riots," *Crisis* 50 (July 1943), 200–201; Fritz Redl, "Zoot Suits: An Interpretation," *Survey Midmonthly* 79 (Oct. 1943), 259–62; Kenneth B. Clark and James Barker, "The Zoot Effect in Personality: A Race Riot Participant," *Journal of Abnormal and Social Psychology* 40 (April 1945), 143–48; Mauricio Mazon, *The Zoot-Suit Riots: The Psychology of Symbolic Annihilation* (Austin: Univ. of Texas Press, 1984), Stuart Cosgrove, "The Zoot Suit and Style Warfare," reprinted in *Radical America* 18, no. 6 (1984), 39–51; Gerald D. Nash, *The American West Transformed: The Impact of the Second World War* (Bloomington: Indiana Univ. Press, 1985), 106–27; "Reflections of a Nisei on Zoot Suits," and "'Zoot Suit Boys' at #Manzanar," in Community Analysis Reports and Community Trend Reports of the War Relocation Authority, 1942–46, #93 and #147 in Manzanar Community Analysis Reports, in Records of the War Relocation Authority, Record Group 210, National Archives.

13. Sanchez, "*Pachucos* in the Making," in Wayne Moquin et al., eds., *A Documentary History of the Mexican Americans* (New York: Praeger, 1971), 317; Walter Davenport, "Race Riots Coming," *Collier's* 112 (Sept. 18, 1943), 11, 79–83.

14. See Carey McWilliams to Katharine F. Lenroot, Oct. 13, 1942, attaching "Testimony of Carey McWilliams, Chief, Division of Immigration and Housing, Department of Industrial Relations, State of California, before the Los Angeles County Grand Jury, Oct. 8, 1942; Rose J. McHugh to Jane M. Hoey, "Report On the Meeting Called by the Co-ordinator of Inter-American Affairs on Problems of Spanish-Speaking Peoples," Dec. 14, 1942; Mrs. Hynning to Muss Colby, "Child-Welfare Services to Spanish-American Groups," Jan. 25, 1943; Katharine F. Lenroot to Children's Bureau Field Staff, "Spanish-American Population," May 26, 1943; all in Records of the Children's Bureau, Record Group 102, National Archives.

15. Letters 277, 332, 426, 437; audiotape by Theresa Negrete-Reyes.

16. Roger Daniels, *Concentration Camps USA: Japanese Americans and World War II* (New York: Holt, Rinehart and Winston, 1971), 104; Michi Weglyn, *Years of Infamy: The Untold Story of America's Concentration Camps* (New York: Morrow, 1976), 88; Bureau of the Census, "Population Statistics: Study 40," in Records of the Bureau of Agricultural Economics, Division of Program Surveys, Project Files, 1940–45, Study 40—Japanese material (observations on West Coast—Feb. 1942), Record Group 83; Thomas James, *Exile Within: The Schooling of Japanese Americans, 1942–1945* (Cambridge: Harvard Univ. Press, 1987), 3, 8, 18, 20, 25, 112, 114, 128–29, 131; *Kansas City Times*, April 21, 1988.

17. Quoted in Fred Barbash, "'Evacuation' of the Japanese Americans," *Washington Post*, Dec. 5, 1982. See also From an Evacuee, "Internment and Family Separation as an Incentive to Repatriation," in Community Analysis Reports and Community Trend Reports of the War Relocation Authority, 1942–1946, #205 in Manzanar Community Analysis Reports, in Record Group 210.

18. Weglyn, *Years of Infamy*, 122.

19. John Tateishi, ed., *And Justice for All: An Oral History of the Japanese American Detention Camps* (New York: Random House, 1984), 15, 24, 104, 124–27, 130, 133–40; Mark Jonathan Harris et al., eds., *The Homefront: America during World War II* (New

York: Putnam's, 1984), 104–13; Dr. W. T. Harrison, Public Health Service, "Tentative Plan for Providing Medical and Hospital Care for Japanese Evacuees," March 28, 1942; Dr. Sappington and Dr. Thompson to Miss Brownlee, "Evacuation of Japanese in California—Request for Children's Bureau Assistance in Medical Care," April 7, 1942; Dr. Thompson to Miss Brownlee, "Medical Care of Women and Children in the Japanese Evacuation Group," April 7, 1942; Helen R. Jeeter to Members of the Family Security Committee, "Japanese Evacuees in Rocky Mountain States," undated; Dr. Sappington to Miss Bloodgood, reports of visits to Japanese assembly centers at Portland, Oregon, and Puyallup, Washington, June 6, 8, 1942; Dr. Martha M. Eliot to Dr. L. R. Coffey, "Visit to Colorado River War Relocation Center, Parker, Arizona," June 27, 1942; Dr. Huse to Dr. Sappington, "Material on the Japanese Assembly Center in Oregon . . . ," Oct. 14, 1942; Mr. Ernst et al., "Topaz, the Japanese Relocation Center at Delta, Utah," Oct. 20, 1942; all in Record Group 210.

20. Miss Brownlee to Miss Atkinson, "Evacuation of Japanese Orphanages," April 8, 1942; Charlotte B. DeForest to Children's Bureau, June 3, 1942; Dr. Sappington to Miss Bloodgood, report of visit to Japanese Assembly Center at Puyallup, Washington, June 8, 1942; Katharine F. Lenroot to Secretary of War, June 10, 1942; Dr. May B. Guy, "Puyallup Washington Japanese Center," undated; Dr. May B. Guy to Dr. T. S. McGowan, June 27, 1942; Mr. Zane to Dr. Eliot and Dr. Daily, July 17, 1942; Dr. Edith P. Sappington, "Japanese Relocation Center at Newell, California," Sept. 16, 1942; Dr. T. G. Ishimaru to Mr. C. R. Zane, Jan. 22, 1943, attaching report on the Children's Village at Manzanar; all in Record Group 210.

21. Letter 97J; Harry H. Matsumoto to Lynn D. Mowat, July 11, 1942, and "The Family in Poston," Aug. 25, 1943, in Community Analysis Reports and Community Trend Reports of the War Relocation Authority, 1942–46, 23 in Colorado River Community Analysis Reports, both in Record Group 210; interview with Helen Kokka Gee. Wartime articles that provide insights into the Japanese-American children's lives are Emory S. Bogardus, "Culture Conflicts in Relocation Centers," *Sociology and Social Research* 27 (May-June 1943), 380–90; Bureau of Sociological Research, Colorado River War Relocation Center, "The Japanese Family in America," *Annals of the American Academy of Political and Social Science* 229 (Sept. 1943), 150–56; Ruth E. Hudson, "Health for Japanese Evacuees," *Public Health Nursing* 35 (Nov. 1943), 615–20; Genevieve W. Carter, "Child Care and Youth Problems in a Relocation Center," *Journal of Consulting Psychology* 8 (July-Aug. 1944), 219–25; Eunice Glenn, "Education Behind Barbed Wire," *Survey Midmonthly* 80 (Dec. 1944), 347–49; Raymond Nathan, "New Neighbors Among Us," *Parents' Magazine* 20 (March 1945), 30–31, 121–24.

22. Letters 97J, 101. There is a multitude of excellent literature on the Japanese-American internment. Helpful for understanding internment's effect of "strengthening peer groups against family groups" is Leonard Broom and John I. Kitsuse, *The Managed Casualty: The Japanese-American Family in World War II* (Berkeley: Univ. of California Press, 1973 ed.). Summaries of the internment appear in Ronald Takaki, *Strangers from a Different Shore: A History of Asian Americans* (Boston: Little, Brown, 1989); Sucheng Chan, *Asian Americans: An Interpretive History* (Boston: Twayne, 1991); Robert A. Wilson and Bill Hosokawa, *East to America: A History of the Japanese in the United States* (New York: Quill ed., 1982).

23. Yarrow et al., "Social Perceptions and the Attitudes of Children," *Genetic Psychology Monographs* 40 (Nov. 1949), 438–39.

24. Allport, *The Nature of Prejudice* (Cambridge: Addison-Wesley, 1954), 291–94, 297–300; Anna W. M. Wolf, "War Begins in the Nursery," *Child Study* 20 (Fall 1942), 35–38; Austin H. MacCormick, Karl Menninger, Mary Shattuck Fisher, and Margaret Mead, "The Roots of War in Human Nature: Aggressiveness and Hate in Childhood and Family Life," *Child Study* 20 (Spring 1943), 72–77, 91; *New York Times*, March 7, 1944; Deborah A. Byrnes, "Children and Prejudice," *Social Education* 52 (April/May 1988), 267–71.

25. Allport, *The Nature of Prejudice*, 297–308.

26. Ibid., 309–11. For a discussion of Piaget's cognitive psychology, see Chapter 7.

27. Letters 24, 72, 97E, 101, 105, 146, 172, 177, 200, 206K, 314, 317, 328.

28. Letters 55G, 206C.

29. Letters 51, 97J, 104, 138, 402.

30. Letters 32, 45, 82, 228, 276, 316, 443; Bernard F. Dick, *The Star-Spangled Screen: The American World War II Film* (Lexington: Univ. Press of Kentucky, 1985), 102–6, 130–31, 137, 179–82, 230–49; Gregory D. Black and Clayton R. Koppes, "The Office of War Information and the Portrayal of Asians in Hollywood Films during World War II" (paper presented at the annual meeting of Asian Studies on the Pacific Coast, 1978).

31. Letters 62, 124, 131, 219, 253, 298.

32. Letters 24, 33, 97J, 294, 314, 412; Harris, *The Homefront*, 87; Lawrence Greenfield, "Leonard Greenfield and Gloria Shostak: Urban Teenage Jews during World War II" (unpub. paper), 91–92; "Button, Button," *Newsweek* 19 (May 25, 1942), 31. On the role of racism in the Pacific war, see John W. Dower, *War Without Mercy: Race and Power in the Pacific War* (New York: Pantheon, 1986).

33. Letters 16, 24, 97J, 154, 220, 270; "How to Tell Your Friends from the Japs," *Time* 38 (Dec. 22, 1941), 11; Shih-Shan Henry Tsai, *The Chinese Experience in America* (Bloomington: Indiana Univ. Press 1986), 116–19.

34. Letters 54, 67, 73, 97A, 97G, 172, 177, 226, 238, 268, 314, 332, 385.

35. Letter 366.

36. Ronald H. Bayor, *Neighbors in Conflict: The Irish, Germans, Jews, and Italians of New York City, 1929–1941* (Baltimore: Johns Hopkins Univ. Press, 1978), xiii; Roger D. Abrahams, ed., *Jump-Rope Rhymes: A Dictionary* (Austin: Univ. of Texas Press for the American Folklore Society, 1969), 168; Arthur Miller, "The Face in the Mirror: Anti-Semitism Then and Now," *New York Times Book Review* (Oct. 14, 1984), 3.

37. Quoted in Gordon W. Allport and Leo Postman, *The Psychology of Rumor* (New York: Henry Holt, 1947), 174.

38. McWilliams, *A Mask for Privilege: Anti-Semitism in America* (Boston: Little, Brown, 1948; rep. ed., Greenwood Press, 1979), 163–64; James N. Gregory, *American Exodus: The Dust Bowl Migration and Okie Culture in California* (New York: Oxford Univ. Press, 1989), 78, 81, 100–113, 179, 208; Lewis M. Killian, "Southern White Laborers in Chicago's West Side" (unpub. Ph.D. diss., Univ. of Chicago, 1949), 118–49.

39. Max Meenes, "A Comparison of Racial Stereotypes of 1935 and 1942," *Journal of Social Psychology* 17 (May 1943), 327–36.

40. Thomas Cripps, *Slow Fade to Black: The Negro in American Film, 1900–1942* (New York: Oxford Univ. Press, 1977); Daniel J. Leab, *From Sambo to Superspade: The Black Experience in Motion Pictures* (Boston: Houghton Mifflin, 1975); Donald Bogle, *Toms, Coons, Mulattoes, Mammies, and Bucks: An Interpretive History of Blacks in American Films* (New York: Viking, 1973); Peter Noble, *The Negro in Films* (New York: Arno Press ed., 1970); Charlotte Ruby Ashton, "The Changing Image of Blacks in American Film: 1944–1973" (unpub. Ph.D. diss., Princeton Univ., 1981); Cripps, "Racial Ambiguities in American Propaganda Films," in K.R.M. Short, ed., *Film & Radio Propaganda in World War II* (London: Croom Helm, 1983), 125–45; Melvin Patrick Ely, *The Adventures of Amos 'n' Andy: A Social History of an American Phenomenon* (New York: Free Press, 1991).

41. Letters 189, 206E; Howard W. Odum, *Race and Rumors of War: Challenge in American Crisis* (Chapel Hill: Univ. of North Carolina Press, 1943), 53–141; Robert H. Knapp, Division of Propaganda Research, Massachusetts Committee on Public Safety, to Robert MacLeod, Bureau of Agricultural Economics, June 1, 1942, enclosing "Rumors in the Boston Region, March 20–May 25, 1942," in Division of Program Surveys, Project Files, 1940–45, S-68, Material on Rumors, Record Group 83; Gloria Brown Melton, "Blacks in Memphis, Tennessee, 1920–1955" (unpub. Ph.D. diss., Washington State Univ., 1982), 203, 209; Dabney, "Nearer and Nearer the Precipice," *Atlantic Monthly* 171 (Jan. 1943), 99; *New York Times*, Sept. 2, 1942; Senter, "Potential Revolution in the Deep South."

42. U.S. Bureau of the Census, *Internal Migration in the United States: April 1940, to April 1947*, Series P-20, No. 14 (Washington: Bureau of the Census, 1948), 1, 6; Reynolds Farley and Walter R. Allen, *The Color Line and the Quality of Life in America* (New York: Oxford Univ. Press ed., 1989), 113; Wynn, *The Afro-American and the Second World War*, 61; Carey McWilliams, *Brothers Under the Skin* (Boston: Little, Brown, 1951), 7–8; Philip M. Hauser, "Population and Vital Phenomena," *American Journal of Sociology* 48 (Nov. 1942), 314.

43. Ralph N. Davis, "The Negro Newspapers and the War," *Sociology and Social Research* 27 (May-June 1943), 373–80; E. Franklin Frazier, "Ethnic and Minority Groups in Wartime, with Special Reference to the Negro," *American Journal of Sociology* 48 (Nov. 1942), 367–77; A. Philip Randolph, "Why Should We March?," *Survey Graphic* 31 (Nov. 1942), 488–89.

44. Jack Yeaman Bryan, "The In-Migrant Menace," *Survey Midmonthly* 81 (Jan. 1945), 9; Arnold R. Hirsch, *Making the Second Ghetto: Race and Housing in Chicago, 1940–1960* (Cambridge: Cambridge Univ. Press, 1983), 45; Lowell Juilliard Carr and James Edson Stermer, *Willow Run: A Study of Industrialization and Cultural Inadequacy* (New York: Harper & Brothers, 1952), 244.

45. United States House of Representatives, 76th and 77th Congs., Select Committee to Investigate the Interstate Migration of Destitute Citizens (later renamed Select Committee Investigating National Defense Migration; hereinafter both committees are referred to as Tolan Committee), *Hearings* (Washington: GPO, 1940–42), part 11, pp. 4498–99, 4577; part 13, p. 5131; part 15, pp. 5897, 6248–53; part 18, pp. 7246–47, 7251–54.

46. Eugene L. Meyer, "A Legacy of Segregation: Woodland Village Seeking Improved Services," *Washington Post*, April 26, 1982; James A. Pooler, "'Shelter' Houses a Lesson in Squalor," *Detroit Free Press*, Nov. 10, 1944; *New York Times*, May 7, 1942,

July 23, 1943, April 9, 1944; U.S. House of Representatives, 78th Cong., 1st Sess., Sub-committee of the Committee on Naval Affairs, *Investigation of Congested Areas* (Washington: GPO, 1943), 661–62, 798–99, 833–34; Hubert Owen Brown, "The Impact of War Worker Migration on the Public School System of Richmond, California, from 1940 to 1945" (unpub. Ph.D. diss., Stanford Univ., 1973), 173–88, 306ff; and the following in Entry 13, PI-128, Records Relating to the Operation of the Area Offices, Folder: Negro Housing, Record Group 212: "Housing for Negroes in Detroit," Sept. 16, 1944; Leroy Peterson to Harry Grayson, "Negro Housing," Sept. 21, 1942.

47. Interviews with "Southerner—Newly-Arrived Factory Worker" and "Negro—Service Worker," 112, 127, in Division of Program Surveys, Project Files, 1940–45, Report on Detroit Study, Record Group 83; House Subcommittee on Naval Affairs, *Investigation of Congested Areas*, 1773; Tolan Committee, *Hearings*, part 15, p. 5947; part 18, pp. 7670–73; part 23, pp. 8762–67; Killian, "Southern White Laborers in Chicago's West Side," 365–83.

48. Letters 57, 72.

49. Letter 206E.

50. Letters 57, 57.5, 123, 206F, 260; audiotape recording by Olivia Longstreet.

51. Stella Chess, "Psychology of Pre-adolescent Children in War Time: IV. War Ideologies of Children," *American Journal of Orthopsychiatry* 13 (July 1943), 508; interview with Lawrence Streeter; *Michigan Ciizen*, April 8–14, 1990; Lawrence Streeter to the author, April 30, 1990; Alan M. Osur, *Blacks in the Army Air Forces During World War II* (Washington, D.C.: Office of Air Force History, 1977), 45–46; Stanley Sandler, *Segregated Skies: The All-Black Combat Squadrons of World War II* (Washington: Smithsonian Institution Press, 1992).

52. Albert Hemingway, *Ira Hayes: Pima Marine* (Lanham, MD: Univ. Press of America, 1988); Nash, *The American West Transformed*, 128–33; Lynne Escue, "Coded Contributions: Navajo Talkers and the Pacific War," *History Today* 41 (July 1991), 13–20.

53. Letters 7.5, 72; John Adair, "The Navajo and Pueblo Veteran: A Force for Culture Change," *The American Indian* 4, no. 1 (1947), 5–11; Momaday, *The Names: A Memoir* (New York: Harper & Row, 1976), 83–91, 99–100; Alice Marriott, "The War Party," in *The Ten Grandmothers* (Norman: Univ. of Oklahoma Press, 1945), 278–84. For the history of Native Americans during the war, see Alison R. Bernstein, *American Indians and World War II: Toward a New Era in Indian Affairs* (Norman: Univ. of Oklahoma Press, 1991); Nash, *The American West Transformed*, 128–47.

54. Erikson, "Childhood and Tradition in Two American Indian Tribes, with Some Reflections on the Contemporary American Scene," in Kluckhohn and Murray, *Personality in Nature, Society, and Culture*, 176–203; Dorothea Leighton and Clyde Kluckhohn, *Children of the People: The Navaho Individual and His Development* (Cambridge: Harvard Univ. Press, 1947), 3–9, 13–64; Robert J. Havighurst and Bernice L. Neugarten, *American Indian and White Children: A Sociopsychological Investigation* (Chicago: Univ. of Chicago Press, 1955); Alex Inkeles and Daniel J. Levinson, "National Character: The Study of Modal Personality and Sociocultural Systems," in Lindzey Gardner and Elliott Aronson, eds., *The Handbook of Social Psychology*, vol. 4 (Reading, MA: Addison-Wesley, second ed., 1969), 462–68; Margaret Connell Szasz, *Education and the American Indian: The Road to Self-Determination, 1928–1973* (Albuquerque: Univ. of New Mexico Press,

1974); Szasz, "Federal Boarding Schools and the Indian Child: 1920–1960," *South Dakota History* 7 (Fall 1977), 371–84; Howard C. Ellis, "From the Battle in the Classroom to the Battle for the Classroom," *American Indian Quarterly* 11 (Summer 1987), 255–65; Nash, *American West Transformed*, 135–42.

55. John Adair and Evon Vogt, "Navaho and Zuni Veterans: A Study of Contrasting Modes of Culture Change," *American Anthropologist* 51 (Oct.-Dec. 1949), 547–61; John Adair, "A Study of Cultural Resistance: The Veterans of World War II at Zuni Pueblo" (unpub. Ph.D. diss., Univ. of New Mexico, 1948); Evon Vogt, *Navaho Veterans: A Study of Changing Values* (Cambridge, MA: Papers of the Peabody Museum of American Archaeology and Ethnology, vol. 41, no. 1, 1951); John H. Bushnell, "From American Indian to Indian American: The Changing Identity of the Hupa," *American Anthropologist* 70 (Dec. 1968), 1112–14.

56. Audiotape recording by Olivia Longstreet; letter 206F.

57. Stephen Fox, *Unknown Internment: An Oral History of the Relocation of Italian Americans during World War II* (Boston: Twayne, 1990), 162–63, 183–87.

58. Letters 72, 183, 206J, 440, 445, 448.

59. Letters 38, 129, 156, 183, 243B, 364, 386, 409, 410, 448, 452.

60. Letters 11, 52, 97I, 123, 184, 206J, 294, 297, 315, 399, 450.

61. Virginia De Carlo to "Dear Mr. President," Feb. 20, 1943, in Record Group 210; Bayor, *Neighbors in Conflict*, xiii; Bruno Bettelheim and Morris Janowitz, *Social Change and Prejudice* (including *Dynamics of Prejudice*) (London: Free Press of Glencoe, 1964), 3–24, 78–83; Daniel Katz and Kenneth W. Braly, "Verbal Stereotypes and Racial Prejudice," in Maccoby et al., *Readings in Social Psychology*, 40–46.

62. Letters 25, 312; Talese, "Wartime Sunday," *Esquire* 107 (May 1987), 100; R. A. Schermerhorn, *These Are Our People: Minorities in American Culture* (Boston: D.C. Heath, 1949), 227–60; Joseph S. Roucek, "Italo-Americans and World War II," *Sociology and Social Research* 29 (July-Aug. 1945), 465–71; Philip A. Bean, "Fascism and Italian-American Identity: A Case Study: Utica, New York," *Journal of Ethnic Studies* 17 (Summer 1989), 101–19.

63. Letters 140, 220.

64. Letters 0.3, 18, 93, 97E, 114, 130, 228, 264, 276, 330, 340, 380, 387, 396, 410, 434.

65. Letters 53, 191, 447, 448; interview with Karolyn Kaufman Zerger, Oral History Collection, Bethel College, KS, with thanks to Rachel Waltner Goossen for bringing this interview to my attention.

66. Interview with Paul Boyer; Mulford Q. Sibley and Philip E. Jacob, *Conscription of Conscience: The American State and the Conscientious Objector, 1940–1947* (Ithaca: Cornell Univ. Press, 1952); Lawrence S. Witner, *Rebels Against War: The American Peace Movement, 1941–1960* (New York: Columbia Univ. Press, 1969), 34–150. For a charming short story about a pacifist homefront boy's life during wartime, see Alfred Habegger, "A Little Spoon," *New Yorker* 57 (Jan. 11, 1982), 34–39.

67. Interview with Paul Boyer; Boyer, *By the Bomb's Early Light: American Thought and Culture at the Dawn of the Atomic Age* (New York: Pantheon, 1985), xvi–xviii.

68. Knapp to MacLeod, enclosing "Rumors in the Boston Region, March 20–May 25, 1942"; Allport and Postman, *The Psychology of Rumor*, 10–16, 21–23, 37, 42, 174; Robert H. Knapp, "A Psychology of Rumor," *Public Opinion Quarterly* 8 (Spring

1944), 23–37; interview with "JEW," 120, in Records of the Bureau of Agricultural Economics, Division of Program Surveys, Project Files, 1940–45, Report on Detroit Study, Record Group 83.

69. Interview with Eric Solomon; Greenfield, "Leonard Greenfield and Gloria Shostak: Urban Teenage Jews During World War II," 83–96; Nat Hentoff, *Boston Boy* (New York: Knopf, 1986), 16–19, 28–31, 68–71, 87, 97–104; Alan Brinkley, *Voices of Protest: Huey Long, Father Coughlin and the Great Depression* (New York: Knopf, 1982), 168, 266–83; Bayor, *Neighbors in Conflict*, 88, 94–97, 105, 112, 196–97; interview with "Southerner—Newly Arrived Factory Worker." Anti-Semitism was also widespread in England; see Tony Kushner, *The Persistence of Prejudice: Antisemitism in British Society During the Second World War* (Manchester, Engl.: Manchester Univ. Press, 1989).

70. Milton Plesur, *Jewish Life in Twentieth-Century America: Challenge and Accommodation* (Chicago: Nelson-Hall, 1982), 106–8; Roth, "My Life as a Boy," *New York Times Book Review* (Oct. 18, 1987), 47. See also Kate Simon, *Bronx Primitive: Portraits in a Childhood* (New York: Viking, 1982), 8, 49.

71. Letters 49, 72, 134, 136, 140, 362, 380, 382, 430; Joseph E. Illick, *At Liberty: The Story of a Community and a Generation: The Bethlehem, Pennsylvania, High School Class of 1952* (Knoxville: Univ. of Tennessee Press, 1989), 76, 316–17.

72. Letters 110, 133, 176, 450.

73. Letters 224, 233, 394, 410.

74. Letters 110, 114, 136, 176, 233, 240D, 275, 362; *New York Times*, Oct. 19, 1988, Sept. 10, 1989. For a collection of essays by Jewish Americans on the meaning of the Holocaust, see David Rosenberg, ed., *Testimony: Contemporary Writers Make the Holocaust Personal* (New York: Times Books, 1989).

75. Letter 275.

76. Letter 131; *New York Times*, Jan. 17, 1943; "Education for Inter-group Co-operation," *The School Review* 52 (Feb. 1944), 67.

77. Shelby M. Harrison, "Some Wartime Social Gains," *Journal of the National Education Association* 32 (May 1943), 131; MacDonald, *Don't Touch That Dial! Radio Programming in American Life from 1920 to 1960* (Chicago: Nelson-Hall, 1979), 346–56; Raymond William Stedman, *The Serials: Suspense and Drama by Installment* (Norman: Univ. of Oklahoma Press, 1977), 329–39.

78. John T. McManus and Louis Kronenberger, "Motion Pictures, the Theater, and Race Relations," *Annals of the American Academy of Political and Social Science* 244 (March 1946), 152–58; K.R.M. Short, "Hollywood Fights Anti-Semitism," in Short, *Film & Radio Propaganda in World War II*, 146–71; Thomas Cripps and David Culbert, "*The Negro Soldier* (1944): Film Propaganda in Black and White," *American Quarterly* 31 (Winter 1979), 616–40; Eleanor Weakley Nolen, "The Colored Child in Contemporary Literature," *The Horn Book* 18 (Sept. 1942), 348–55; Clara E. Breed, "Books That Build Better Racial Attitudes," *The Horn Book* 21 (Jan.-Feb. 1945), 55–61; "Down-to-Earth Experiment," *Time* 44 (July 24, 1944), 46; M. F. Ashley Montagu, "Democracy, Education, and Race," *School and Society* 59 (April 1, 1944), 227–29; "Education for Intergroup Co-operation," 67–68.

79. "Education for Internal and International Peace," *The School Review* 53 (Feb. 1945), 67–74; Gleason is quoted in Roger Daniels, *Coming to America: A History of Immigration and Ethnicity in American Life* (New York: HarperCollins, 1990), 305; see also

Philip Gleason, "American Identity and Americanization," in Stephan Thernstrom, ed., *Harvard Encyclopedia of American Ethnic Groups* (Cambridge: Harvard Univ. Press, 1980), 47–58.

80. Agnes Bendict, "Towards Democracy," *Parents' Magazine* 19 (May 1944), 30–31, 128–29.

81. Letters 38, 53, 129, 156, 183, 243B, 364, 386, 409, 410, 448, 452; Helen Steers Burgess, "Facing War with Our Children," *Parents' Magazine* 17 (March 1942), 19, 36; "Japanese Incarceration," *Spectrum* 6 (May 1983), unpaginated; *New York Times*, June 25, 1989.

82. Badener is quoted in *New York Times Book Review* (Feb. 12, 1989), 3.

83. Allport, *The Nature of Prejudice*, 303–5; letter 30.

Chapter 11. Children's Health and Welfare

1. "Infantile Paralysis," *Life* (July 31, 1944), 25–28.

2. Ibid.; Don W. Gudakunst, "Fighting Infantile Paralysis," *Survey Midmonthly* 80 (Sept. 1944), 254–55; Myrtle Ellen LaBarr, "How One State Met Its Polio Epidemic," *Independent Woman* 24 (Jan. 1945), 14–15, 27–29.

3. "Infantile Paralysis," 25–28.

4. Letters 114, 167, 168, 345; Bundesen, "Learn the Truth—Avoid 'Polio Panic,'" *Ladies Home Journal* 62 (Jan. 1945), 86; Howe, "Are You Afraid of Polio?," *Harper's* 190 (June 1945), 646–53; "Infantile Paralysis," *Good Housekeeping* 121 (Aug. 1945), 42.

5. "Tonsils and Polio," *Time* 40 (Sept. 14, 1942), 69; Leonard M. Folkers, "Tonsillectomy Ahead?," *Parents' Magazine* 19 (Aug. 1944), 10; "Take These Simple Precautions Against Polio," *Parents' Magazine* 20 (Aug. 1945), 101.

6. Fay S. Copellman, "Follow-up of One Hundred Children with Poliomyelitis," *The Family* 25 (Dec. 1944), 289–97; Alice A. Grant, "Medical Social Work in an Epidemic of Poliomyelitis," *Journal of Pediatrics* 24 (June 1944), 691–723.

7. Letters 167, 225, 243B, 268; Copellman, "Follow-up of One Hundred Children with Poliomyelitis," 292, 294, 296; Don W. Gudakunst, "Today's Attack on Infantile Paralysis," *Education for Victory* 2 (May 3, 1944), 3; "Polio, 1943," *Time* 42 (Aug. 23, 1943), 68.

8. Letters 0.2, 167, 168, 206I, 225, 243B, 268, 309, 312; U.S. Census Bureau, *Historical Statistics of the United States, Bicentennial Edition*, part 1, p. 77; "Health of the United States at War," 1281; Don W. Gudakunst, "Infantile Paralysis—1943," *Public Health Nursing* 35 (Nov. 1943), 611; *Science News Letter* 44 (July 17, Aug. 7, 14, 21, Sept. 4, Oct. 2, 1943), 37, 83, 92, 106, 120, 152, 221; "Polio, 1943," 68; "Infantile Paralysis Week," *School and Society* 59 (Jan. 1, 1944), 7; A. L. Van Horn and A. J. Lesser, "Combating Epidemics of Poliomyelitis," *The Child* 8 (April 1944), 157–59; Coralynn A. Davis, "Meeting an Epidemic of Infantile Paralysis," *Public Health Nursing* 36 (July 1944), 332–35; Jessie L. Stevenson, "Public Health Nursing in the 1943 Polio Epidemic," *Public Health Nursing* 36 (July 1944), 336–39; *New York Times*, July 27, Aug. 29, 31, Oct. 6, 1943, Jan. 14, 1944.

9. U.S. Bureau of the Census, *Historical Statistics of the United States, Colonial Times to the Present, Bicentennial Edition* (Washington: GPO, 1975), part 1, p. 77; "Nerves

and Polio," *Newsweek* 26 (July 23, 1945), 78; A. L. Van Horn, "Planning to Meet Polio-myelitis," *The Child* 10 (July 1945), 6; *Science News Letter* 46 (July 22, Aug. 5, 19, 26, Sept. 23, Oct. 14, 1944), 61, 89, 120, 139, 201, 249; "Polio's Battleground," *Newsweek* 24 (July 24, 1944), 78; *American Journal of Nursing* 44 (Sept. 1944), 900; Marion Williamson, "Review of a Polio Epidemic," *Public Health Nursing* 37 (June 1945), 310–13.

10. *New York Times*, Jan. 30, 1944; letter 0.2.

11. *New York Times*, Jan. 30, July 4, 14, Aug. 1, 5, 12, 18, 21, 22, 25, 28, 29, 30, Sept. 1, 17, 19, 1944; letter 167.

12. *New York Times*, Jan. 22, 31, Feb. 23, 28, June 2, 28, 1945.

13. Ibid., July 28, 31, Aug. 4, 5, 6, 8, 10, 11, 24, 27, 28, 29, 30, 31, Sept. 1, 6, 14, 18, 20, Oct. 25, 1945; *Science News Letter* 47 (April 21, June 30, 1945), 248, 404; 48 (July 14, 21, Aug. 4, 11, 18, Sept. 1, 1945), 24, 43, 68, 86, 102, 137; "Polio Report," *Time* 46 (Sept. 10, 1945), 70; U.S. Census Bureau, *Historical Statistics of the United States, Bicentennial Edition*, part 1, p. 77.

14. "Polio Spread by Humans . . . ," *Science News Letter* 41 (April 18, 1942), 243–44; "Overlooked Disease Cause," *Science News Letter* 47 (June 9, 1945), 354; "Link Paralysis to Food Intake," *Science Digest* 13 (May 1943), 44; "Salt and Polio," *Consumers' Research Bulletin* 15 (May 1945), 25; "Infantile Paralysis Virus May Enter Body through Mouth," *Science News Letter* 40 (July 12, 1941), 20; "Cavities and Polio," *Science News Letter* 47 (May 19, 1945), 307; "Teeth and Polio," *Newsweek* 25 (May 28, 1945), 100.

15. Letter 322; J. R. Paul et al., "The Detection of Poliomyelitis Virus in Flies," *Science* 94 (Oct. 24 1941), 395–96; "Infantile Paralysis Carried by Flies," *Science News Letter* 40 (Oct. 25, 1941), 259; Sabin and Ward, "Flies As Carriers of Poliomyelitis Virus in Urban Epidemics," *Science* 94 (Dec. 19, 1941), 590–91; "Insects As Carriers of Polio-myelitis Virus," *Science* 95 (Feb. 13, 1942), 169–70; "Sanitary Privies May Help Solve Polio Problem," *Science News Letter* 42 (Nov. 7, 1942), 301; Walter Adams, "DDT," *Better Homes & Gardens* 22 (May 1944), 19, 100–103; Robert Ward et al., "Poliomyelitis Virus in Fly-Contaminated Food Collected at an Epidemic," *Science* 101 (May 11, 1945), 491–93; "New Polio Weapon," *Science News Letter* 48 (Aug. 25, 1945), 114; J. D. Ratcliff, "New Knowledge about Polio," *Hygeia* 20 (Oct. 1942), 799; *New York Times*, Aug. 5, 14, 20, 21, Sept. 5, 10, 1945.

16. Ratcliff, "New Knowledge about Polio," 799; Albert Milzer and Sidney O. Levinson, "Recent Research in Infantile Paralysis," *Public Health Nursing* 34 (Aug. 1942), 434–37; Margaret S. Arey, "Infantile Paralysis: What Can Be Done about It?," *The American Home* 30 (Sept. 1943), 80–81; *New York Times*, July 29, 1944; Bundesen, "Learn the Truth—Avoid 'Polio Panic,'" 87; "Take These Simple Precautions against Polio," *Parents' Magazine* 20 (Aug. 1945), 101; "Polio, 1943," 68.

17. Elizabeth Kenny, *The Treatment of Infant Paralysis in the Acute Stage* (Minneapolis: Bruce, 1941); Kenny and John F. Pohl, *The Kenny Concept of Infantile Paralysis and Its Treatment* (Minneapolis: Bruce, 1943); Kenny and Martha Otenso, *And They Shall Walk: The Life Story of Sister Elizabeth Kenny* (New York: Dodd, Mead, 1943); Kenny, *My Battle and Victory* (London: Hall, 1955); "Sister Kenny," *Life* (Sept. 28, 1942), 73, 75, 77; Inez L. Armstrong, "How We Met the Poliomyelitis Epidemic," *American Journal of Nursing* 44 (June 1944), 529–32; "Pain of Polio Relieved," *Science News Letter* 46 (Sept. 2, 1944), 146; "When Polio Strikes," *Newsweek* 23 (Feb. 7, 1944), 93; "Co-operation in Fighting Polio," *American Journal of Nursing* 44 (Oct. 1944), 922; Don W.

Gudakunst, "It's a Fifty-Fifty Chance," *Parents' Magazine* 19 (July 1944), 50; Ephraim Fischoff and Don W. Gudakunst, "The Fight against Infantile Paralysis Continues," *American Journal of Nursing* 44 (June 1944), 535–36; Jessie Stevenson, "The Kenny Method," *American Journal of Nursing* 42 (Aug. 1942), 904–10; "Polio Polemic," *Time* 28 (Sept. 27, 1943), 58, 60.

18. U.S. Census Bureau, *Historical Statistics of the United States, Bicentennial Edition*, part 1, p. 77; Lawrence, KS, *Journal-World*, Aug. 14, 1983, April 7, 1985; Jane S. Smith, *Patenting the Sun: Polio and the Salk Vaccine* (Garden City, NY: Doubleday, 1990).

19. Letter 206A; Leonard Kriegel, "Taking It," *New York Times Magazine* (June 9, 1985), 80; Joy Horowitz, "Polio's Painful Legacy," *New York Times Magazine* (July 7, 1985), 16ff.; "A New Scare for Polio Victims," *Newsweek* 103 (April 23, 1984), 83.

20. Carpenter, "Children: Social Priority Number One," *Survey Midmonthly* 77 (Dec. 1941), 347–50; Dorothy B. Nyswander, "Preventing Child Casualties on the Home Front," *Public Health Nursing* 35 (Sept. 1943), 493–96; Dorothy W. Baruch, "Child Care Centers and the Mental Health of Children in the War," *Journal of Consulting Psychology* 7 (Nov.-Dec. 1943), 252–66. A study of the relationship between the lives of German children during the First World War homefront privations and their later Nazism is Peter Loewenberg, "The Psychohistorical Origins of the Nazi Youth Cohort," in *Decoding the Past: The Psychohistorical Approach* (Berkeley: Univ. of California Press ed., 1985), 240–83.

21. "Scarlet Fever Epidemic," *Science News Letter* 45 (Feb. 19, 1944), 116–17; "Influenza and Pneumonia Excess Mortality at Specific Ages in the Epidemic of 1943–44, with Comparative Data for Preceding Epidemics," *Public Health Reports* 60 (July 20, 27, 1945), 821–35, 853–63; U.S. Census Bureau, *Historical Statistics of the United States, Bicentennial Edition*, part 1, p. 58; U.S. House of Representatives, 78th Cong., 1st Sess., Subcommittee of the Committee on Naval Affairs, *Investigation of Congested Areas*, part 3 (Washington: GPO, 1943), 653–54, 659; "Increase in Infant Mortality and Infant Diarrhea in San Francisco, Calif.," *Public Health Reports* 58 (June 11, 1943), 917; U.S. House of Representatives, 77th Cong., 1st and 2nd Sess., Select Committee Investigating National Defense Migration (Tolan Committee), *Hearings* (Washington: GPO, 1941–42), Parts 12, 17–19, 25, pp. 4340–78, 4571–79, 4896 ff., 6982–87, 7552–56, 7791–94, 9836. For weekly data on contagious diseases, see the U.S. Public Health Service's *Public Health Reports*.

22. "Nursery Killer," *Newsweek* 20 (Nov. 2, 1942), 68.

23. "Health of the United States at War," *Public Health Reports* 58 (Aug. 20, 1943), 1281–82; "Provisional Mortality Rates for the First Half of 1944," *Public Health Reports*, 60 (Feb. 9, 1945), 153–61; U.S. Senate, 78th Cong., 1st Sess., Subcommittee of the Committee on Education and Labor (Pepper Committee), *Wartime Health and Education* (Washington: GPO, 1944), part 5, p. 1857; *New York Times*, Feb. 8, Oct. 8, 1942, Jan. 7, 1944.

24. *New York Times*, May 31, 1945; U.S. Children's Bureau Publication 311, *Childhood Mortality from Accidents: By Age, Race, and Sex, and by Type of Accident* (Washington: GPO, 1945); E. Laurence Palmer, "Commando Training Just Outside His Own Door," *The American Home* 32 (Oct. 1944), 42, 44, 46; U.S. Dept. of Interior, Bureau of Mines Information Circular, D. Harrington and R. G. Warncke, *Accidents to Children from Blasting Caps* (Washington: mimeo., 1944).

25. "The Disturbing Return of a Long-Gone Disease," *Newsweek* 108 (Dec. 22, 1986), 60; Fishbein, "Rheumatic Fever," *Hygeia* 22 (May 1944), 333.

26. Shultz, "Protect Your Family from Rheumatic Fever," *Better Homes & Gardens* 22 (Feb. 1944), 32; U.S. Children's Bureau Publication 308, *Proceedings of Conference on Rheumatic Fever, Washington, D.C., Oct. 5–7, 1943* (Washington: GPO, 1943).

27. Martha M. Eliot, "Baby Care Today and Tomorrow," *Parents' Magazine* 19 (Jan. 1944), 25; U.S. Children's Bureau Publication 272, *White House Conference on Children in a Democracy: Final Report* (Washington: GPO, 1942), 25–26; Herbert J. Sommers, "Infant Mortality in Rural and Urban Areas," *Public Health Reports* 57 (Oct. 2, 1942), 1494–1501; "Mr. and Mrs. America 1970," *Consumers' Guide* 12 (Jan. 1946), 8–10.

28. U.S. Children's Bureau, *White Conference on Children in a Democracy*, 149–96; "Mr. and Mrs. America 1970," *Consumers Guide* 12 (Jan. 1946), 8. The best contemporary study of childhood poverty is Justine Wise Polier's *Everyone's Children, Nobody's Child: A Judge Looks at Underprivileged Children in the United States* (New York: Scribner's, 1941).

29. Marjorie Gooch, "Maternal and Infant Mortality in the United States, 1942," *The Child* 8 (June 1944), 180, 183–85.

30. Ibid.; U.S. Children's Bureau, *White House Conference on Children in a Democracy*, 286–311; Pepper Committee, *Wartime Health and Education*, 814, 816, 840–41, 848–49; "1943 Touches New Low in Infant Mortality," *American Journal of Public Health* 34 (Sept. 1944), 988; "Ten-Year Maternal and Infant-Mortality Record Points to Unfinished Business in Saving Mothers' and Babies' Lives," *The Child* 10 (Oct. 1945), 56–59; Marjorie Gooch, "Ten Years of Progress in Reducing Maternal and Infant Mortality: With Figures Showing Changes in the Rates between the 2 Years 1942–43," *The Child* 10 (Nov. 1945), 77–83; *New York Times*, May 7, 1942, March 13, 1945.

31. "Ten-Year Maternal and Infant Mortality Record Points to Unfinished Business . . . ," 56–59; *New York Times*, Dec. 7, 1945, March 20, April 5, 1985. Forty years after the Second World War, children's health in America had improved markedly; but the infant mortality rate for blacks was still twice that for whites, 19.6 black infant deaths per 1000 live births compared with 10.1 for whites.

32. *Congressional Record* 91 (July 26, 1945), 8054; Martha M. Eliot, "Women and Children in Wartime: British Experience and American Plans," *Survey Graphic* 31 (March 1942), 117–18; Eliot, "Baby Care Today and Tomorrow," 25; "Preliminary Statement" to White House Conference, quoted in Polier, *Everyone's Children, Nobody's Child*, 273; *New York Times*, July 11, 22, Oct. 25, 1944.

33. Gordon, *Heroes of Their Own Lives: The Politics and History of Family Violence: Boston 1880–1960* (New York: Viking, 1988), 23, 155–59; Kempe et al., "The Battered-Child Syndrome," *Journal of the American Medical Association* 181 (July 1962), 17–24; Stephen J. Pfohl, "The 'Discovery' of Child Abuse," *Social Problems* 24 (Feb. 1977), 310–23; "Skeletons in the Closet," *Survey Midmonthly* 80 (Feb. 1944), 52–53; Austin E. Smith, "The Beaten Child," *Hygeia* 22 (May 1944), 386–88. Helpful introductions to the subject of child abuse are David G. Gil, *Violence Against Children: Physical Child Abuse in the United States* (Cambridge: Harvard Univ. Press, 1970); John Demos, "Child Abuse in Context: An Historian's Perspective," in Demos, *Past, Present, and Personal: The Family and the Life Course in American History* (New York: Oxford Univ. Press, 1986), 68–91; C. Henry Kempe and Ray E. Helfer, eds., *The Battered Child* (Chicago: Univ. of Chicago Press, fourth ed., 1987); Elizabeth H. Pleck, *Domestic Tyranny: The Making of Social Policy*

Against Family Violence from Colonial Times to the Present (New York: Oxford Univ. Press, 1987).

34. Letters 75, 77A, 83H, 115, 331, 346, 355, 392; Pepper Committee, *Wartime Health and Education*, 488–89; Donald A. Hansen and Reuben Hill, "Families under Stress," in Harold T. Christensen, ed., *Handbook of Marriage and the Family* (Chicago: Rand McNally, 1964), 808–9.

35. "The Darkest Secret," *People Weekly* 35 (June 10, 1991), 88 ff. See also Ellen Bass and Louise Thornton, eds., *I Never Told Anyone: Writings by Women Survivors of Child Abuse* (New York: Harper & Row, 1983); Judith Herman and Lisa Hirschman, "Father-Daughter Incest," in Elizabeth Abel and Emily K. Abel, eds., *The Signs Reader: Women, Gender & Scholarship* (Chicago: Univ. of Chicago Press, 1983), 257–78; W. Arens, *The Original Sin: Incest and Its Meaning* (New York: Oxford Univ. Press, 1986); Diana E. H. Russell, *The Secret Trauma: Incest in the Lives of Girls and Women* (New York: Basic, 1986); James B. Twitchell, *Forbidden Partners: The Incest Taboo in Modern Culture* (New York: Columbia Univ. Press, 1987); John Crewdson, *By Silence Betrayed: Sexual Abuse of Children in America* (Boston: Little, Brown, 1988).

36. The author of "Riddle" has requested anonymity.

37. Paul H. Jacobson, "Cohort Survival for Generations since 1840," *Milbank Memorial Fund Quarterly* 42 (July 1964), 36–51; Eileen M. Crimmins, "The Changing Pattern of American Mortality Decline, 1940–1977, and Its Implications for the Future," *Population and Development Review* 7 (June 1981), 229–54; Gladys Denny Shultz, "More Babies Are Alive," *Better Homes & Gardens* 20 (May 1942), 46, 85. According to Crimmins's analysis, the period from 1940 to 1977 was marked by three distinct phases. First, "from 1940 to the mid-1950's mortality declined at a pace unprecedented in American history." Then, "a marked slowing in the rate of decline occurs after 1954, presumably because the diffusion of antibiotics has been largely accomplished by that date." Third, beginning in 1968, "a new decline in mortality set in at rates close to those of the 1940's and early 1950's." In the third period, there occurred "a new trend—a sharp decline in mortality due to cardiovascular diseases," especially for people over thirty-five.

38. See also the Children's Bureau, Statistical Series No. 1, *Maternal and Infant Mortality in 1944* (Washington: GPO, 1947), 10–17; No. 2, *Deaths of Premature Infants in the United States* (Washington: GPO, 1947), 3–9; No. 15, *Main Causes of Infant, Childhood and Maternal Mortality 1939–1949* (Washington: GPO, 1953), passim.

39. Gooch, "Ten Years of Progress in Reducing Maternal and Infant Mortality," 77; J. Yerushalmy, "Births, Infant Mortality, and Maternal Mortality in the United States—1942," *Public Health Reports* 59 (June 23, 1944), 797–811; "Provisional Mortality Rates for the First Half of 1944," *Public Health Reports* 60 (Feb. 9, 1945), 153–61. For maternal mortality, see also *New York Times*, Feb. 6, 1943, Oct. 1, 1944; Martha M. Eliot, "Making Childbirth Safer in Wartime," *Ladies' Home Journal* 60 (Aug. 1943), 6, 56; Pepper Committee, *Wartime Health and Education*, part 3, pp. 1856–57; U.S. House of Representatives, 78th Cong., 2nd Sess., Subcommittee of the Committee on Appropriations, *Department of Labor–Federal Security Agency Appropriation Bill, 1945* (Washington: GPO, 1944), 282–83; "More Women Live through Childbirth," *Science Digest* 16 (Dec. 1944), 51; "Mortality in Childbirth Cut by One-Third," *Science News Letter* 48 (Aug. 25, 1945), 117–18; Meyer F. Nimkoff, "The Family," *American Journal of Sociology* 47 (May 1942), 869–75; *New York Times*, Oct. 25, Nov. 5, 17, 1944, Jan. 6, 18, 1945.

A general study is Judith Walzer Leavitt, *Brought to Bed: Child-bearing in America, 1750–1950* (New York: Oxford Univ. Press, 1986). Great Britain's wartime history of maternal and infant mortality was similar to that of the United States; see J. M. Winter, "The Demographic Consequences of the Second World War for Britain," Occasional Paper 34, Office of Population Censuses and Surveys and British Society for Population Studies, *Measuring Socio-Demographic Change* (published conference papers, Univ. of Sussex, 1985), unpaginated.

40. John L. Springer, "The Battle for Child Health," *Parents' Magazine* 18 (Jan. 1943), 30; Carpenter, "Children: Social Priority Number One," 348; "The Children's Bureau Looks Ahead," *Social Service Review* 15 (March 1941), 129–32.

41. "Allowances for Servicemen's Dependents, 1942," *Monthly Labor Review* 55 (Aug. 1942), 226–28; Marietta Stevenson, "New Governmental Services for People in Wartime," *Social Service Review* 16 (Dec. 1942), 595–602; Denzel C. Cline, "Allowances to Dependents of Servicemen in the United States," *Annals of the American Academy of Political and Social Science* 227 (May 1943), 1–8; Helen Tarasov, "Family Allowances: An Anglo-American Contrast," *Annals of the American Academy of Political and Social Science* 227 (May 1943), 9–21; Robert E. Bondy, "Special Welfare Services to Families of Men in Service," *National Conference of Social Work Proceedings* 70 (1943), 76–79.

42. *Congressional Record* 88, pp. A3010–11, 3119–21; 89, pp. A4341–42, 4824–25; "Liberalized Allowances for War Service Dependents," *Monthly Labor Review* 58 (Jan. 1944), 67–69; "Military Family Allowances in the United States, Canada, and Great Britain, 1945," *Monthly Labor Review* 63 (July 1946), 94–97; John Modell and Duane Steffey, "Waging War and Marriage: Military Service and Family Formation, 1940–1950" (unpub. paper), 5–7.

43. Harry Henderson and Sam Shaw, "Pay for Soldier's Families," *Collier's* 111 (May 22, 1943), 18, 76; Harry Grossman, "Administration of the Servicemen's Dependents Allowance Act of 1942," *Social Security Bulletin* 6 (July 1943), 21–24; *New York Times*, May 5, 1944.

44. Phyllis Aronson, "The Adequacy of the Family Allowance System As It Affects the Wives and Children of Men Drafted into the Armed Forces" (unpub. M.S.W. thesis, Wayne Univ., 1944), 87; Harold N. Gilbert, "Green Checks for the Folks Back Home," *American Magazine* 137 (Jan. 1944), 94; *New York Times*, Feb. 16, 1944.

45. Furnas, "They Get 'Em Paid," *Saturday Evening Post* (June 5, 1943), 105.

46. Harold N. Gilbert, "Bigger Checks for Dependents," *Christian Science Monitor*, Jan. 15, 1944; Gretta Palmer, "The Army's Problem Wives," *Reader's Digest* 45 (July 1944), 66–68; "Checks by the Million," *Fortune* 30 (Dec. 1944), 151–54, 194, 196.

47. Eliot, "Maternity Care for Service Men's Wives," *Survey Midmonthly* 79 (April 1943), 113–14.

48. U.S. Census Bureau, *Historical Statistics of the United States, Bicentennial Edition*, part 2, p. 1140; Pepper Committee, *Wartime Health and Education*, part 5, pp. 1624–35.

49. Nathan Sinai and Odin W. Anderson, *EMIC (Emergency Maternity and Infant Care): A Study of Administrative Experience* (Ann Arbor: School of Public Health, Univ. of Michigan, 1948), 17–20, 22–29; *The Child* 7 (Nov. 1942, April 1943), 64, 148; 8 (Oct. 1943), 55; 10 (July 1945), 11; U.S. Children's Bureau Publication 302, Edward E. Schwartz and Eloise R. Sherman, *Community Health and Welfare Expenditures in Wartime: 1942 and 1940—30 Urban Areas* (Washington: GPO, 1944), 18–22; Mary Zahrobsky,

"Health and Medical Care of School Children Legislation in Illinois," *Social Service Review* 19 (Sept. 1945), 343–51; U.S. Children's Bureau Defense of Children Series No. 5, *Are We Defending Their Right to Health?* (Washington: GPO, n.d.); Stevenson, "New Governmental Services for People in Wartime," 603–4; Anna R. Moore and Marie Chard, "Private Don Jones' Baby," *American Journal of Nursing* 43 (Jan. 1943), 46–50; Bondy, "Special Welfare Services to Families of Men in Service," 81–84; "Uncle Sam Provides Obstetric Care," *American Journal of Nursing* 43 (May 1943), 470–71; "From Far and Near," *Public Health Nursing* 35 (May 1943), 293–94; Stuart W. Adler, "Medical Care for Dependents of Men in Military Service," *American Journal of Public Health* 33 (June 1943), 645–50; "Congress Votes Maternity Care Funds for Servicemen's Wives," *Social Service Review* 17 (June 1943), 216–17; Charles P. Taft, "Public Health and the Family in World War II," *Annals of the American Academy of Political and Social Science* 229 (Sept. 1943), 148–49; Amy Porter, "Babies for Free," *Collier's* 116 (Aug. 4, 1945), 18–19, 28; U.S. Children's Bureau Publication 294, *Facts about Child Health 1946* (Washington: GPO, 1946), 10–18.

50. Eliot, "Maternity Care for Service Men's Wives," 113–14; Eliot, "Baby Care Today and Tomorrow," 61.

51. Porter, "Babies for Free," 19; *New York Times*, Nov. 1, 1942, June 25, July 2, Aug. 3, Sept. 30, Oct. 2, 1943, April 28, 1944; U.S. House of Representatives, 78th Cong., 1st Sess., Subcommittee of the Committee on Appropriations, *Department of Labor–Federal Security Agency Appropriation Bill, 1944* (Washington: GPO, 1943), 218–19, 241–56; U.S. House, *Department of Labor–Federal Security Agency Appropriation Bill, 1945*, 342–43, 356–61, 370–73, 444–70, U.S. Children's Bureau, EMIC Information Circular No. 1, *Administrative Policies: Emergency Maternity and Infant Care Program* (Washington: GPO, 1943); Sinai and Anderson, *EMIC*, 29–79; "The Emergency Maternity and Infant Care Program," *American Journal of Nursing* 44 (March 1944), 242–45; Martha M. Eliot, "Experience with the Administration of a Medical Care Program for Wives and Infants of Enlisted Men," *American Journal of Public Health* 34 (Jan. 1944), 34–39; Porter, "Babies for Free," 19; "Medical Care for the Wives and Infants of Servicemen," *American Journal of Public Health* 34 (May 1944), 529–31.

52. *Social Security Bulletin* 7 (Sept. 1944), 38–39; 9 (March 1946), 47; "Effect of the Increased Birth Rate on Maternal and Child Health Problems," *American Journal of Public Health* 35 (Feb. 1945), 140–42; "Emergency Maternity and Infant-Care Program," *The Child* 10 (July 1945), 11; Sinai and Anderson, *EMIC*, 54, 81–88, i-x; U.S. House of Representatives, 79th Cong., 1st Sess., Subcommittee of the Committee on Appropriations, *Department of Labor–Federal Security Agency Appropriation Bill for 1946* (Washington: GPO, 1945), 265–70, 326–28, 333–34; *New York Times*, July 15, 27, Nov. 22, Dec. 24, 1944, Jan. 20, March 18, Nov. 11, 1945.

53. U.S. Children's Bureau Publication 294, *Facts about Child Health 1943* (Washington: GPO, 1945), 10–12; U.S. Children's Bureau Statistical Series Number 4, *Further Progress in Reducing Maternal and Infant Mortality*, 22–27.

54. See the table printed earlier in this chapter (*Infant Death Rates by Race: United States, 1933–1945*); U.S. Children's Bureau, *Further Progress in Reducing Maternal and Infant Mortality*, 19, 22–27; Allan M. Brandt, *No Magic Bullet: A Social History of Venereal Disease in the United States since 1880* (New York: Oxford Univ. Press, 1987 ed.), 157, 169–70; James H. Jones, *Bad Blood: The Tuskegee Syphilis Experiment* (New York:

Free Press, 1981), esp. 16–29; U.S. Children's Bureau, *Facts about Child Health 1943*, 10–12.

55. *New York Times*, Aug. 14, 1945.

56. "Mr. and Mrs. America 1970," 10.

57. "President Truman Calls Health of Children a Public Responsibility," *The Child* 10 (Feb. 1946), 124; Eliot, "Baby Care Today and Tomorrow," 61; "Maternity and Infant Care," *Elementary School Journal* 44 (April 1944), 446; Pepper Committee, *Wartime Health and Education*, part 5, pp. 1657, 1856–62.

58. *New York Times*, July 27, 1945; *Washington Post* quoted in *Congressional Record* 91, p. 8052–58, A3714; Porter, "Babies for Free," 28.

59. Sinai and Anderson, *EMIC*, 37; "G.I. Babies," *Time* 41 (Oct. 4, 1943), 46; U.S. House, *Department of Labor—Federal Security Agency Appropriation Bill, 1945*, 312–23, 488–517, 525–29, 567–70; "Medical Care for the Wives and Infants of Servicemen," *American Journal of Public Health* 34 (May 1944), 529–30; *New York Times*, Aug. 3, 1944; "Reductio ad absurdum," *Survey Midmonthly* 80 (Sept. 1944), 260; "Next Steps in Maternal and Child Health," *American Journal of Public Health* 35 (Feb. 1945), 122.

60. *New York Times*, Oct. 3, 25, 1944, July 27, 1945.

61. For the political history of Medicare and Medicaid, see Richard Harris, *A Sacred Trust* (New York: New American Library, 1966); Monte M. Poen, *Harry S. Truman Versus the Medical Lobby: The Genesis of Medicare* (Columbia: Univ. of Missouri Press, 1979). Also helpful are Robert J. Myers, *Medicare* (Homewood, IL: R. D. Irwin, 1970); Theodore R. Marmor, *The Politics of Medicare* (Chicago: Aldine, 1973).

62. Lawrence, KS, *Journal-World*, March 1, 1987; Kevin Phillips, *The Politics of Rich and Poor: Wealth and the American Electorate in the Reagan Aftermath* (New York: Random House, 1990); Thomas Byrne Edsall and Mary D. Edsall, *Chain Reaction: The Impact of Race, Rights, and Taxes on American Politics* (New York: Norton, 1991).

Chapter 12. "Daddy's Coming Home!"

1. Letters 42, 83E, 438.

2. Letters 27, 244.

3. Letters 82, 243, 286, 316, 332, 348, 390.

4. Letters 282, 320, 352, 408.

5. Letters 82A, 172, 254, 440.

6. Letters 350, 358A, 382.

7. Letters 312, 366.

8. Letter 332.

9. Letters 41, 137, 154.

10. Letters 2, 76, 244; U.S. Bureau of the Census, *Historical Statistics of the United States, Colonial Times to 1970, Bicentennial Edition* (Washington: GPO, 1975), part 2, p. 1140.

11. Letter 435.

12. Letter 243; Census Bureau, *Historical Statistics of the United States, Bicentennial Edition*, part 2, p. 1141; Army Service Forces, Office of Dependency Benefits, *Annual Report for the Fiscal Year 1944* (Newark, NJ: Office of Dependency Benefits, 1945), 72–73; Army Service Forces, Office of Dependency Benefits, *Annual Report for the Fiscal Year*

1945 (Newark, NJ: Office of Dependency Benefits, 1946), 15; Steven Mintz and Susan Kellogg, *Domestic Revolutions: A Social History of American Family Life* (New York: Free Press, 1988), 167, 173–74.

13. Wylie, *Generation of Vipers* (New York: Rinehart, 1942), 185, 203.

14. U.S. Census Bureau, *Historical Statistics of the United States, Bicentennial Edition*, part 2, p. 1140; Myra MacPherson, *Long Time Passing: Vietnam and the Haunted Generation* (Garden City, NY: Doubleday, 1984), 197–205. A study of "combat stress" during the Second World War is Samuel A. Stouffer et al., *The American Soldier*, vol. 2: *Combat and Its Aftermath* (Princeton: Princeton Univ. Press, 1949). See also Ernest R. Mowrer and Harriet R. Mowrer, "The Disabled Veteran in the Family," *Annals of the American Academy of Political and Social Science* 239 (May 1945), 150–59; Reuben Hill, *Families under Stress: Adjustment to the Crises of War Separation and Reunion* (New York: Harper & Brothers, 1949); Robert J. Havighurst et al., *The American Veteran Back Home: A Study of Veteran Readjustment* (New York: Longmans, Green, 1951), esp. 23–36; Roy Richard Grinker and John P. Spiegel, *Men under Stress* (Philadelphia: Blakiston, 1945); Grinker and Spiegel, *War Neuroses* (Philadelphia: Blakiston, 1945); James H. S. Bossard, "What War Is Still Doing to the Family," in Howard Becker and Reuben Hill, eds., *Family, Marriage and Parenthood* (Boston: D.C. Heath, 1946), 728–29; Ruth Shonle Cavan, *The American Family* (New York: Thomas Y. Crowell, 1953), 558–59; Herbert C. Archibald et al., "Gross Stress Reaction Following Combat: A 15-Year Follow-Up," *American Journal of Psychiatry* 119 (1962), 317–22; Archibald and Read D. Tuddenham, "Persistent Stress Reaction after Combat: A 20-Year Follow-Up," *Archives of General Psychiatry* 12 (May 1965), 475–81; Glen H. Elder, Jr., and Elizabeth C. Clipp, "Combat Experience, Comradeship, and Psychological Health" (unpub. paper, 1986). Valuable insights into stigmatization can be found in Beatrice A. Wright, "Attitudes and the Fundamental Negative Bias," in H. E. Yuker, ed., *Attitudes toward Persons with Disabilities* (New York: Springer, 1988), 3–21.

15. "The Courage of the Men: An Interview with John Huston"; and Huston, "Let There Be Light/*The Script*," both in Robert Hughes, ed., *Film: Book 2: Films of Peace and War* (New York: Grove Press, 1962), 30–35, 205–33; Erik Barnouw, *Documentary: A History of the Non-Fiction Film* (New York: Oxford Univ. Press, rev. ed., 1983), 163–64.

16. *Diagnostic and Statistical Manual of Mental Disorders: DSM-III-R* (Washington: American Psychiatric Association, rev. third ed., 1987), 247–51; Motti Mark, "The Trauma of War" (address to Faculty Forum, Univ. of Kansas, April 10, 1991). See also Robert Jay Lifton, *Home from the War: Vietnam Veterans: Neither Victims nor Executioners* (New York: Simon & Schuster, 1973); Charles R. Figley and Seymour Leventman, eds., *Strangers at Home: Vietnam Veterans since the War* (New York: Prager, 1980); MacPherson, *Long Time Passing*, 177–96; Bessel Van der Kolk, ed., *Post-Traumatic Stress Disorder: Psychological and Biological Sequelae* (Washington: American Psychiatric Press, 1984); Arthur Egendorf, *Healing from the War: Trauma & Transformation after Vietnam* (Boston: Houghton Mifflin, 1985); William E. Kelley, ed., *Post-Traumatic Stress Disorder and the War Veteran Patient* (New York: Brunner/Mazel, 1985); Glen H. Elder, Jr., and Elizabeth C. Clipp, "Combat Experience, Comradeship, and Psychological Health," in John P. Wilson et al., eds., *Human Adaptation to Extreme Stress: From the Holocaust to Vietnam* (New York: Plenum, 1988), 131–56; Paul Fussell, *Wartime: Understanding and Behavior in the Second World War* (New York: Oxford Univ. Press, 1989); U.S. Bureau of the Census,

Estimated Number of Veterans of World War II in Continental United States, by States: April 1, 1947, Series P-25, No. 5 (Washington: Bureau of the Census, 1947), 1–2; Administrator of Veterans Affairs, *Annual Report . . . 1946* (Washington: GPO, 1947), 104–5.

17. Letters 15, 203, 232, 240B, 371, 377.

18. Letters 119, 213, 352, 354.

19. Letters 12A, 23, 72, 83H, 92, 97D, 97F, 127, 176, 206, 208, 243, 255, 276, 316, 352, 406; interview with Jay W. Baird, Aug. 1989; "Children Must Face the Facts of Disability," *Hygeia* 22 (Nov. 1944), 818–19, 872–74; Mark Jonathan Harris et al., eds., *The Homefront: America During World War II* (New York: Putnam's, 1984), 199, 223.

20. Letters 23, 40, 92, 94, 97F, 165, 249, 255, 346, 413, 416, 432.

21. Letter 165; Arnold is quoted in MacPherson, *Long Time Passing,* 183; "The Second Battleground," *Newsweek* 117 (May 6, 1991), 61.

22. Letter 125; McPherson, *Long Time Passing,* 204.

23. U.S. Census Bureau, *Historical Statistics of the United States, Bicentennial Edition,* part 2, p. 49; Mintz and Kellogg, *Domestic Revolutions,* 171–73; *New York Times,* Jan. 8, 1945; letters 64, 94, 97A, 97D, 156, 165, 170, 249, 255, 294, 308, 416.

24. Letters 97D, 127, 255, 291, 354; Otto Butz, ed., *The Unsilent Generation: An Anonymous Symposium in Which Eleven College Seniors Look at Themselves and Their World* (New York: Rinehart, 1958), 45–46.

25. Letters 12A, 249, 380.

26. Therese Benedek, *Insight and Personality Adjustment: A Study of the Psychological Effects of War* (New York: Ronald, 1946), 232; letter 413; Frank O'Connor, "My Oedipus Complex," in O'Connor's *Collected Stories* (New York: Knopf, 1981), 282–92; Mary M. Leichty, "The Effect of Father-Absence During Early Childhood upon the Oedipal Situation as Reflected in Young Adults," *Merrill-Palmer Quarterly* 6 (July 1960), 212–17.

27. Letters 78, 97D.

28. Bach, "Father-Fantasies and Father-Typing in Father-Separated Children," *Child Development* 17 (March-June 1946), 78; John F. Day, "While Dad's Away," *Life* (Aug. 13, 1945), 13–14.

29. Letters 145, 291, 371; Marshall L. Hamilton, *Father's Influence on Children* (Chicago: Nelson-Hall, 1977), 1–18; Ross D. Parke, *Fathers* (Cambridge: Harvard Univ. Press, 1981), 43–46, 56–63; Hill, *Families Under Stress,* 58–63; L. Pearl Gardner, "A Survey of the Attitudes and Activities of Fathers," *Journal of Genetic Psychology* 63 (Sept. 1943), 15–53; Evelyn Millis Duvall, "Soldier Come Home: An Annotated Bibliography," *Marriage and Family Living* 7 (1945), 61–63, 72; Bossard, "What War Is Still Doing to the Family," 724–25; Cavan, *The American Family,* 556–57; Havighurst et al., *The American Veteran Back Home,* 83–85; John Demos, "The Changing Faces of Fatherhood," in *Past, Present, and Personal: The Family and the Life Course in American History* (New York: Oxford Univ. Press, 1986), esp. 60–64.

30. Letters 97D, 232, 249; interview with Jacqueline Dowd Hall; Harris et al., *The Homefront,* 231–32.

31. Rautman, "Children of War Marriages," *Survey Midmonthly* 80 (July 1944), 198–99; Harris et al *The Homefront,* 231–32; Rautman, St. Petersburg, FL, to the author, March 18, 1990.

32. *New York Times,* Jan. 9, 1945; Stolz et al., *Father Relations of War-Born Children* (Stanford: Stanford Univ. Press, 1954), 27–51, 169–80, 316–26; Stolz, "The Effect of Mobilization and War on Children," *Social Casework* 32 (April 1951), 146.

33. Rautman to the author, March 18, 1990; Rautman, "Children of War Marriages," 198–99; Phillips Cutright, "Illegitimacy in the United States: 1920–1968," and Kingsley Davis, "The American Family in Relation to Demographic Change," both in Charles F. Westoff and Robert Parke, Jr., eds., Commission on Population Growth and the American Future, Research Reports, vol. 1: *Demographic and Social Aspects of Population Growth* (Washington: GPO, 1972), 383–91, 250–54; U.S. Census Bureau, *Historical Statistics of the United States, Bicentennial Edition,* part 1, p. 52; *New York Times,* Dec. 6, 1945; U.S. House of Representatives, 79th Cong., 1st Sess., Subcommittee of the Committee on Appropriations, *Department of Labor–Federal Security Agency Appropriation Bill for 1946* (Washington: GPO, 1945), 270–71, 292–93; Maud Morlock, "Wanted: A Square Deal for the Baby Born Out of Wedlock," *The Child* 10 (May 1946), 167–69; Bossard, "What War Is Still Doing to the Family," 716; Lottie S. Levi, "Unmarried Mothers in Wartime," in Child Welfare League of America, *The Impact of War on Children's Services* (New York: Child Welfare League of America, 1943), 4–8.

34. Sams, with Lynn Z. Bloom, ed., *Forbidden Family: A Wartime Memoir of the Philippines, 1941–1945* (Madison: Univ. of Wisconsin Press, 1989), 48–103.

35. Ibid., 103–86.

36. Ibid., 278–300.

37. Ibid., 23, 307–8.

38. Anna W. M. Wolf and Irma Simonton Black, "What Happened to the Younger People," in Jack Goodman, ed., *While You Were Gone: A Report on Wartime Life in the United States* (New York: Simon and Schuster, 1946), 71.

39. Virginia M. Moore, "When Father Comes Marching Home," *Parents' Magazine* 20 (Jan. 1945), 16–17, 112; Catherine Mackenzie, "Fathers Home from the War," *New York Times,* Feb. 4, 1945; John F. Cuber, "Family Readjustment of Veterans," *Marriage and Family Living* 7 (Spring 1945), 28–30; Reuben Hill, "The Returning Father and His Family," *Marriage and Family Living* 7 (Spring 1945), 31–34; J. H. S. Bossard, "Family Problems of the Immediate Future," *Journal of Home Economics* 37 (Sept. 1945), 383–87; Whitman M. Reynolds, "When Father Comes Home Again," *Parents' Magazine* 20 (Oct. 1945), 28 ff.; and the entire issue of *American Journal of Sociology* 51 (March 1946).

40. "The Menace of the Maternal Father," *Hygeia* 20 (June 1942), 468–70. Apparently complicating the homecoming of some fathers were two other factors. Some fathers who returned having had homosexual experiences were determined to repress their sexual orientation; see Allan Berube, *Coming Out under Fire: The History of Gay Men and Women in World War II* (New York: Free Press, 1990), esp. 175–254. Second, some fathers returned as confirmed misogynists. According to an observational study done by a soldier, "Fundamentally, the G.I. did not like or desire women other than as means of gratifying . . . his primitive sexual desire. He commonly referred to women by the profane term for vagina . . ." (Henry Elkin, "Aggressive and Erotic Tendencies in Army Life," *American Journal of Sociology* 51 [March 1946], 413).

41. Gregory P. Stone, "The Play of Little Children," in R. E. Herron and Brian Sutton-Smith, eds., *Child's Play* (New York: John Wiley, 1971), 11.

42. Carlsmith, "Effect of Early Father Absence on Scholastic Aptitude," *Harvard Educational Review* 34 (Winter 1964), 3–21.

43. Ibid., 17.

44. Chodorow, *The Reproduction of Mothering: Psychoanalysis and the Sociology of Gender* (Berkeley: Univ. of California Press, 1978), 169; Gilligan, *In a Different Voice: Psychological Theory and Women's Development* (Cambridge: Harvard Univ. Press, 1982); Mary Field Belensky et al., *Women's Ways of Knowing: The Development of Self, Voice, and Mind* (New York: Basic Books, 1986); Mary M. Brabeck, ed., *Who Cares? Theory, Research, and Educational Implications of the Ethic of Care* (New York: Praeger, 1989).

45. Carlsmith, "Effect of Early Father Absence on Scholastic Aptitude," 10; Carlsmith, "Some Personality Characteristics of Boys Separated from Their Fathers during World War II," *Ethos* 1 (Winter 1973), 470–73.

46. Carlsmith, "Effect of Early Father Absence on Scholastic Aptitude," 3–4, 16. Carlsmith borrowed her explanatory framework from John W. M. Whiting's theory of cross-sex identification; see Whiting, "The Absent Father and Cross-Sex Identity," *Merrill-Palmer Quarterly* 7 (April 1961), 85–95.

There is no doubt that standardized national tests, including the SAT, have historically been culture-bound, having been written essentially for white, middle-class males. The 1960 SAT test booklet itself, the one used by these seventeen- and eighteen-year-old men, stated: "In general girls do less well than boys on the Mathematical parts of the test and should not be surprised if their Mathematical scores are noticeably lower than their Verbal." Clearly, this attitude was not a scientific fact, but a social and cultural construct, which threatened to become a self-fulfilling prophecy. See Gita Z. Wilder and Kristin Powell, *Sex Differences in Test Performance: A Survey of the Literature* (New York: College Entrance Examination Board [Report No. 89-3], 1989).

In recent years, however, the gaps between men and women on mathematical and verbal problems have been disappearing. In 1989 Marcia C. Linn and Janet S. Hyde argued that as early as 1974 the reported gender differences in math were "so small as to be negligible." See Alan Feingold, "Cognitive Gender Differences Are Disappearing," *American Psychologist* 43 (Feb. 1988), 95–103.

The text of this book is not the best place for a lengthy discussion of this topic. And yet, since at least the outlines of this topic should be on the record and understood, a brief summary is included here. Closely related to Carlsmith's work is Edward A. Nelsen and Eleanor E. Maccoby, "The Relationship between Social Development and Differential Abilities on the Scholastic Aptitude Test," *Merrill-Palmer Quarterly* 12 (Oct. 1966), 269–84. Reporting on males in Stanford University's classes of 1963 and 1964, they found that "a high-verbal, low-mathematics pattern was associated with reports of father absence, punishment exclusively by the mother, fear of father, and reports of having been a 'mamma's boy' or 'daddy's boy.'" See also David B. Lynn, "A Note on Sex Differences in the Development of Masculine and Feminine Identification," *Psychological Review* 66 (March 1959), 126–35.

Importantly, Beatrice B. Whiting points out that there is an exception to these findings, observing that in some cases father absence produces "protest masculinity" (compulsive reaction formation), explaining that violent and aggressive behavior plus concern over masculinity validates the status-envy hypothesis of identification: "Sex Identity Conflict and Physical Violence: A Comparative Study," *American Anthropologist* 67 (Dec.

1965), 123–40. Finally, cognitive gender differences from father absence have narrowed in recent years; see Marybeth Shinn, "Father Absence and Children's Cognitive Development," *Psychological Bulletin* 85 (March 1978), 295–324.

47. Parke, *Fathers*, 57.

48. See Donald B. Rinsley, "The Adolescent, the Family, and the Culture of Narcissism: A Psychosocial Commentary," *Adolescent Psychiatry* 13 (1986), 24.

49. Natalie Angier, "The Biology of What It Means To Be Gay," *New York Times*, Sept. 1, 1991. At all points along the nature-nurture continuum, there is disagreement about homosexuality. An example is female homosexuality. At one time, psychologists asserted that this resulted from a girl's overly strong identification with her father. But an English study comparing homosexual and heterosexual women found that lesbians, in referring to their fathers, "used items expressing hostility and disapproval more often, saw the father as more frightening and described the father as lower in competence and strength of personality" than the mother. See Paul Mussen and Eldred Rutherford, "Parent-Child Relations and Parental Personality in Relation to Young Children's Sex-Role Preferences," *Child Development* 34 (Sept. 1963), 589–607; Eva Bene, "On the Genesis of Female Homosexuality," *British Journal of Psychiatry*, 111 (Sept. 1965), 815–21.

50. D'Emilio, *Sexual Politics, Sexual Communities: The Making of a Homosexual Community, 1940–1970* (Chicago: Univ. of Chicago Press, 1983), 233–37; Berube, *Coming Out Under Fire*, 201–54.

51. Letter 97F.

52. Robert R. Sears et al., "Effect of Father Separation on Preschool Children's Doll Play Aggression," *Child Development* 17 (Dec. 1946), 242; Robert R. Sears, Stanford Univ., to the author, April 19, 1989.

53. In addition to the paucity of studies of father-daughter relations, another obstacle is that the studies done contradict each other. Some contend that the fathers of highly feminine girls are more masculine than the fathers of less feminine girls, since the masculine fathers encourage what they presume to be sex-appropriate behavior. Another concludes that "fathers who are highly nurturant enhance masculinity in sons and femininity in daughters." Yet other studies assert the opposite, contending that girls who describe themselves as self-confident identify with the "instrumental" side of the father whereas "the expressive mother-identified females" describe themselves "as considerate, fearful, gentle, silent, submissive, and trusting." See Hamilton, *Father's Influence on Children*, 79; Alfred B. Heilbrun, Jr., "Sex Differences in Identification Learning," *Journal of Genetic Psychology* 106 (June 1965), 185–93; Heilbrun; "Sex Role, Instrumental-Expressive Behavior, and Psychopathology in Females," *Journal of Abnormal Psychology* 73 (April 1968), 131–36; Heilbrun, "Sex-Role Identity in Adolescent Females: A Theoretical Paradox," *Adolescence* 3 (Spring 1968), 79–88; David B. Lynn, "Fathers and Sex-Role Development," *Family Coordinator* 25 (Oct. 1976), 403–10; Michael E. Lamb, "Paternal Influences and the Father's Role: A Personal Perspective," *American Psychologist* 34 (Oct. 1979), 938–43.

54. Edward F. Zigler et al., *Socialization and Personality Development* (New York: Oxford Univ. Press, 1982), 60.

55. Letters 27, 83, 97D, 160, 232, 249, 371, 373, 376, 380, 403, 436; Hartmann, "Prescriptions for Penelope: Literature on Women's Obligations to Returning World War II Veterans," *Women's Studies* 5 (1978), 236; Mona Gardner, "Has Your Husband Come

Home to the Right Woman?," *Ladies' Home Journal*, 62 (Dec. 1945), 41 ff.; Eli Ginzberg, "The Changing Pattern of Women's Work: Some Psychological Correlates," *American Journal of Orthopsychiatry* 28 (April 1958), 313–16.

56. Lerner, "The Feminists: A Second Look," *Columbia Forum* 13 (Fall 1970), 26.

57. Letters 95, 101, 243, 413. An important article is E. Mavis Hetherington's "Effects of Father Absence on Personality Development in Adolescent Daughters," *Developmental Psychology* 7 (Nov. 1972), 313–26.

58. Letter 65.

59. Letter 70.

Chapter 13. Confronting War's Enormity, Praising Its Glory

1. Miller, "Report from the Inferno," *Newsweek* 98 (Sept. 7, 1981), 72; Committee for the Compilation of Materials on Damage Caused by the Atomic Bombs in Hiroshima and Nagasaki, *Hiroshima and Nagasaki: The Physical, Medical, and Social Effects of the Atomic Bombings* (New York: Basic, 1981).

2. Mailer, "The White Negro (Superficial Reflections on the Hipster)," *Dissent* 4 (Summer 1957), 276–77. For a later statement that echoes some of these points, see Robert Jay Lifton and Eric Markusen, *The Genocidal Mentality: Nazi Holocaust and Nuclear Threat* (New York: Basic, 1990).

3. Letters 12B, 38, 42, 83A, 97E, 150, 178, 182, 187, 206D, 208, 222, 276, 312, 316, 360, 404, 405, 412, 431, 442B, 450.

4. Letters 30, 49, 72, 362, 416, 426.

5. Letters 38, 206F, 360.

6. Letters 16, 102, 408.

7. Burr Leyson, "Packaged Devastation," *Boys' Life* (Nov. 1945), 3, 40–42; letters 28, 97E 184, 206P, 264, 446.

8. Letters 83B, 97K, 270, 324F, 385, 422; Les K. Adler and Thomas G. Paterson, "Red Fascism: The Merger of Nazi Germany and Soviet Russia in the American Image of Totalitarianism, 1930s–1950s," *American Historical Review* 75 (April 1970), 1046–64; Paul Boyer, *By the Bomb's Early Light: American Thought and Culture at the Dawn of the Atomic Age* (New York: Pantheon, 1985); Spencer R. Weart, *Nuclear Fear: A History of Images* (Cambridge: Harvard Univ. Press, 1988); and the film *The Atomic Cafe* (1982).

9. Terr, *Too Scared to Cry: Psychic Trauma in Childhood* (New York: Harper & Row, 1990), ix–x.

10. Letters 66, 121, 162, 206, 213, 255, 272, 399, 403, 438, 454; J. Louise Despert, *Preliminary Report on Children's Reactions to the War, Including a Critical Survey of the Literature* (New York: Josiah Macy, Jr., Foundation, 1942), 73.

11. Letters 82A, 97C, 119, 140, 189, 222, 352, 400, 412, 433.

12. Letters 0.3, 4, 12, 24, 27, 46, 62, 165, 168, 336; Harris et al., *The Homefront*, 73.

13. Letters 10, 32, 57, 70, 87, 204, 228, 270, 286, 322, 362.

14. Letters 24, 49, 156B, 228, 240B, 426; U.S. Census Bureau, *Historical Statistics of the United States, Bicentennial Edition*, part 1, p. 262.

15. Letters 16, 32, 62, 238.

Chapter 14. Age, Culture, and History

1. For an essay dealing in greater depth with the relationship between individual development and historical change, see my "America's Home-front Children in World War II," in Glen H. Elder, Jr., et al., eds., *Children in Time and Place: The Intersection of Developmental and Historical Perspectives* (Cambridge: Cambridge Univ. Press, 1993), 27–46. Several extended essays have been helpful in elaborating my ideas about age, culture, and history as the essential elements, which, in configuration, both determine individual development and shape social change. See George A. De Vos and Arthur A. Hippler, "Cultural Psychology: Comparative Studies of Human Behavior," and Alex Inkeles and Daniel J. Levinson, "National Character: The Study of Modal Personality and Sociocultural Systems," both in Gardner Lindzey and Elliott Aronson, eds., *Handbook of Social Psychology*, vol. 4 (Reading, MA: Addison-Wesley, second ed., 1968–69), 323–506, esp. 361–63, 419, 422, 463.

Also informative are these essays on socialization: Robert A. LeVine, "Culture, Personality, and Socialization: An Evolutionary View," Alex Inkeles, "Social Structure and Socialization," and Boyd R. McCandless, "Childhood Socialization," all in David A. Goslin, ed., *Handbook of Socialization Theory and Research* (Chicago: Rand McNally, 1969), 503–41, 615–32, 791–819; Edward Zigler and Irvin L. Child, "Socialization," in Lindzey and Aronson, *Handbook of Social Psychology*, vol. 3, pp. 450–589; John A. Clausen, ed., *Socialization and Society* (Boston: Little, Brown, 1968), 3–71.

2. Labouvie and Nesselroade, "Age, Period, and Cohort Analysis and the Study of Individual Development and Social Change," in Nesselroade and Alexander Von Eye, eds., *Individual Development and Social Analysis* (New York: Academic Press, 1985), 208–9; Bronfenbrenner, Kessel, Kessen, and White, "Toward a Critical Social History of Developmental Psychology: A Propaedeutic Discussion," *American Psychologist* 41 (Nov. 1986), 1218–30.

3. Anna W. M. Wolf and Irma Simonton Black, "What Happened to the Younger People," in Jack Goodman, ed., *While You Were Gone: A Report on Wartime Life in the United States* (New York: Simon and Schuster, 1946), 65.

4. Some scholars contend that there is another element in the configuration: sociobiology resulting from genetics. Although these human biological variables are beyond the province of an historian, working with retrospective materials, to measure, see Carl N. Degler, *In Search of Human Nature* (New York: Oxford Univ. Press, 1991); Edward O. Wilson, *Sociobiology: The New Synthesis* (Cambridge: Harvard Univ. Press, 1975); Wilson, *On Human Nature* (Cambridge: Harvard Univ. Press, 1978).

Helpful critiques of sociobiology are Jill S. Quadagno, "Paradigms in Evolutionary Theory: The Sociobiological Model of Natural Selection," *American Sociological Review* 44 (Feb. 1979), 100–109; Richard C. Lewontin et al., *Not in Our Genes: Biology, Ideology, and Human Nature* (New York: Pantheon, 1984); Philip Kitcher, *Vaulting Ambition: Sociobiology and the Quest for Human Nature* (Cambridge: MIT Press, 1985). Feminist scholars have also criticized sociobiology; see Anne Fausto-Sterling, *Myths of Gender: Biological Theories About Men and Women* (New York: Basic Books, 1986).

5. Erikson, *Childhood and Society* (New York: Norton, 1950), 219–34, 244–83; Robert Coles, *Erik H. Erikson: The Growth of His Work* (Boston: Little, Brown, 1970), 105–6, 122–23, 168–69. Erikson is also sensitive to the significance of culture to human development. Moreover, while Freudians tend to focus on the parents as the totality of the child's external world, Erikson's view is broader. For him, the family is the central influence on the child, particularly in the early stages of development, but it is not the exclusive influence; Erikson also addresses the roles played by the neighborhood or tribe, the school, and society at large. Erikson's emphasis on ego development offers a challenging perspective on the history of a generation of children. Being psychosocial, cultural, and historical, Eriksonian theory deals with categories historians can use and understand; and stage theory would seem to be compatible with historians' notions of development, whether individual or societal.

6. Two historians of childhood, John Demos and David Hunt, have separately praised Erikson for the wisdom and clarity of his stages, singling out his second stage, the formation of autonomy, as especially pertinent to the historical study of childhood. See Hunt, *Parents and Children in History: The Psychology of Family Life in Early Modern France* (New York: Basic Books, 1970), 191–96; Demos, "Developmental Perspectives on the History of Childhood," *Journal of Interdisciplinary History* II (Autumn 1971), 321–27; Erikson, *Identity and the Life Cycle* (New York: Norton, 1980); 70–71.

7. Helpful in constructing this table were Jonas Langer, *Theories of Development* (New York: Holt, Rinehart and Winston, 1969); Henry W. Maier, *Three Theories of Child Development* (New York: Harper & Row, third ed., 1978); R. Murray Thomas, *Comparing Theories of Child Development* (Belmont, CA: Wadsworth, 1979).

8. Erikson, "Problems of Infancy and Early Childhood," in Gardner Murphy and A. J. Bachrach, eds., *An Outline of Abnormal Psychology* (New York: Modern Library 1954), 24; Coles, *Erik H. Erikson: The Growth of His World* (Boston: Little, Brown, 1970), 81–82. See also by Erikson, *Identity, Youth and Crisis* (New York: Norton, 1968); *Dimensions of a New Identity* (New York: Norton, 1974); *Life History and the Historical Moment* (New York: Norton, 1975); and by Allen Wheelis, *The Quest for Identity* (New York: Norton, 1958). On ego resiliency, see Emmy E. Werner and Ruth S. Smith, *Vulnerable but Invincible: A Longitudinal Study of Resilient Children and Youth* (New York: McGraw-Hill, 1982).

9. For cultural diversity in American childhood during the war, see Chapter 6, on ethnic and class differences in child-rearing. For helpful introductions to the literature on socialization, see the suggestions in note 1 of this chapter.

10. Letter 174.

11. Elder, "Social History and Life Experience," in Dorothy Eichorn et al., eds., *Present and Past in Middle Life* (New York: Academic, 1981), 3; Elder, "The Timing of Lives," (unpub. ms.), 1; Elder and Avshalom Caspi, "Human Development and Social Change: An Emerging Perspective on the Life Course," in Nial Bolger et al., eds., *Persons in Context: Developmental Processes* (New York: Cambridge Univ. Press, 1988), 77–113.

12. Elder, "Families and Lives: Some Developments in Life-Course Studies," in Tamara Hareven and Andrejs Plakans, eds., *Family History at the Crossroads: A Journal of Family History Reader* (Princeton: Princeton Univ. Press, 1987), 179–99; Elder, "Family History and the Life Course," in Tamara K. Hareven, ed., *Transitions: The Family and the Life Course in Historical Perspective* (New York: Academic, 1978), 21; Elder, "Perspectives

on the Life Course," in Elder, ed., *Life Course Dynamics: Trajectories and Transitions, 1968–1980* (Ithaca: Cornell Univ. Press, 1985), 23–49; Elder and Caspi, "Human Development and Social Change," 2–3.

13. Elder, "Perspectives on the Life Course," in Elder, ed., *Life Course Dynamics: Trajectories and Transitions, 1968–1980* (Ithaca: Cornell Univ. Press, 1985), 23–49; Elder and Caspi, "Human Development and Social Change," 2–3. For helpful discussions of the relevance of a child's developmental stage during the war, see Francis E. Merrill, *Social Problems on the Home Front: A Study of War-time Influences* (New York: Harper & Brothers, 1948), 55–59; Lois Meek Stolz, "The Effect of Mobilization and War on Children," *Social Casework* 32 (April 1951), 143–49.

14. Letters 7.5, 130, 146, 184, 330, 353, 442B; *New York Times*, Dec. 12, 1943, Aug. 6, 22, 27, Oct. 11, 1944; Stein, *Wars I Have Seen* (New York: Random House, 1945), 7; John Modell, *Into One's Own: From Youth to Adulthood in the United States, 1920–1975* (Berkeley: Univ. of California Press, 1989), 165–70; Katharine F. Lenroot, "Let's Send Them Back to School," *The Rotarian* 65 (Aug. 1944), 7; "'Go-to-School' Drive Opens," *Education for Victory* 3 (Aug. 3, 1944), 1–4; "'To Serve Your Country, Go to School,'" *Education for Victory* 3 (Sept. 4, 1944), 1–2; "Rightful Heritage of Every Child," *Education for Victory* 3 (Sept. 20, 1944), 1–3; "National Go-to-School Drive," *The Child* 9 (Sept. 1944), 42–43; "High-School Attendance," *School Life* 29 (Oct. 1946), 23.

15. Ernest R. Groves and Gladys Hoagland Groves, *The Contemporary American Family* (Chicago: Lippincott, 1947), 200.

16. Letters 248, 270.

17. Letters 62, 174, 228, 248.

18. Letters 79, 97A, 102, 111, 171, 176, 251, 322, 332, 334, 405, 408, 438.

19. Letters 206E, 240C, 318.

20. Letters 2, 8, 72, 156, 178, 220, 294.

21. Letters 76, 206O.

22. Letters 206P, 345.

23. Letter 388.

24. Letters 24, 30, 80, 127, 128, 332.

25. Letter 422.

26. Letters 148, 370; Ann Landers, "Who Had It Rougher? Old-Timers or Teens," *Lawrence Journal-World*, June 25, 1989.

27. Landon Y. Jones, *Great Expectations: America and the Baby Boom Generation* (New York: Coward, McCann & Geoghegan, 1980), 3–4; Arthur J. Norton and Jeanne E. Moorman, "Marriage and Divorce Patterns of U.S. Women in the 1980s" (unpub. paper, Aug. 25, 1986), 18; Elinor Lenz, "The Generation Gap: From Persephone to Portnoy," *New York Times Book Review* (Aug. 30, 1987), 1, 36–37; Lewis S. Feuer, *The Conflict of Generations: The Character and Significance of Student Movements* (New York: Basic Books, 1969), 527–31; David I. Kertzer, "Generation as a Sociological Problem," *Annual Review of Sociology* 9 (1983), 125–49. Helpful for understanding the historical concept of a generation are Alan B. Spitzer, "The Historical Problem of Generations," *American Historical Review* 78 (Dec. 1973), 1353–85; Morton Keller, "Reflections on Politics and Generations in America," *Daedalus* 107 (Fall 1978), 123–35; Robert Wohl, *The Generation of 1914* (Cambridge: Harvard Univ. Press, 1979).

28. Conroy, "The Fifties: America in a Trance," *Esquire* 57 (June 1983), 116, 118;

Howard Schuman and Jacqueline Scott, "Generations and Collective Memories," *American Sociological Review* 54 (June 1989), 359.

29. Loewenberg, *Decoding the Past: The Psychohistorical Approach* (Berkeley: Univ. of California Press ed., 1985), 245–46. What makes Mannheim's essay important, Loewenberg writes, is that it is "the seminal conceptual formulation of the generation as a force acting in history. . . ." According to Mannheim, "The human consciousness . . . is characterized by a particular inner 'dialectic.' It is of considerable importance for the formation of the consciousness which experiences happen to make those all-important 'first impressions,' 'childhood experiences'—and which follow to form the second, third, and other 'strata.' Conversely, in estimating the biographical significance of a particular experience, it is important to know whether it is undergone by an individual as a decisive childhood experience, or later in life, superimposed upon other basic and early impressions. Early impressions tend to coalesce into a *natural view* of the world. All later experiences then tend to receive their meaning from this original set, whether they appear as that set's verification and fulfillment or as its negation and antithesis. . . . Mental data are of sociological importance not only because of their actual content, but also because they cause the individuals sharing them to form a group—they have a socializing effect." Mannheim, "The Problem of Generations," in Mannheim, *Essays on the Sociology of Knowledge*, edited by Paul Kecskemeti (New York: Oxford Univ. Press, 1952), 298, 304.

30. Mannheim, "The Problem of Generations," 302–12; Loewenberg, *Decoding the Past*, 246; Warren Dunham, "War and Mental Disorder: Some Sociological Considerations," *Social Forces* 22 (Oct. 1943), 141.

31. In a sense, however, the homefront girls and boys have gained the status of a generation somewhat by default. While this cohort has received little scholarly attention, those preceding and succeeding it have. Sociologist Glen H. Elder, Jr., has studied children born in the early 1920s who became adolescents in the 1930s in his *Children of the Great Depression: Social Change in Life Experience* (Chicago: Univ. of Chicago Press, 1974). And there have been studies of the baby boom generation, depicted as beginning in 1946, such as Landon Jones's *Great Expectations* and Elaine Tyler May's *Homeward Bound: American Families in the Cold War Era* (New York: Basic Books, 1988). By bracketing the homefront children in this way, these other studies have helped to create an in-between generation.

32. Ortega y Gasset is quoted in an insightful essay on generations by Peter N. Carroll, in *The Other Samuel Johnson: A Psychohistory of Early New England* (Rutherford, NJ: Fairleigh Dickinson Univ. Press, 1978), 17–30; interview with Sharyn Brooks Katzman; Reuben Hill, "The American Family of the Future," *Journal of Marriage and the Family* 26 (Feb. 1964), 20–28.

33. Schaller, *Reflections of a Contrarian: Second Thoughts on the Parish Ministry* (Nashville, TN: Abingdon Press, 1989), 65–95; *Lawrence Journal-World*, March 9, 1989; Schaller to the author, June 22, July 5, 1989; U.S. Bureau of the Census, *Historical Statistics of the United States, Colonial Times to 1970, Bicentennial Edition* (Washington: GPO, 1975), part 1, pp. 49–55.

34. Schaller, *Reflections of a Contrarian*, 65–69; Annie Gottlieb, *Do You Believe in Magic? The Second Coming of the Sixties Generation* (New York: Times Books, 1987); Jones, *Great Expectations*, 5–6; Richard Stengel, "Lost in Time: Too Young for the '60s, Too Old for the '80s," *San Francisco Chronicle*, July 15, 1987.

35. "Harvard Marxism of John Womack, Jr.," *Washington Post*, Jan. 1, 1983. On the children of the homefront children, see Leon Botstein, "The Children of the Lonely Crowd," *Change* 10 (May 1978), 16–20, 54; Peter N. Stearns, "The Fading of Youth," *Newsweek* 94 (July 23, 1979), 15; David Leavitt, "The New Lost Generation: It's Post-Sixties, Pre-Eighties and Forever in Between," *Esquire* 103 (May 1985), 85ff; Daniel Schnur, "Lament of the Puppies," *Newsweek on Campus* (Nov. 1985), 56; Dick Dahl, "Kids Just Wanna Have Bucks," *Utne Reader* (June/July 1986), 71–73; Michael Moffatt, *Coming of Age in New Jersey: College and American Culture* (New Brunswick, NJ: Rutgers Univ. Press, 1989); Felicity Barringer, "What IS Youth Coming To?," *New York Times*, Aug. 19, 1990.

36. In "a visual culture like ours," Jon Pareles of the *New York Times* has written, "MTV has amplified the importance of image over sound," not to mention the importance of image over substance, such as the written word or informed political debate ("As MTV Turns 10, Pop Goes the World," *New York Times*, July 7, 1991).

37. Pioneering economists Joseph Schumpeter and Simon Kuznets, as well as more recent scholars, have written of periodic economic swings of fifteen to twenty years in length. In this view, as the economic historian Michael A. Bernstein has written, there were three separate cycles at work during the 1930s, when the latency-age homefront children were born: ". . . (1) the Kondratieff, a wave of fifty or more years associated with the introduction and dispersion of major inventions, (2), the Juglar, a wave of approximately ten years' duration that appeared to be linked with population movements, and (3) the Kitchin, a wave of about forty months' length that had the appearance of a typical inventory cycle." See Bernstein, *The Great Depression: Delayed Recovery and Economic Change in America, 1929–1939* (Cambridge: Cambridge Univ. Press, 1987), 11–13.

38. The Project on the Vietnam Generation, *Report* 2 (Dec. 1986), 2; Fernand Braudel, *Civilization & Capitalism 15th–18th Century*, vol. 1: *The Structures of Everyday Life: The Limits of the Possible* (New York: Harper & Row, 1979), 70–92; Jung, *Synchronicity: An Acausal Connecting Principle* (Princeton: Princeton Univ. Press, 1973 ed.), 104. For sharing his knowledge of, insights into, and materials about cycles, I wish to thank Dennis E. Domer.

39. Schlesinger, *Paths to the Present* (New York: Macmillan, 1949), 77–92. See also Arthur M. Schlesinger, Jr., *The Cycles of American History* (Boston: Houghton Mifflin, 1986), especially 23–48.

40. Norman B. Ryder, "The Cohort As a Concept in the Study of Social Change," *American Sociological Review* 30 (Dec. 1965), 845. I wish to thank Professor Sanford Dornbusch of Stanford Univ. who, at an early point in my research, suggested that I should view this cohort in terms of its internal divisions occurring about every four years.

41. Baker, *Growing Up* (New York: Plume ed., 1983), 190; Elder, *Children of the Great Depression*, 283–91; Elder and Richard C. Rockwell, "The Depression Experience in Men's Lives," in Allan J. Lichtman and Joan R. Challinor, eds., *Kin and Communities: Families in America* (Washington, D.C.: Smithsonian Institution Press, 1979), 97.

42. David Shribman, "A Closer Look at the Hart Generation," *New York Times Magazine* (May 27, 1984); Evans, "A New Generation Takes Over," *Newsweek* 104 (July 9, 1984), 13; *New York Times*, May 17, 1987.

43. Arthur Prager, *Rascals at Large, or The Clue in the Old Nostalgia* (Garden City, NY: Doubleday, 1971), 8; Evans, "A New Generation Takes Over," 13.

44. Letters 32, 197, 345, 413; Michael Zuckerman, "Dr. Spock: The Confidence Man," in Charles E. Rosenberg, ed., *The Family in History* (Philadelphia: Univ. of Pennsylvania Press, 1975), 192–96; Daniel R. Miller and Guy E. Swanson, *The Changing American Parent: A Study in the Detroit Area* (New York: Wiley, 1958).

45. Letters 119, 133, 156, 210, 228, 324E; Evans, "A New Generation Takes Over," 13.

46. Letters 97E, 119, 120; Linda Dittmar and Gene Michaud, "America's Vietnam War Films: Marching Toward Denial," in Dittmar and Michaud, eds., *From Hanoi to Hollywood: The Vietnam War in American Film* (New Brunswick, NJ: Rutgers Univ. Press, 1990), 4–6; "Refugio," *U.S. News & World Report* 107 (Dec. 18, 1989), 51.

47. Letter 240.

48. Jones, *Great Expectations*, 2; Phillip Longman, "Justice Between Generations," *Atlantic Monthly* 255 (June 1985), 73–81.

49. Wangh, "A Psychogenetic Factor in the Recurrence of War," *International Journal of Psycho-Analysis* 49 (Parts 2 & 3, 1968), 319–23; letter 440. I wish to thank Dr. Nanette C. Aurerhahn not only for pointing out this article but for familiarizing me with psychoanalytic insights into generations.

Chapter 15. The Homefront Children at Middle Age

1. Howard Schuman and Jacqueline Scott, "Generations and Collective Memories," *American Sociological Review* 54 (June 1989), 359–81; William L. O'Neill, *American High: The Years of Confidence, 1945–1960* (New York: Free Press, 1986); John Patrick Diggins, *The Proud Decades: America in War and Peace, 1941–1960* (New York: Norton, 1988); Marty Jezer, *The Dark Ages: Life in the United States, 1945–1960* (Boston: South End Press, 1982).

2. Hodgson, *America in Our Time* (New York: Random House Vintage ed., 1978), 12, 75. On the importance of commemorating patriotism in postwar America, see John Bodnar, *Remaking America: Public Memory, Commemoration, and Patriotism in the Twentieth Century* (Princeton: Princeton Univ. Press, 1992); Karal Ann Marling and John Wetenhall, *Iwo Jima: Monuments, Memories, and the American Hero* (Cambridge: Harvard Univ. Press, 1991).

3. Charles Poore, "Shades of the Roaring Twenties," *New York Times Magazine* (April 28, 1946), 22–23ff.; Oren Root, Jr., "Youth Will Not Permit Another 1920," *New York Times Magazine* (June 9, 1946), 18–19; "Preview of the Postwar Generation," *Fortune* 27 (March 1943), 116–17ff.; William F. Ogburn, "Who Will Be Who in 1980," *New York Times Magazine* (May 30, 1948), 23, 34–35.

4. Samuelson, "The Great Postwar Prosperity," *Newsweek* 105 (May 20, 1985), 57; U.S. Bureau of the Census, *Historical Statistics of the United States, Colonial Times to 1970, Bicentennial Edition* (Washington: GPO, 1975), part 1, p. 224.

5. Letter 54. Some psychologists have contended that the "Baby Bust" children have benefited significantly in another way compared with the baby boom: They were raised in smaller families. "On a number of independent measures," the psychologist John C. Wright has explained, "the intellectual and educational development and attainments of children decline statistically with birth order, such that first-born . . . children achieve more than later-born children. Obviously cohorts that are crowded (high birth rate) also

experience a higher percentage of second- and later-born family members": Wright to William Tuttle, March 11, 1980. See R. B. Jajonc and Gregory B. Markus, "Birth Order and Intellectual Development," *Psychological Review* 82 (Jan. 1975), 74–88; Jajonc, "Family Configuration and Intelligence," *Science* 192 (April 16, 1976), 227–36.

6. *New York Times*, March 28, 1943; June 23, 1944; Phil Perdue, "Miracle Bean," *Collier's* 111 (Jan. 23, 1943), 14, 60.

7. Landon Y. Jones, *Great Expectations: America and the Baby Boom Generation* (New York: Coward, McCann & Geoghegan, 1980), 19–35; U.S. Census Bureau, *Historical Statistics of the United States, Bicentennial Edition*, part 1, p. 49; Richard A. Easterlin, "The American Baby Boom in Historical Perspective," *American Economic Review* 51 (Dec. 1961), 869–911; Porter, "Babies Equal Boom," *Reader's Digest* 59 (Aug. 1951), 5–6; "Rocketing Births: Business Bonanza," *Life* 44 (June 16, 1958), 83–87. The connection between babies and business was a frequent topic in *U.S. News and World Report*: see "More Babies—More Business," 32 (April 18, 1952), 51; "37 Million Babies—Key to Business Future," 39 (July 29, 1955), 30–32; "What 4 Million Babies Mean to Business," 42 (March 29, 1957), 39.

8. Richard A. Easterlin, *Birth and Fortune: The Impact of Numbers on Personal Welfare* (New York: Basic Books, 1980), 54; Lyle E. Schaller, "Were You Born in the 1930s?," *The Parish Paper* (Yokefellow Institute) 16 (Feb. 1987), unpaginated; Susan Householder Van Horn, *Women, Work, and Fertility, 1900–1986* (New York: New York Univ. Press, 1988), 82–123; Hugh Carter and Paul Glick, *Marriage and Divorce: A Social and Economic Study* (Cambridge: Harvard Univ. Press, rev. ed., 1976), 40–111.

9. Newman, *Falling from Grace: The Experience of Downward Mobility in the American Middle Class* (New York: Free Press, 1988), 238.

10. Bailey, *From Front Porch to Back Seat: Courtship in Twentieth-Century America* (Baltimore: Johns Hopkins Univ. Press, 1988), 49; Ellen K. Rothman, *Hands and Hearts: A History of Courtship in America* (New York: Basic Books, 1984), 285–311; John Modell, *Into One's Own: From Youth to Adulthood* (Berkeley: Univ. of California Press, 1989), 233–43.

11. May, *Homeward Bound: American Families in the Cold War Era* (New York: Basic Books, 1988), 3–5.

12. Conroy, "The Fifties: America in a Trance," *Esquire* 57 (June 1983), 116, 118; Pogrebin, "Consequences," *New York Times Magazine* (June 2, 1991), 20, 22; Ronald Steel, "Life in the Last Fifty Years," *Esquire* 57 (June 1983), 23–24; Jib Fowles, "The 1950s Revisited . . . ," *New York Times*, June 5, 1983; Benita Eisler, *Private Lives: Men and Women of the Fifties* (New York: Franklin Watts, 1986), 76–158; Wini Breines, *Young, White, and Miserable: Growing Up Female in the Fifties* (Boston: Beacon, 1992). For youth in the 1950s, see James Gilbert, *A Cycle of Outrage: America's Reaction to the Juvenile Delinquent in the 1950s* (New York: Oxford Univ. Press, 1986); and William Graebner, *Coming of Age in Buffalo: Youth and Authority in the Postwar Era* (Philadelphia: Temple Univ. Press, 1990). Two revealing books about college students in the 1950s (based on Princeton's classes of 1954 and 1957) are Roy Heath, *The Reasonable Adventurer* (Pittsburgh: Univ. of Pittsburgh Press, 1964); and Otto Butz, ed., *The Unsilent Generation* (New York: Rinehart, 1958). See also David Riesman, "The Found Generation," *American Scholar* 25 (Autumn 1956), 421–36.

13. May, *Homeward Bound*, 113–34; Barbara Ehrenreich, "The Male Revolt,"

Mother Jones 8 (April 1983), 26ff.; Ehrenreich, *Hearts of Men: American Dreams and the Flight from Commitment* (Garden City, NY: Anchor Books ed., 1984), 1–28.

14. Steel, "Life in the Last Fifty Years," 23.

15. Letters 20, 129, 176, 187, 228, 240B, 288, 367.

16. Letters 148, 195, 362, 370, 456.

17. Letter 383; Mills, "Lasting Commitment," *New York Times Magazine* (Oct. 28, 1990), 28, 56.

18. Norman B. Ryder and Charles F. Westoff, "The United States: The Pill and the Birth Rate, 1960–1965," *Studies in Family Planning*, no. 20 (June 1967), 1–3; Loren Baritz, *The Good Life: The Meaning of Success for the American Middle Class* (New York: Harper & Row Perennial Library ed., 1990), 229–30, 238–39, 241.

19. Glen H. Elder, Jr., *Children of the Great Depression: Social Change in Life Experience* (Chicago: Univ. of Chicago Press, 1974), 64–82, 271–97; Alice Rossi, "Lifespan Theories and Women's Lives," *Signs* 6 (Autumn 1980), 31; Larry Hirschhorn, "Social Policy and the Life Cycle: A Developmental Perspective," *Social Service Review* 51 (Sept. 1977), 434–50; Hirschhorn, "Post-Industrial Life: A U.S. Perspective," *Futures* 11 (Aug. 1979), 287–98; Dennis P. Hogan, *Transitions and Social Change: The Early Lives of American Men* (New York: Academic Press, 1981), 1–13, 35–64, 209–19; Hogan, "Reintroducing Culture in Life Course Research," *Contemporary Sociology* 20 (Jan. 1991), 1–4; Howard Chudacoff, *How Old Are You? Age Consciousness in American Culture* (Princeton: Princeton Univ. Press, 1989), 161–72; Leroy Ashby, "Partial Promises and Semi-Visible Youths: The Depression and World War II," in Joseph M. Hawes and N. Ray Hiner, eds., *American Childhood: A Research Guide and Historical Handbook* (Westport, CT: Greenwood, 1985), 510–16. See also Daniel Bell's *The Coming of Post-Industrial Society: A Venture in Social Forecasting* (New York: Basic Books, 1973); and Joseph Veroff et al., *The Inner American: A Self-Portrait from 1957 to 1976* (New York: Basic Books, 1981), which surveys attitudes on issues regarding marriage, family, and work.

20. Letters 64, 144, 162, 208, 210, 322, 332, 375, 398.

21. Letters 206H, 244, 444.

22. Letters 83C, 237, 358B.

23. Letters 32, 38, 156, 200A, 210, 228, 324E, 350.

24. Especially helpful are two books by Naomi Golan, *Passing Through Transitions: A Guide for Practitioners* (New York: Free Press, 1981), and *The Perilous Bridge: Helping Clients Through Mid-Life Transitions* (New York: Free Press, 1986). See also Steel, "Life in the Last Fifty Years," 24; John Demos, "Toward a History of Mid-Life: Preliminary Notes and Reflections, in *Past, Present, and Personal: The Family and the Life Course in American History* (New York: Oxford Univ. Press, 1986), 114–34.

25. Erikson, "Eight Stages of Man," in *Childhood and Society* (New York: Norton, 1950), 231–33; John Kotre, *Outliving the Self: Generativity and the Interpretation of Lives* (Baltimore: Johns Hopkins Univ. Press, 1984); Lawrence S. Wrightsman, *Personality Development in Adulthood* (Newbury Park, CA: Sage Publications, 1988), 55–76. For a historian's application of Erikson's concept of generativity, see Peter Charles Hoffer, *Revolution and Regeneration: Life Cycle and the Historical Vision of the Generation of 1776* (Athens: Univ. of Georgia Press, 1983).

Erikson's model fails to account for gender differences; for example, it is inapplicable to numerous middle-aged women, who—having already raised their children—are now pursuing their careers. Life-course researchers have concluded that while men in this stage tend to undergo "a shift . . . from dominance to nurturance (agency to affiliation)," the reverse is true for women: Rossi, "Lifespan Theories and Women's Lives," 4–7, 9–10, 19–21; David Gutmann, "Parenthood: A Key to the Comparative Study of the Life Cycle," in Nancy Datan and Leon H. Ginsberg, eds., *Life Span Developmental Psychology: Normative Life Crises* (New York: Academic Press, 1975), 167–84; Marjorie Fiske Lowenthal et al., *Four Stages of Life* (San Francisco: Jossey-Bass, 1975), 223–45; Lillian B. Rubin, *Women of a Certain Age: The Midlife Search for Self* (New York: Harper & Row, 1979).

26. Csikszentmihalyi, *Flow: The Psychology of Optimal Experience* (New York: Harper & Row, 1990), 132; letter 270; K. Robert Schwarz, "For Steve Reich, War and Rediscovery," *New York Times*, May 18, 1989. Not all of the homefront children who have worked artistically to capture the era's essence are well known. I wish to thank June Oakley (Gonzalez) for sending me her record album entitled *Dec. 7, 1941* (Nashville: Wild Hare Productions).

27. Letters 97M, 255, 358, 442H.

28. Letter 97J.

NOTES ON SOURCES

The most valuable sources for this book are the letters written to me by more than 2500 homefront children. These letters, which document both the scope of the children's experiences and their intensity, are on deposit in the Kansas Collection, Spencer Research Library, University of Kansas, and can be used there by researchers.

Periodicals, both popular magazines such as *Life* and *Newsweek* and scores of special-focus publications, were also essential sources. My research began with a thorough review of all pertinent articles listed in the *Reader's Guide to Periodical Literature*, the *International Index to Periodicals*, and other indexes. There were hundreds of articles on children's health, education, and recreation. Articles on America's wartime schools, for example, appeared in the *Elementary School Journal, Journal of Educational Sociology, Journal of the National Education Association, The School Review, School and Society*, and other journals. Filled with information on the homefront children's health were articles in the *American Journal of Nursing, Hygeia, Journal of Pediatrics, Science Digest*, and *Science News Letter*. Children's emotional health was an important subject in the *American Journal of Ortho-psychiatry* and *Mental Hygiene*. In addition, many of America's leading magazines had experts giving advice on children's lives. Three of the best were Catherine Mackenzie of the *New York Times Magazine*, Clara Savage Littledale of *Parents' Magazine*, and Gladys Denny Shultz of *Better Homes & Gardens*; relevant articles about children's wartime concerns also appeared in *American Home, Good Housekeeping, House & Garden, Ladies' Home Journal*, and *Woman's Home Companion*. Last, children's own magazines were important, such as *Boys' Life*.

Government documents were invaluable in a variety of ways. Voluminous testimony about social conditions on the homefront was heard by committees of Congress; the most helpful of these hearings were:

U.S. House of Representatives, 76th and 77th Congs., Select Committee to Investigate the Interstate Migration of Destitute Citizens (later renamed Select Committee Investigating National Defense Migration). *Hearings*, Parts 1–34. Washington: GPO, 1940–1942.

U.S. House of Representatives, 77th Cong., 2nd Sess., through 79th Cong., 1st Sess., Subcommittee of the Committee on Appropriations. *Department of Labor–Federal Security Agency Appropriation Bill for 1943* [and *1944, 1945, 1946*]. Washington: GPO, 1942 [and 1943, 1944, 1945].

U.S. House of Representatives, 78th Cong., 1st Sess., Subcommittee of the Committee on Naval Affairs. *Investigation of Congested Areas*. Washington: GPO, 1943.

U.S. Senate, 78th Cong., 1st Sess., Committee on Education and Labor. *Wartime Care and Protection of Children of Employed Mothers*. Washington: GPO, 1943.

U.S. Senate, 78th Cong., 1st Sess., Subcommittee of the Committee on Education and Labor. *Wartime Health and Education*. Washington: GPO, 1943–44.

Too numerous to list are all the governmental publications that proved helpful, including the *Congressional Record*, the bulletins of the Women's Bureau and the Children's Bureau,

Infant Care, Education for Victory, Public Health Reports, Monthly Labor Review, Social Security Bulletin, and numerous reports issued by the Bureau of the Census.

Documents of a more primary nature were located in the National Archives in Washington, D.C. The archival holdings were strong in some areas where documentation is otherwise weak—for example, African-American children. Also in the archives are letters which mothers wrote to Eleanor Roosevelt and which children wrote to Franklin D. Roosevelt; and there are surveys of wartime rumors and of racial and ethnic stereotypes. Most helpful were the records of the Bureau of Agricultural Economics, the Children's Bureau, the Committee for Congested Production Areas, the Provost Marshall General, and the War Relocation Authority. Also consulted but less helpful were the records of the Federal Works Agency, Office of Community War Services, Office of Education, Public Health Service, War Manpower Commission, War Production Board, and Women's Bureau.

Although I have included a bibliography of selected books, I would suggest that the footnote citations are the best place to look for helpful articles and books. In the footnotes I have tried to provide detailed information for readers who want to know more about children's development, both historically and psychologically. The following list focuses on general surveys of the homefront, oral histories and specialized studies of homefront life, and psychological books that also are pertinent to research in the humanities.

Abrahams, Roger D., ed. *Jump-Rope Rhymes: A Dictionary.* Austin: Univ. of Texas Press for the American Folklore Society, 1969.

Allport, Gordon W. *The Nature of Prejudice.* Cambridge: Addison-Wesley, 1954.

Allport, Gordon W., and Leo Postman. *The Psychology of Rumor.* New York: Henry Holt, 1947.

Anderson, Karen. *Wartime Women: Sex Roles, Family Relations, and the Status of Women During World War II.* Westport, CT: Greenwood Press, 1981.

Bailey, Beth L. *From Front Porch to Back Seat: Courtship in Twentieth-Century America.* Baltimore: Johns Hopkins Univ. Press, 1988.

Baker, M. Joyce. *Images of Women in Film: The War Years, 1941–1945.* Ann Arbor, MI: UMI Research Press, 1981.

Barker, Roger G. *Ecological Psychology: Concepts and Methods for Studying the Environment of Human Behavior.* Stanford, CA: Stanford Univ. Press, 1968.

Barnouw, Erik. *The Golden Web: A History of Broadcasting in the United States,* vol. 2: *1933 to 1953.* New York: Oxford Univ. Press, 1968.

Barnouw, Victor. *Culture and Personality.* Homewood, IL: Dorsey, third ed., 1979.

Baruch, Dorothy W. *You, Your Children, and the War.* New York: D. Appleton-Century, 1942.

Basinger, Jeanine. *The World War II Combat Film.* New York: Columbia Univ. Press, 1986.

Bayor, Ronald H. *Neighbors in Conflict: The Irish, Germans, Jews, and Italians of New York City, 1929–1941.* Baltimore: Johns Hopkins Univ. Press, 1978.

Benedek, Therese. *Insight and Personality Adjustment: A Study of the Psychological Effects of War.* New York: Ronald, 1946.

Berube, Allan. *Coming Out under Fire: The History of Gay Men and Women in World War II.* New York: Free Press, 1990.

Blum, John Morton. *V Was for Victory: Politics and American Culture During World War II*. New York: Harcourt Brace Jovanovich, 1976.

Boyer, Paul. *By the Bomb's Early Light: American Thought and Culture at the Dawn of the Atomic Age*. New York: Pantheon, 1985.

Brady, Alice. *Children Under Fire*. Los Angeles: Columbia Publishing, 1942.

Broom, Leonard, and John I. Kitsuse. *The Managed Casualty: The Japanese-American Family in World War II*. Berkeley: Univ. of California Press, 1973 ed.

Campbell, D'Ann. *Women at War with America: Private Lives in a Patriotic Era*. Cambridge: Harvard Univ. Press, 1984.

Capeci, Dominic J., Jr. *The Harlem Riot of 1943*. Philadelphia: Temple Univ. Press, 1977.

——. *Race Relations in Wartime Detroit: The Sojourner Truth Housing Controversy of 1942*. Philadelphia: Temple Univ. Press, 1984.

Capeci, Dominic J., Jr., and Martha Wilkerson. *Layered Violence: The Detroit Rioters of 1943*. Jackson: Univ. Press of Mississippi, 1991.

Carr, Lowell Juilliard, and James Edson Stermer. *Willow Run: A Study of Industrialization and Cultural Inadequacy*. New York: Harper & Brothers, 1952.

Chafe, William H. *The Paradox of Change: American Women in the 20th Century*. New York: Oxford Univ. Press, 1991.

Child, Irvin L. *Italian or American? The Second Generation in Conflict*. New Haven: Yale Univ. Press, 1943.

Child Study Association of America. *The Child, the Family, the Community: A Classified Booklist*. New York: Child Study Association of America, undated.

Child Welfare League of America, Inc. *The Impact of War on Children's Services*. New York: Child Welfare League of America, 1943.

Chodorow, Nancy. *The Reproduction of Mothering: Psychoanalysis and the Sociology of Gender*. Berkeley: Univ. of California Press, 1978.

Clive, Alan. *State of War: Michigan in World War II*. Ann Arbor: Univ. of Michigan Press, 1979.

Cohen, Ronald D. *Children of the Mill: Schooling and Society in Gary, Indiana, 1906–1960*. Bloomington: Indiana Univ. Press, 1990.

Costello, John. *Virtue Under Fire: How World War II Changed Our Social and Sexual Attitudes*. New York: Fromm, 1987.

Cremin, Lawrence A. *The Transformation of the School: Progressivism in American Education, 1876–1957*. New York: Knopf, 1961.

Daniels, Roger. *Coming to America: A History of Immigration and Ethnicity in American Life*. New York: HarperCollins, 1990.

——. *Concentration Camps USA: Japanese Americans and World War II*. New York: Holt, Rinehart and Winston, 1971.

D'Emilio, John. *Sexual Politics, Sexual Communities: The Making of a Homosexual Community, 1940–1970*. Chicago: Univ. of Chicago Press, 1983.

Demos, John. *Past, Present, and Personal: The Family and The Life Course in American History*. New York: Oxford University Press, 1986.

Despert, J. Louise. *Preliminary Report on Children's Reactions to the War, Including a Critical Survey of the Literature*. New York: Josiah Macy, Jr., Foundation, 1942.

Dick, Bernard F. *The Star-Spangled Screen: The American World War II Film.* Lexington: Univ. of Kentucky Press, 1985.

Elder, Glen H., Jr., et al., eds. *Children in Time and Place: The Intersection of Developmental and Historical Perspectives.* Cambridge: Cambridge Univ. Press, 1993.

——. *Children of the Great Depression: Social Change in Life Experience.* Chicago: Univ. of Chicago Press, 1974.

Elshtain, Jean Bethke. *Women and War.* New York: Basic Books, 1987.

Ely, Melvin Patrick. *The Adventures of Amos 'n' Andy: A Social History of an American Phenomenon.* New York: Free Press, 1991.

Erikson, Erik. *Childhood and Society.* New York: Norton, 1950.

——. *Dimensions of a New Identity.* New York: Norton, 1974.

——. *Identity, Youth and Crisis.* New York: Norton, 1968.

——. *Life History and the Historical Moment.* New York: Norton, 1975.

Filene, Peter G. *Him/Her/Self: Sex Roles in Modern America.* Baltimore: Johns Hopkins Press, second ed., 1986.

Ford, Thomas R., ed. *The Southern Appalachian Region: A Survey.* Lexington: Univ. of Kentucky Press, 1962.

Fox, Stephen. *Unknown Internment: An Oral History of the Relocation of Italian Americans during World War II.* Boston: Twayne, 1990.

Freud, Anna, and Dorothy Burlingham. *Infants Without Families.* New York: International Univ. Press, 1944.

——. *War and Children.* New York: Medical War Books, 1943.

Fussell, Paul. *Wartime: Understanding and Behavior in the Second World War.* New York: Oxford Univ. Press, 1989.

Gans, Herbert J. *The Urban Villagers: Group and Class in the Life of Italian-Americans.* Glencoe: Free Press, 1962.

Geertz, Clifford. *The Interpretation of Cultures: Selected Essays.* New York: Basic Books, 1973.

Gesell, Arnold, and Frances L. Ilg. *The Child from Five to Ten.* New York: Harper & Brothers, 1946.

Gesell, Arnold, Frances L. Ilg, and Louise Bates Ames. *Youth: The Years from Ten to Sixteen.* New York: Harper & Brothers, 1956.

Gesell, Arnold, et al. *The First Five Years of Life.* New York: Harper & Brothers, 1940.

Gilligan, Carol. *In a Different Voice: Psychological Theory and Women's Development.* Cambridge: Harvard Univ. Press, 1982.

Gluck, Sherna Berger. *Rosie the Riveter Revisited: Women, the War, and Social Change.* Boston: Twayne, 1987.

Goodman, Jack, ed. *While You Were Gone: A Report on Wartime Life in the United States.* New York: Simon and Schuster, 1946.

Gordon, Linda. *Heroes of Their Own Lives: The Politics and History of Family Violence: Boston 1880–1960.* New York: Viking, 1988.

Gorham, Ethel. *So Your Husband's Gone to War!* Garden City, NY: Doubleday, Doran, 1942.

Greenstein, Fred I. *Children and Politics.* New Haven: Yale Univ. Press, rev. ed., 1969.

Gregory, James N. *American Exodus: The Dust Bowl Migration and Okie Culture in California*. New York: Oxford Univ. Press, 1989.

Gruenberg, Sidonie Matsner, ed., *The Family in a World at War*. New York: Harper & Brothers, 1942.

Hareven, Tamara K., ed. *Transitions: The Family and the Life Course in Historical Perspective*. New York: Academic, 1978.

Harris, Mark Jonathan, et al., eds. *The Homefront: America During World War II*. New York: Putnam's, 1984.

Hartmann, Susan M. *The Home Front and Beyond: American Women in the 1940s*. Boston: Twayne, 1982.

Havighurst, Robert J., et al. *The American Veteran Back Home: A Study of Veteran Readjustment*. New York: Longmans, Green, 1951.

Havighurst, Robert J., with Hugh Gerthon Morgan. *The Social History of a War-Boom Community*. New York: Longmans, Green, 1951.

Hawes, Joseph M., and N. Ray Hiner, eds. *American Childhood: A Research Guide and Historical Handbook*. Westport, Conn.: Greenwood, 1985.

Herron, R. E., and Brian Sutton-Smith, eds. *Child's Play*. New York: Wiley, 1971.

Higonnet, Margaret Randolph, et al., eds. *Behind the Lines: Gender and the Two World Wars*. New Haven: Yale Univ. Press, 1987.

Hill, Reuben, et al. *Families under Stress: Adjustment to the Crises of War Separation and Reunion*. New York: Harper & Brothers, 1949.

Hoopes, Roy. *Americans Remember the Home Front: An Oral Narrative*. New York: Hawthorn Books, 1977.

Hutchinson, E. P. *Immigrants and Their Children, 1850–1950*. New York: Wiley, 1956.

James, Thomas. *Exile Within: The Schooling of Japanese Americans 1942–1945*. Cambridge: Harvard Univ. Press, 1987.

Jones, Jacqueline. *Labor of Love, Labor of Sorrow: Black Women, Work, and the Family from Slavery to the Present*. New York: Basic Books, 1985.

Jones, Landon Y. *Great Expectations: America and the Baby Boom Generation*. New York: Coward, McCann & Geoghegan, 1980.

Kenney, William. *The Crucial Years, 1940–1945*. New York: Macfadden, 1962.

Kesselman, Amy. *Fleeting Opportunities: Women Shipyard Workers in Portland and Vancouver During World War II and Reconversion*. Albany: State Univ. of New York Press, 1990.

Kessler-Harris, Alice. *Out to Work: A History of Wage-Earning Women in the United States*. New York: Oxford Univ. Press, 1982.

Killian, Lewis M. *White Southerners*. Amherst: Univ. of Massachusetts Press, rev. ed., 1985.

Kinsey, Alfred C., et al. *Sexual Behavior in the Human Female*. Philadelphia: W. B. Saunders, 1953.

———. *Sexual Behavior in the Human Male*. Philadelphia: W. B. Saunders, 1948.

Komarovsky, Mirra. *Women in the Modern World: Their Education and Their Dilemmas*. Boston: Little, Brown, 1953.

Koppes, Clayton R., and Gregory D. Black. *Hollywood Goes to War: How Politics,*

Profits, and Propaganda Shaped World War II Movies. New York: Free Press, 1987.

Leaf, Munro. *A War-time Handbook for Young Americans.* Philadelphia: Frederick A. Stokes, 1942.

Lee, Alfred McClung, and Norman Daymond Humphrey. *Race Riot.* New York: Dryden Press, 1943.

Leighton, Dorothea, and Clyde Kluckhohn. *Children of the People: The Navaho Individual and His Development.* Cambridge: Harvard Univ. Press, 1947.

Lewis, Claudia. *Children of the Cumberland.* New York: Columbia Univ. Press, 1946.

Lingeman, Richard R. *Don't You Know There's a War On? The American Home Front, 1941–1945.* New York: Putnam's (Capricorn ed.), 1976.

Litoff, Judy Barrett, and David C. Smith, eds. *Dear Boys: World War II Letters from a Woman Back Home.* University: Univ. Press of Mississippi, 1991.

———. *Since You Went Away: World War II Letters from American Women on the Home Front.* New York: Oxford Univ. Press, 1991.

Litoff, Judy Barrett, David C. Smith, Barbara Wooddall Taylor, and Charles E. Taylor. *Miss You: The World War II Letters of Barbara Wooddall Taylor and Charles E. Taylor.* Athens: Univ. of Georgia Press, 1990.

MacDonald, J. Fred. *Don't Touch That Dial: Radio Programming in American Life, 1920–1960.* Chicago: Nelson-Hall, 1980.

McWilliams, Carey. *A Mask for Privilege: Anti-Semitism in America.* Boston: Little, Brown, 1948.

———. *North from Mexico: The Spanish-Speaking People of the United States.* Philadelphia: Lippincott, 1949.

Manvell, Roger. *Films and the Second World War.* South Brunswick, NJ: A. S. Barnes, 1974.

Mazon, Mauricio. *The Zoot-Suit Riots: The Psychology of Symbolic Annihilation.* Austin: Univ. of Texas Press, 1984.

May, Elaine Tyler. *Homeward Bound: American Families in the Cold War Era.* New York: Basic Books, 1988.

Menefee, Selden. *Assignment: U.S.A.* New York: Reynal & Hitchcock, 1943.

Merrill, Francis E. *Social Problems on the Home Front: A Study of War-time Influences.* New York: Harper & Brothers, 1948.

Meyer, Agnes. *Journey through Chaos.* New York: Harcourt, Brace, 1944.

Mikesh, Robert C. *Japan's World War II Balloon Bomb Attacks on North America.* Washington, D.C.: Smithsonian Institution Press, 1990.

Milkman, Ruth. *Gender at Work: The Dynamics of Job Segregation by Sex during World War II.* Urbana: Univ. of Illinois Press, 1987.

Miller, Marc Scott. *The Irony of Victory: World War II and Lowell, Massachusetts.* Urbana: Univ. of Illinois Press, 1988.

Mintz, Steven, and Susan Kellogg. *Domestic Revolutions: A Social History of American Family Life.* New York: Free Press, 1988.

Modell, John. *Into One's Own: From Youth to Adulthood in the United States, 1920–1975.* Berkeley: Univ. of California Press, 1989.

Mussen, Paul H., ed. *Handbook of Child Psychology.* (fourth ed.); 4 vols. New York: Wiley, 1983.

Nash, Gerald D. *The American West Transformed: The Impact of the Second World War.*
Bloomington: Indiana Univ. Press, 1985.

Patri, Angelo. *Your Children in Wartime.* Garden City, NY: Doubleday, Doran, 1943.

Perrett, Geoffrey. *Days of Sadness, Years of Triumph: The American People, 1939–1945.*
New York: Coward, McCann & Geoghegan, 1973.

Piaget, Jean. *The Origins of Intelligence in Children.* New York: Norton, second ed.,
1963.

———. *Play, Dreams and Imitation in Childhood.* New York: Norton, 1962.

Piaget, Jean, and Barbel Inhelder. *The Psychology of the Child.* New York: Basic Books,
1969.

Polenberg, Richard. *War and Society: The United States, 1941–1945.* Philadelphia:
Lippincott, 1972.

Polier, Justin Wise. *Everyone's Child, Nobody's Child: A Judge Looks at Underprivileged
Children in the United States.* New York: Scribner's, 1941.

Preston, Ralph C. *Children's Reactions to a Contemporary War Situation.* New York:
Teachers College, Bureau of Publications, 1942.

Riley, Denise. *War in the Nursery: Theories of Child and Mother.* London: Virago,
1983.

Rosenberg, David, ed. *Testimony: Contemporary Writers Make the Holocaust Personal.*
New York: Times Books, 1989.

Rothman, Ellen K. *Hands and Hearts: A History of Courtship in America.* New York:
Basic Books, 1984.

Rothman, Sheila M. *Woman's Proper Place: A History of Changing Ideals and Practices,
1870 to the Present.* New York: Basic Books, 1978.

Rupp, Leila J. *Mobilizing Women for War: German and American Propaganda, 1939–
1945.* Princeton: Princeton Univ. Press, 1978.

Sandler, Stanley. *Segregated Skies: The All-Black Combat Squadrons of World War II.*
Washington: Smithsonian Institution Press, 1992.

Satterfield, Archie, ed. *The Home Front: An Oral History of the War Years in America.*
New York: Playboy Press, 1981.

Savage, William W., Jr. *Comic Books and America, 1945–1954.* Norman: Univ. of
Oklahoma Press, 1990.

Sears, Robert R., et al. *Identification and Child Rearing.* Stanford: Stanford Univ. Press,
1965.

———. *Patterns of Child Rearing.* Evanston: Row, Peterson, 1957.

Shindler, Colin. *Hollywood Goes to War: Films and American Society, 1939–1952.*
London: Routledge & Kegan Paul, 1979.

Short, K.R.M., ed. *Film & Radio Propaganda in World War II.* London: Croom Helm,
1983.

Sinai, Nathan, and Odin W. Anderson. *EMIC (Emergency Maternity and Infant Care): A
Study of Administrative Experience.* Ann Arbor: School of Public Health, Univ. of
Michigan, 1948.

Stolz, Lois Meek. *Our Changing Understanding of Young Children's Fears, 1920–1960.*
New York: National Association for the Education of Young Children, 1964.

Stolz, Lois Meek, et al., *Father Relations of War-Born Children.* Stanford: Stanford Univ.
Press, 1954.

Stouffer, Samuel A., et al. *The American Soldier*, vol. 2: *Combat and Its Aftermath*. Princeton: Princeton Univ. Press, 1949.

Szasz, Margaret Connell. *Education and the American Indian: The Road to Self-Determination, 1928–1973*. Albuquerque: Univ. of New Mexico Press, 1974.

Tateishi, John, ed. *And Justice for All: An Oral History of the Japanese American Detention Camps*. New York: Random House, 1984.

Terkel, Studs. *"The Good War": An Oral History of World War Two*. New York: Pantheon, 1984.

Tolley, Howard, Jr. *Children and War: Political Socialization to International Conflict*. New York: Teachers College Press, 1973.

Uslan, Michael, ed. *America at War: The Best of DC War Comics*. New York: Simon and Schuster, 1979.

Van Horn, Susan Householder. *Women, Work, and Fertility, 1900–1986*. New York: New York Univ. Press, 1988.

Vogt, Evon Z., and Ethel M. Albert, eds. *People of Rimrock: A Study of Values in Five Cultures*. Cambridge: Harvard Univ. Press, 1966.

Weglyn, Michi. *Years of Infamy: The Untold Story of America's Concentration Camps*. New York: Morrow, 1976.

Weibel, Kathryn. *Mirror Mirror: Images of Women Reflected in Popular Culture*. Garden City, NY: Anchor, 1977.

West, Elliott, and Paula Petrik, eds. *Small Worlds: Children and Adolescents in America, 1850–1950*. Lawrence: Univ. Press of Kansas, 1992.

Whyte, William Foot. *Street Corner Society: The Social Structure of an Italian Slum*. Chicago: Univ. of Chicago Press, second ed., 1955.

Wolfenstein, Martha. *Children's Humor: A Psychological Analysis*. Glencoe: Free Press, 1954.

Wolfenstein, Martha, and Nathan Leites. *Movies: A Psychological Study*. Glencoe, IL: Free Press, 1950.

Wynn, Neil A. *The Afro-American and the Second World War*. London: Paul Elek, 1976.

INDEX

Abbett, Barbara, 39
Abell, Irvin, 198
Abortion, 79, 80
Adler, Pat, 37
Adultery, 28, 220, 223–24
Advertising, 151–52
African Americans: child-rearing practices,
105; children's view of death, 45; civil
rights movement, 229; displacement of,
58; and diversity, 188; fear of, 175, 177;
feelings of inferiority, 103–4; health of,
176, 200-01; conflicted self-identity, 294,
n46; income, 93; latchkey children, 78;
living conditions, 176; migration of, 59–
60, 165, 175; mortality rates, 208, 211,
324 n.31; prejudice against, 60, 78, 99,
164–66, 174–77, 180; pride, 177–78; and
racial intermarriage, 99–100; and racial
violence, 165–66; and social class, 105,
200; sympathy for, 177, 187; working
women, 78
Age: effect on children's experiences, 16–17,
149, 170–71, 214, 235, 240–41; influence
on child development, 226–27, 237–39,
240–41
Air raid drills: advice on, 13–14; children's
reactions to alarms and blackouts, 5–6, 7–
9; with gas masks, 8; parents and, 7, 8; at
schools, 5–6, 7
Alcoholism, 218–19
Alcock, A. T., 13
Aldrich, C. Andrews, 107–8
Aldrich, Mary M., 107–8
Alexander, Paul, 131
Allen, David, 211
Allen, Woody, 262
Allport, Gordon W., 116–17, 170, 183, 188
Amaro, Concepcion, 38, 213
American Academy of Pediatrics, 210
American Committee on Maternal Welfare,
13-14

American Medical Association, 210
Anderson, Charlotte, 12
Anglea, Ruby, 30
Anti-Semitism. *See* Jewish Americans
Anzovino, Susan, 142
Appalachian migrants, 49, 60–61, 62–63
Applebury, Sally, 38
Arendt, Hannah, xiii
Arnheim, Rudolph, 309 n.14
Arnold, Arthur, 219
Atomic bombs, children' reaction to, 214,
231, 232–33

*Babies Are Human Beings: An Interpretation of
Growth* (Aldrich), 107–8
Baby boom: during the war, 24–28; postwar,
25, 27
Bach, George R., 221
Badanes, Jerome, 188
Bailey, Beth L., 24, 257
Baker, Russell, 249
Balch, Barbara, 42
Baldwin, Helen, 164
Barnhouse, Richard, R., 10
Barnouw, Victor, 97
Baruch, Dorothy W., 135
Bateman, Robert, 67
Bauer, Larry Paul, 77, 119, 136, 139;
143
Bayor, Ronald H., 181, 183
Behrens, Ruth, 218
Berlin, Sir Isaiah, xi
Berman, Judy Milano, 181
Bernal, Diana, 130, 166–67
Berner, Nancy, 189
Bernstein, Michael A., 339 n.37
Betti, Kathy Lynn, 40
Beydler, Jean, 36–37, 212
Bigotry. *See* Prejudice; Racism; Stereotypes
Bingham, Henrietta, 47–48, 215
Birth order, 104, 340-41 n.5

353